Lecture Notes in Mechanical Engineering

Lecture Notes in Mechanical Engineering (LNME) publishes the latest developments in Mechanical Engineering—quickly, informally and with high quality. Original research reported in proceedings and post-proceedings represents the core of LNME. Volumes published in LNME embrace all aspects, subfields and new challenges of mechanical engineering.

To submit a proposal or request further information, please contact the Springer Editor of your location:

Europe, USA, Africa: Leontina Di Cecco at Leontina.dicecco@springer.com
China: Ella Zhang at ella.zhang@springer.com
India: Priya Vyas at priya.vyas@springer.com
Rest of Asia, Australia, New Zealand: Swati Meherishi at swati.meherishi@springer.com

Topics in the series include:

- Engineering Design
- Machinery and Machine Elements
- Mechanical Structures and Stress Analysis
- Automotive Engineering
- Engine Technology
- Aerospace Technology and Astronautics
- Nanotechnology and Microengineering
- Control, Robotics, Mechatronics
- MEMS
- Theoretical and Applied Mechanics
- Dynamical Systems, Control
- Fluid Mechanics
- Engineering Thermodynamics, Heat and Mass Transfer
- Manufacturing
- Precision Engineering, Instrumentation, Measurement
- Materials Engineering
- Tribology and Surface Technology

Indexed by SCOPUS, EI Compendex, and INSPEC

All books published in the series are evaluated by Web of Science for the Conference Proceedings Citation Index (CPCI)

To submit a proposal for a monograph, please check our Springer Tracts in Mechanical Engineering at https://link.springer.com/bookseries/11693

Harish Kumar · Prashant K. Jain · Saurav Goel
Editors

Recent Advances in Intelligent Manufacturing

Select Proceedings of ICAME 2022

Springer

Editors
Harish Kumar
Department of Mechanical Engineering
National Institute of Technology Delhi
Delhi, India

Prashant K. Jain
Design and Manufacturing
Indian Institute of Information Technology
Jabalpur, India

Saurav Goel
London South Bank University
London, UK

ISSN 2195-4356 ISSN 2195-4364 (electronic)
Lecture Notes in Mechanical Engineering
ISBN 978-981-99-1307-7 ISBN 978-981-99-1308-4 (eBook)
https://doi.org/10.1007/978-981-99-1308-4

This Springer imprint is published by the registered company Springer Nature Singapore Pte Ltd.
The registered company address is: 152 Beach Road, #21-01/04 Gateway East, Singapore 189721, Singapore

Contents

About the Editors

Dr. Harish Kumar is currently working as Associate Professor at National Institute of Technology Delhi. He has more than 18 years of research and academic experience and served as Scientist at different grades in CSIR—National Physical Laboratory, India (NPLI). He has been Active Researcher in the area of mechanical measurement and metrology. He has worked as Guest Researcher at National Institute of Standards and Technology, USA, in 2016. He has been instrumental in ongoing redefinition of kilogram in India. He has authored more than 100 publications in peer-reviewed journals and conferences. He is Active Researcher in the area of additive manufacturing. He has served as Guest Editor of different journals and conference proceedings volumes including Mapan—Journal of Metrology Society of India, Materials Today: Proceedings (Elsevier), Lecture Notes in Mechanical Engineering (Springer), Smart Innovation, Systems and Technologies (Springer), Advanced Science, Engineering and Medicine (American Scientific Publishers), etc. He has been serving as Associate Editor of the Mapan—Journal of Metrology Society of India.

Dr. Prashant K. Jain is currently Professor at Department of Mechanical Engineering, Indian Institute of Information Technology, Design and Manufacturing, Jabalpur. He obtained his B.E. (Mechanical Engineering) from Dr. H. S. Gour University, Sagar, M.E. (Advanced Production Systems) from SATI, Vidisha, and Ph.D. from Indian Institute of Technology Delhi, Delhi. He also has served at IIT Delhi as Project Scientist and at Delhi College of Engineering, Delhi (now Delhi Technological University, Delhi) as Lecturer. His major research interests include rapid prototyping/additive manufacturing, geometric modeling, CAD/CAM integration, computational geometry, rapid prototyping, and tooling. He has more than 85 publications to his credit, published in international peer-reviewed journals, and national and international conferences in India and abroad.

Dr. Saurav Goel FHEA, CEng, is a Professor at London South Bank University (LSBU) and Associate Director of the EPSRC NetworkPlus in Digitalised Surface Manufacturing (EP/S036180/1). He is Visiting Professor at IIT Guwahati and Shiv Nadar University in India. He is also Fellow of the Higher Education Academy and

Chartered Engineer. His thematic research on "Material Oriented Manufacturing" at the confluent of experiments and atomic modeling focuses on creating socially responsible STEM professionals and to deploy the use of science for the benefit of society. He started working toward this mission by implementing new EDI policies in the Centre of Doctoral Training in Ultra-Precision (EP/K503241/1 and EP/L016567/1) at Cranfield University. He is currently a Co-I on three EPSRC research projects, one of which aims to generate affordable healthcare solutions by developing novel heterostructure scintillators as next-generation radiation detectors for the TOF-PET imaging systems (EP/S013652/1). He is also looking at addressing social issues such as gender imbalance in an ODA country like India via a Royal Academy of Engineering (IAPP18-19\295) UK-India collaborative project and has secured the Prestigious Newton Fellowship Award from the Royal Society (NIF\R1\191571) to model the "science of interfaces" in surface manufacturing. More recently, he joined hands with NHS in developing COVID-19-resistant biosafe surfaces for which he secured an Engineering Pandemic Preparedness Award from ^2500 citations (Google Scholar). His research is strongly supported by various industries like Airbus and NPL who have provided him with an EPSRC I Case studentship as well as research organizations like STFC who have part-funded his doctorate students. He is currently maintaining a research group of 10 people (4 PDRA's, 6 Ph.D. students) with 50% gender balance.

Exploring the Constructs and Measures of Innovation Management in Indian MSMEs

Bisma Mannan, Sonal Khurana, and Abid Haleem

Abstract In this global era, Innovation Management is referred to as a fundamental requirement for organizational survival, success and competitiveness. There is an imperative need for innovation management. The global marketplace is like a war zone; if the firm does not manage the innovation, the firm will eventually die. Therefore, this study endeavors to explore the constructs and measures of innovation management in Indian MSMEs. The constructs and their respective measures from the past literature were identified, and then the hypotheses are developed. Based on the developed hypotheses, a questionnaire was formulated. Obtained response on more than 400 questionnaires through various sources, out of these 379 questionnaires were filtered and rest were discarded, and then on the filtered data Exploratory factor analysis (EFA) was applied using SPSS. From the past literature, identified ten constructs and 39 measures. From the pattern matrix, it is found out that all the identified 39 measures belong to their respective constructs and are highly loaded on them. The results of the correlation matrix show that the top management is positively correlated with all the constructs and influence significantly the communication and organization culture and also the financial resources construct has a high correlation with collaboration. This paper helps in exploring the significant constructs of innovation management and their respective measures and also gives the clarity on which measures facilitates which constructs of innovation management in Indian MSMEs.

Keywords Constructs · EFA (Exploratory Factor Analysis) · Innovation · Innovation management · Measures · MSMEs (Micro Small & Medium Enterprises)

B. Mannan (✉)
University of Plymouth, Plymouth, UK
e-mail: bismamannan@gmail.com

S. Khurana
Department of Engineering, Vivekananda Institute of Professional Studies-Technical Campus, Delhi, India

A. Haleem
Department of Mechanical Engineering, Jamia Millia Islamia, Delhi, India

H. Kumar et al. (eds.), *Recent Advances in Intelligent Manufacturing*,
Lecture Notes in Mechanical Engineering,
https://doi.org/10.1007/978-981-99-1308-4_1

1 Introduction

Innovation is one of the most discussed topics amongst the academician and practitioners in this global era. Innovation is typically defined as a new or improved product, method, service, or technology in research. The globe, not just India, is facing new problems that call for creative answers. Growth and greater job prospects are mostly driven by innovation. For the past few decades, innovation has been regarded as a prerequisite for the success and competitiveness of associations or organizations. Innovation has grown increasingly important in the current business environment as it is more difficult to survive in competitive market. Organizations must put more of an emphasis on the managerial aspects of innovation that create a sustainable competitive advantage because of the increased pace of technological development and competition. In developing country like India where employment is the prime issue for the youth, Micro, Small and medium-sized enterprises (MSMEs) become more and more the center of attention and the focal point of the policymakers. Most of the commercial structures mainly possess of MSMEs, and regardless of the presence of large organizations most of the employees get employment in these MSMEs. Especially, MSMEs is considered to be of great importance for job creation and economic development.

The research team led by the National Knowledge Commission (NKC) on "Entrepreneurship in India" indicated that large organizations in India still have the outlook of innovation which is incremental in contrast with the breakthrough innovation. Large organizations innovate to increase competitiveness in the market whereas MSMEs advance and innovate to improve the share in the market. Past studies were specific to large organization mostly manufacturing and automotive sector. So, there is a need to identify innovation management practices related to MSMEs. This study further identified the related constructs and measures of innovation management. This study felt a need to perform a comprehensive exploratory factor analysis so that one can correctly identify which measures are essential, the loading of factors, and correlation of constructs and undertake the associated analysis for some meaningful outcome.

1.1 Objectives of the Study

The innovation helps the organization to survive in the competitive market but the literature presents in the context of a large organization of UK and China. There is not much literature available in the Indian MSMEs context. So, this study endeavors to answers the research objectives given below:

- Identification of significant constructs that govern innovation management in Indian MSMEs
- Identification of measure that helps in measuring the construct of innovation management

- Developed a hypothesis to find out which measure significantly affect the construct
- Measuring the values of construct and measures using EFA and its validation
- Analyzing the results of EFA and discuss it in detail

2 Constructs and Measures of Innovation Management

This adopted constructs of innovation management and MSMEs mostly from Adams, Bessant & Phelps [1] and Hoffman et al. [2] which have 1201 and 754 citations respectively. After that, identified the measures concerning the constructs from the various articles.

2.1 Knowledge Management

The notion of knowledge was always a prime factor, but in the recent decade, it has grown much consideration (e.g. Nonaka et al. [3], Page and Schirr [4]). From the early of time, knowledge has always played an affirmed part in the advancement of innovation [5]. Knowledge management deals with interactive thoughts, ideas, information and imparting knowledge that motivates innovation abilities and helps in networking, knowledge generation, knowledge absorption, knowledge transfer, and knowledge sharing. Knowledge management also deals with the management of the flow of implicit and explicit knowledge seized by the individual, body or association [6, 3, 7] and besides the process of accumulating and utilizing information.

H1a: Knowledge creation has a significant effect on Knowledge Management

H1b: Knowledge absorption has a significant effect on Knowledge Management

H1c: Knowledge sharing has a significant effect on Knowledge Management

H1d: Knowledge transfer has a significant effect on Knowledge Management

2.2 Organization Culture

Hamel and Getz [8] advocate that it is a prerequisite for the organization to incorporate innovation as a culture. The administration should also have confidence in the idea that innovation may come from any level of hierarchy within the organization not only from the R&D department. Subsequently, it is essential to free the organization employees to entirely involved in innovation practices and policies [8].

Innovation nowadays becomes a core value, and the organizations encourage and provide incentives to innovative employees. Cassell and Symon [9] argues that the organization culture is vital for the innovation management practices but also said that the employees want the balanced work atmosphere for innovation.

H2a: Employ involvement has a significant effect on Organization Culture

H2b: Cross culture has a significant effect on Organization Culture
H2c: Provincialism has a significant effect on Organization Culture
H2d: cultural Issues has a significant effect on Organization Culture

2.3 Diffusion and Adoption

Past literature establishes diffusion and Adoption as one of the most important constructs of innovation which has four measures Innovativeness, Customer responsiveness, social system and timescale.

H3a: Innovativeness has a significant effect on Diffusion and adoption
H3b: Customer responsiveness has a significant effect on Diffusion and adoption
H3c: The Social system has a significant effect on Diffusion and adoption
H3d: Timescale has a significant effect on Diffusion and adoption

2.4 Communication

Clear and understandable communication in the organization and outside it is essential for innovation. For proper diffusion and adoption of innovation need proper communication to customers through advertisements and digital marketing.

H4a: Mode of Communication has a significant effect on Communication
H4b: Understandable communication has a significant effect on Communication
H4c: Communication channels have a significant effect on Communication
H4d: Networking has a significant effect on Communication

2.5 Collaboration

Presently, the influence of innovation development and university collaboration on the output of the MSMEs is the first apprehension of innovation strategies and technology in developing countries. Nevertheless, innovation development and university collaboration are highly intricate which require structured and systematic approach [10] that include noteworthy non-codified and tacit components of knowledge,it helps in knowledge mobility and networking [11, 12]. We observe that policy and decision makers are more and more inspired by open innovation, this further upsurge local growth and economic development [13]. Research has produced abstruse outcomes regarding the result of collaboration. Lee et al. [14] findings suggest that there is no direct relation between industries working with universities and the market grows, while analysis shows that MSMEs technological capabilities of high levels may get advantage from such collaboration. So, the influence of universities on the growth of MSMEs requires careful interpretation.

H5a: Incubation centre has a significant effect on Collaboration
H5b: Open Innovation has a significant effect on Collaboration
H5c: Trust between partners has a significant effect on Collaboration
H5d: Knowledge mobility has a significant effect on Collaboration

2.6 *Benchmarking*

Innovation in every developing country is extremely dependent on the external environment and competition. Organizations who have been capable of making benchmarking in their internal systems and are capable of working with the shortfalls of the external situations or market competition are effective and successful, whereas others remain insignificant [15].

H6a: Business strategy has a significant effect on Benchmarking
H6b: Competitiveness has a significant effect on Benchmarking
H6c: Protocol has a significant effect on Benchmarking
H6d: Internal and external assessment has a significant effect on Benchmarking

2.7 *Top Management*

Tidd [16] found the conduct of top management is the most influential construct. However, top management follows the path of innovation to get the things done and have a future vision for operational and directional change in the organization than the organization can easily survive in this competitive environment [17].

H7a: Transformational leadership has a significant effect on Top Management
H7b: Transactional leadership has a significant effect on Top Management
H7c: Top management reflexivity has a significant effect on Top Management
H7d: Motivation & incentives have a significant effect on Top Management

2.8 *Portfolio Management*

Portfolio management emerges as a key factor for successful innovation more used explicitly in product innovation in the recent literature. It is critical due to the quickness at which assets are used in the process of innovation and the requirement for these assets to be relocated [18], [19].

H8a: Risk management has a significant effect on Portfolio Management
H8b: Resource allocation has a significant effect on Portfolio Management
H8c: IT infrastructure has a significant effect on Portfolio Management

2.9 Government Policies

SMEs have many constrained in exporting their products and services to the global market. Government meddling and bureaucracy frightens the investors of the global market in the developing country like India which develops uncertainty and risk. Thus, Innovation in India is sometimes negatively related with the government policies and intrusion [20], [21].

H9a: Legislation, norms, regulations, standards &Taxation has a significant effect on Government policies

H9b: Policy on patents and licenses has a significant effect on Government policies

H9c: Export & import has a significant effect on Government policies

H9d: Subsidiaries has a significant effect on Government policies

2.10 Financial Resources

Understanding and identifying the credit risk of MSMEs ought to be different from the large organizations. MSMEs are Owners directly and significantly influence the MSME by their owner [22]. Bank policies on credit and subsidiaries may help in the effective management of finances in MSMEs [23].

H10a: Financial means has a significant effect on financial resources

H10b: Bank policies on credit has a significant effect on financial resources

H10c: Interest rates have a significant effect on financial resources

H10d: R & D investment has a significant effect on financial resources (Table 1)

3 Methodology

After developing the hypotheses with the help of a literature review, one of the most challenging tasks of this study is to develop and fill out the questionnaire from the MSMEs employees and industrial experts. We have collected the data from middle-level workers and managers from Delhi/ NCR MSMEs. It was the most time-consuming task of the entire study.

Bentler and Chou [64] provided the thumb rule for sample size. It says, when there are several measures of the constructs and when the data follow the normal distribution, and there are large factor loadings, the ratio between the size of the sample and the number of constructs must be at least 5:1 to get a reliable estimation of different parameters. They further recommend that the ratios would be greater than 10:1 to acquire results. However, in some studies, the number of respondents per estimated construct surges to 15:1. Moreover, if the sample size increases beyond 400 to 500 results become "too sensitive" [65]. Therefore, 400 questionnaires have

Table 1 Construct, measure and their references

Constructs	Measures	Authors
Knowledge Management (KM)	– Knowledge creation (KM_1) – Knowledge absorption (KM_2) – Knowledge sharing (KM_3) – Knowledge transfer (KM_4)	Nonaka et al. [3], Blackler [5], McAdam [24], Ambrosini, and Bowman [25], Adams, Bessant, and Phelps [1], Tsai [7], Zemaitis [26]
Organisation Culture (OC)	– Employ involvement (OC_1) – Cross culture (OC_2) – Provincialism (OC_3) – Cultural Issues (OC_4)	Ekvall [27, 28], Mathisen and Einarsen [29], Dougherty [30], Mannan, Khan and Khurana [31]
Diffusion and adoption (DA)	– Innovativeness (DA_1) – Customer responsiveness (DA_2) – Social system (DA_3) – Timescale (DA_4)	Zikmund et al. [7], Kinnear [32, 33], Cooper [18], Tidd et al. [34], Kotler et al. [35], Roger [36], Winer [37]
Communication (COM)	– Mode of Communication (COM_1) – Understandable communication (COM_2) – Communication channels (COM_3) – Networking (COM_4)	Gopalakrishnan and Damanpour [38], Gopalakrishnan and Bierly [39], Hameed et al. [40], Mannan, Khan and Khurana [41], Dahnil et al. [42, 43, 44], Mannan, Khurana, and Haleem [45]
Collaboration (COL)	– Incubation centre (COL_1) – Open Innovation (COL_2) – Trust between partners (COL_3) – Knowledge mobility (COL_4)	[10], Robson and Bennett [46], Lee et al. [14, 11], Pickernell et al. [47], Packham et al. [13], Khurana, Mannan, and Haleem [48]

(continued)

Table 1 (continued)

Constructs	Measures	Authors
Benchmarking (BM)	– Business strategy (BM_1) – Competitiveness (BM_2) – Protocol (BM_3) – Internal and external assessment (BM_4)	Kovacic [49], Gomes, and Yasin [50], Giovannoni, and Pia Maraghini [15], Madsen, Slåtten, and Johanson [51]
Top Management (TM)	– Transformational leadership (TM_1) – Transactional leadership (TM_2) – Top management reflexivity (TM_3) – Motivation & incentives (TM_4)	Birkinshaw [52], Vaccaro et al. [53], Mihalache [54], Mannan Jameel and Haleem [55]
Portfolio Management (PM)	– Risk management (PM_1) – Resource allocation (PM_2) – IT infrastructure (PM_3)	Cooper et al. [56], Adams, Bessant, and Phelps [1], Faems, Janssens, and Neyens [19], Wadhwa, Phelps, and Kotha [57]
Government policies (GOVT)	– Legislation, norms, regulations, standards & Taxation (GOVT_1) – Policy on patents and licenses (GOVT_2) – Export & import (GOVT_3) – Subsidiaries (GOVT_3)	García-Teruel and Martínez-Solano [58], Escribá-Esteve; García-Teruel and Martínez-Solano [23], Sharma, et al. [59], Newman, Gunessee and Hilton [60]
Financial resources (FR)	– Financial means (FR_1) – Bank policies on credit (FR_2) – Interest rates (FR_3) – R & D investment (FR_4)	Newman, Gunessee and Hilton [60], Bhunia [61], Xia [62], Mannan, Khurana, and Haleem [63]

been filled by the respondent. After data filtering process, EFA is used only on 379 questionnaire responses with the help of IBM-SPSS.

Exploratory Factor Analysis (EFA) is a technique which uses statistical estimation for finding the correlation between the measures in the given dataset. This approach of estimation results in a factor structure (forming a group of measures) which rely on the strong correlations. In most cases, an EFA is used to select the measures which help in developing a cleaner SEM. To organise and group the new data one should always perform EFA. The benefit of an EFA when compare to CFA (confirmatory factor analysis) is that researcher can analyse without knowing any previous concept about which measures belong to which constructs. That means the EFA is more capable than CFA to recognise the problematic measures without any difficulty.

3.1 EFA (Exploratory Factor Analysis)

In EFA, first Kaiser–Meyer–Olkin (KMO) estimates the adequacy of samples. KMO shows that EFA can be performed on a data sample as it can group the measures into the smaller set of factors which is known as constructs. KMO ranges from 0 to 1, and it must be equal to or higher than 0.60 to proceed in EFA analysis. If the KMO value is lesser than 0.50, the outcome of EFA possibly would not be very beneficial.

The below Table 2 shows the KMO value for this study:

It is the Kaiser criterion or K1 rule in which eigenvalue should higher than 1.0. Henry F. Kaiser gives this rule. This K1 rule for extraction of constructs/factors shows in the below Table 3.

It is found out that eigenvalues are more significant than 1 for ten factors. That means, all the 39 measures are grouped into ten factors/constructs. The values of % of Variance in ranges from 19.297 to 3.437 which gives the total Cumulative % of the variance is 84.699%. After ten factors, there is not much change in the eigenvalues (Fig. 1).

Scree plot is used to visualise the factor extraction using eigenvalues. This graph is a plot on the based on the above table using the first two columns. One can see that there is a visible change in eigenvalue till factor 10 after that it almost becomes constant. So, it shows that all the 39 measures grouped into ten constructs.

3.2 Pattern Matrix

Next, factor loading in done in EFA. Factor loading in EFA can be assessed using the pattern matrix. Pattern matrix is like a result of EFA which shows loading of measures on constructs. The pattern matrix represents the correlations between the measures

Table 2 Kaiser–Meyer–Olkin measure of sampling adequacy

KMO and Bartlett's Test		
Kaiser–Meyer–Olkin measure of sampling adequacy		0.829
Bartlett's test of sphericity	Approx. Chi-square	15,051.193
	Df	741
	Sig.	0.000

Table 3 Eigenvalues and percentage of variance

Total variance explained							
Factor	Initial eigenvalues			Extraction sums of squared loadings			Rotation sums of squared loadings
	Total	% of Variance	Cumulative %	Total	% of Variance	Cumulative %	Total
1	7.514	19.267	19.267	5.512	14.133	14.133	4.150
2	4.889	12.537	31.804	5.535	14.192	28.326	3.927
3	4.548	11.662	43.466	3.815	9.783	38.109	4.721
4	3.561	9.130	52.596	3.633	9.316	47.425	3.959
5	2.801	7.183	59.779	2.900	7.435	54.860	4.637
6	2.465	6.321	66.100	2.357	6.042	60.903	3.749
7	2.197	5.634	71.734	2.280	5.847	66.750	4.796
8	1.991	5.105	76.838	1.923	4.932	71.682	3.387
9	1.725	4.424	81.262	1.545	3.961	75.643	4.400
10	1.340	3.437	84.699	1.536	3.937	79.580	3.370
11	0.524	1.345	86.044				
12	0.442	1.134	87.178				
13	0.429	1.100	88.278				
14	0.406	1.040	89.318				
15	0.358	0.917	90.236				
16	0.348	0.893	91.128				
17	0.286	0.733	91.861				
18	0.282	0.724	92.585				
19	0.278	0.712	93.298				
20	0.253	0.649	93.946				
21	0.236	0.605	94.552				
22	0.206	0.529	95.081				
23	0.188	0.482	95.563				
24	0.177	0.454	96.017				
25	0.169	0.433	96.450				
26	0.156	0.399	96.850				
27	0.152	0.390	97.239				
28	0.136	0.349	97.588				
29	0.129	0.330	97.918				
30	0.120	0.308	98.226				
31	0.102	0.262	98.489				
32	0.098	0.252	98.741				
33	0.090	0.230	98.971				

(continued)

Table 3 (continued)

Total variance explained							
34	0.084	0.215	99.186				
35	0.078	0.200	99.387				
36	0.073	0.186	99.573				
37	0.069	0.176	99.749				
38	0.056	0.143	99.892				
39	0.042	0.108	100.000				

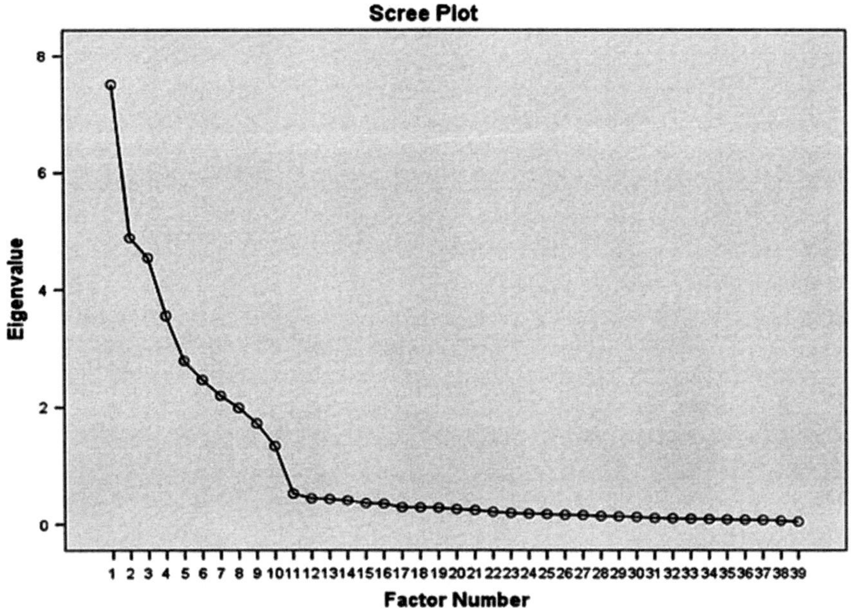

Fig. 1 Scree plot using eigenvalues of constructs

and the constructs, and it also contains the coefficients for the linear combination of the measures.

From the Table 4, it is observed that all the measures are highly loaded only on one construct, and each measure has a higher value than 0.4. Therefore, all the 39 measures of innovation management in Indian MSMEs belongs to 10.

Table 4 Pattern matrix of EFA

	Constructs									
	1	2	3	4	5	6	7	8	9	10
GOVT_1	0.958									
GOVT_4	0.946									
GOVT_3	0.942									
GOVT_2	0.911									
FR_3		0.981								
FR_4		0.973								
FR_2		0.927								
FR_1		0.856								
TM_3			0.984							
TM_1			0.932							
TM_2			0.894							
TM_4			0.849							
DA_4				0.953						
DA_1				0.914						
DA_3				0.836						
DA_2				0.827						
KM_4					0.950					
KM_3					0.942					
KM_1					0.845					
KM_2					0.795					
COL_4						0.924				
COL_1						0.911				
COL_2						0.888				
COL_3						0.782				
COM_2							0.931			
COM_1							0.891			
COM_4							0.822			
COM_3							0.740			
BM_4								0.957		
BM_1								0.889		
BM_3								0.812		
BM_2								0.702		
OC_1									0.909	
OC_2									0.876	
OC_3									0.826	

(continued)

Table 4 (continued)

	Constructs									
	1	2	3	4	5	6	7	8	9	10
OC_4									0.695	
PM_3										1.010
PM_2										0.922
PM_1										0.705

3.3 Correlation Matrix

The above table shows the correlation of constructs with other constructs. It shows that top management is positively correlated with all the constructs and significantly influence the communication and organisation culture. Financial resources are highly correlated with collaboration. Diffusion and adoption are highly correlated with portfolio management and communication whereas organisation culture is highly correlated with knowledge management (Table 5).

3.4 Reliability of the Constructs

Here reliability signifies to the internal consistency of the measure-level error within the single construct. In EFA, reliability refers to a set of reliable measures which are consistently loaded on the same construct. The check the reliability in the EFA is to estimate the value Cronbach's alpha for each construct. The value of Cronbach's alpha should be above 0.7 for internal consistency.

Although, the Cronbach's alpha will generally increase for constructs having more measures and decrease for construct having a fewer measure. For this reason, each construct should at-least have three measures, although two measures are permissible sometimes. Table 6 shows the Cronbach's alpha values for each construct. All the values are above 0.7 and range from 0.897 to 0.968 that implies all the constructs are reliable.

4 Results and Its Discussion

Various constructs control the ability of organisations to innovate and determine by multiple measures, and the primary objective of this paper is to identify and explore the constructs and measures of innovation management in MSMEs in the Indian context. After that, 39 measures are identified. Most of the construct have four measures, and only portfolio management has three measures which form 39 hypotheses, i.e. from H1 to H39. Exploratory Factor Analysis (EFA) is applied to

Table 5 Factor correlation matrix

Constructs	1(GOVT)	2 (FR)	3 (TM)	4 (DA)	5 (KM)	6 (COL)	7 (COM)	8 (BM)	9 (OC)	10(PM)
1 (GOVT)	1.000	0.115	0.104	−0.014	0.190	0.232	0.110	0.030	0.242	−0.024
2 (FR)	0.115	1.000	0.030	0.067	0.120	0.275	0.033	−0.103	0.061	−0.076
3 (TM)	0.104	0.030	1.000	0.230	0.230	0.088	0.463	0.022	0.283	0.108
4 (DA)	−0.014	0.067	0.230	1.000	0.151	0.035	0.243	0.067	0.113	0.336
5 (KM)	0.190	0.120	0.230	0.151	1.000	−0.013	0.316	−0.183	0.439	−0.166
6 (COL)	0.232	0.275	0.088	0.035	−0.013	1.000	0.091	0.187	0.043	0.018
7 (COM)	0.110	0.033	0.463	0.243	0.316	0.091	1.000	0.017	0.373	0.220
8 (BM)	0.030	−0.103	0.022	0.067	−0.183	0.187	0.017	1.000	−0.001	0.306
9 (OC)	0.242	0.061	0.283	0.113	0.439	0.043	0.373	−0.001	1.000	0.079
10 (PM)	−0.024	−0.076	0.108	0.336	−0.166	0.018	0.220	0.306	0.079	1.000

Extraction Method: Maximum Likelihood
Rotation Method: Promax with Kaiser Normalization

Table 6 Cronbach's alpha values for constructs

Constructs	No. of measures	Cronbach's alpha (α)
Knowledge Management	4	0.939
Organisation Culture	4	0.897
Diffusion and adoption	4	0.937
Communication	4	0.913
Collaboration	4	0.931
Benchmarking	4	0.908
Top Management Support	4	0.956
Portfolio Management	3	0.915
Government policies	4	0.968
Financial resources	4	0.965

analyse the 39 factors Exploratory Factor Analysis (EFA) is a statistical approach to determine the correlation between the measures in a dataset. Kaiser–Meyer–Olkin (KMO) value for the data sample is 0.829, and it is also found out that the eigenvalues are greater than 1 for ten factors. That means, all the 39 measures are grouped into ten constructs. The values of % of Variance in ranges from 19.297 to 3.437 which gives the total Cumulative % of the variance is 84.699%. After ten factors, there is not much change in the eigenvalues. From the pattern matrix, it is found out that all the measures are highly loaded on their respective constructs and having a minimum value of 0.702 that means all the measures significantly affect their construct and all the hypothesis which was formulated are accepted.

The correlation matrix shows that top management is positively correlated with all the constructs and significantly influence the communication and organisation culture. Financial resources are highly correlated with collaboration. Diffusion and adoption are highly correlated with portfolio management and communication. Organisation culture is highly correlated with knowledge management. The last part of analysis shows that the minimum value of Cronbach's alpha is 0.897 which is a reliability parameter that means all the results and the data fed to the software for EFA analysis is highly reliable.

5 Limitations and Future Scope

This study is focused on innovation management in MSMEs which does not consider the parameter of sustainability. So, the future study may consider the sustainability aspects of innovation management. From the past literature, the ten most critical constructs and 39 measures have been identified and analysed in this study. Thus, one can consider more constructs for future study. The questionnaire was mostly filled by the Delhi-NCR MSMEs employees. So, for future study, another region in India can be considered.

6 Conclusion

The main aim is to identify and understand the constructs and measures of innovation management related to MSMEs in the Indian context and then to validate them using EFA. There are ten significant constructs which are identified from the literature review of innovation management and then 39 measures are identified. Using these measures and construct 39 hypotheses were formed and a questionnaire was developed. After that Exploratory Factor Analysis (EFA) is done on the 379 datasets. Kaiser–Meyer–Olkin (KMO) value for the data sample is 0.829, and it is also found out that the eigenvalues are more significant than 1 for ten factors. That means, all the 39 measures are grouped into ten factors/constructs. Ultimately, pattern matrix in EFA gives a loading of measures on constructs and shows that all the measures are highly loaded on constructs and having a minimum value of 0.702, which implies that all the measures significantly affect their respective constructs, and all the hypothesis which were formulated are accepted. This study also shows the correlation between the constructs that help in developing strategic and tactical plans which eventually helps in achieving the relative advantages over competitors and gives a competitive edge to the Indian MSMEs.

References

1. Adams R, Bessant J, Phelps R (2006) Innovation management measurement: a review. Int J Manag Rev 8(1):21–47. https://doi.org/10.1111/j.1468-2370.2006.00119.x
2. Hoffman K, Parejo M, Bessant J, Perren L (1998) Small firms, R&D, technology and innovation in the UK: a literature review. Technovation 18(1):39–55. https://doi.org/10.1016/s0166-4972(97)00102-8
3. Nonaka I, Takeuchi H (1995) The knowledge creating company: how Japanese companies create the dynamics of innovation. Oxford University Press, New York
4. Page AL, Schirr GR (2008) Growth and development of a body of knowledge: sixteen years of new product development research, 1989–2004. J Prod Innov Manag 5(3):233–248
5. Blackler F (1995) Knowledge, knowledge work and organizations: an overview and interpretation. Organ Stud 16:1021–1046

6. Davis MC (1998) Knowledge management. Inf Strat Execut J 15:11–22
7. Tsai W (2001) Knowledge transfer i nextraorganizationall networks: effects of network position and absorptive capacity on business unit innovation and performance. Acad Manag J 44:996–1004
8. Hamel G, Getz G (2004) Funding growth in an age of austerity. Harv Bus Rev 82(7/8):76–84
9. Cassell C, Symon G (2006) Taking qualitative methods in organization and management research seriously. Qualit Res Organiz Manag 1(1):4–12
10. Autio E (1997) New, technology-based firms in innovation networks: symplectic and generative impacts. Res Policy 26(3):263–281
11. Agrawal A (2006) Engaging the inventor: exploring licensing strategies for university inventions and the role of latent knowledge. Strateg Manag J 27(1):63–79
12. Bougrain F, Haudeville B (2002) Innovation collaboration and SMEs internal research capacities. Res Policy 31:735–747
13. Packham G, Pickernell D, Brooksbank D (2010) Introduction: the changing role funiversiies in nknowelge egeneratono dissemination and commercialisation. Int J Entrep Innov 11(4):261–263
14. Lee J, Win HN (2004) Technology transfer between university research centers and industry in Singapore. Technovation 24(5):433–442
15. Giovannoni E, Pia Maraghini M (2013) The challenges of integrated performance measurement systems. Acc Aud Account J 26(6):978–1008. https://doi.org/10.1108/aaaj-04-2013-1312
16. Tidd J (2010) Gaining momentum. Imperial College Press, London
17. Cooper R (2005) Product leadership. Basic Books, New York
18. Cooper R (2001) Excelling under pressure: increasing your energy for leadership and innovation in a world of stress, change and unprecedented opportunities. Strat Lead 29(4):15–20. https://doi.org/10.1108/eum0000000005751
19. Faems D, Janssens M, Neyens I (2012) Alliance portfolios and innovation performance. Group Org Manag 37(2):241–268. https://doi.org/10.1177/1059601112441246
20. Eriksson K, Johanson J, Majkgård A, Sharma DD (1997) Experiential knowledge and cost in the internationalization process. J Int Bus Stud 28(2):337–360
21. Knight GA, Cavusgil ST (1996) The born global firm: a challenge to traditional internationalization theory, in ST Cavusgil, TK Madsen. (Eds), 26. http://dx.doi.org/https://doi.org/10.1016/j.sbspro.2014.07.025
22. Rutherford MW, Muse LA, Oswald SL (2006) A new perspective on the developmental model for family business. Fam Bus Rev 19(4):317–333
23. García-Teruel PJ, Martínez-Solano P (2010) Determinants of trade credit: a comparative study of European SMEs. Int Small Bus J 28:215–233
24. McAdam RMS (1999) The process of knowledge management within organizations: a critical assessment of both theory and practice. Knowl Process Manag 6:101–111
25. Ambrosini V, Bowman C (2001) Tacit knowledge: some suggestions for operationalisation. J Manage Stud 38:811–829
26. Zemaitis E (2014) Knowledge management in open innovation paradigm context: high tech sector perspective. Proc Soc Behav Sci 110:164–173
27. Ekvall G (1996) Organizational climate for creativity and innovation. Eur J Work Organ Psy 5:105–123
28. Amabile TM, Conti R, Coon H, Lazenby J, Herron M (1996) Assessing the work environment for creativity. Acad Manag J 39:1154–1184
29. Mathisen GE, Einarsen S (2004) A review of instruments assessing creative and innovative environments within organizations. Creat Res J 16:119–140
30. Dougherty D (1992) A practice-centered model of organizational renewal through product innovation. Strat Manag J Sum Spec Iss 13:77–92
31. Mannan B, Khan J, Khurana S (2012) Information technology: a green supply chain enable emerging paradigms in marketing. In National conference on emerging paradigms in marketing. New Delhi: JMI. http://dx.doi.org/https://doi.org/10.13140/RG.2.1.3556.1205

32. Kinnear TC, Bernhardt KL, Krentler KA (1995) Principles of marketing, 4th edn. Harper Collins, New York
33. Baker M (1999) The IEBM encyclopedia of marketing. International Thomson Business, London
34. Tidd J (2001) Innovation managemen tn ncotnex ftenvironmentlaaorganisation and performance. Int J Manag Rev 3(3):169–183. https://doi.org/10.1111/1468-2370.00062
35. Khurana S, Khan J, Mannan B (2012) Enablers and barriers for implementing technology transfer projects: a study of SMEs in India. Emerging paradigms in marketing. In National conference on emerging paradigms in marketing. New Delhi: JMI. http://dx.doi.org/https://doi.org/10.13140/RG.2.1.1458.9686
36. Rogers E (2003) Diffusion of innovations. Free Press, New York
37. Winer (2007) Marketing management (3rd ed.). Prentice Hall
38. Gopalakrishnan S, Damanpour F (1994) Patterns of generation and adoption of innovation in organisations: contingency models of innovation attributes. J Eng Tech Manage 11(2):95–116. https://doi.org/10.1016/0923-4748(94)90001-9
39. Gopalakrishnan S, Bierly P (2001) Analyzing innovation adoption using a knowledge-based approach. J Eng Tech Manage 18(2):107–130. https://doi.org/10.1016/s0923-4748(01)00031-5
40. Hameed M, Counsell S, Swift S (2012) A conceptual model for the process of IT innovation adoption in organisations. J Eng Tech Manage 29(3):358–390. https://doi.org/10.1016/j.jengtecman.2012.03.007
41. Mannan B, Khan J, Khurana S (2013) Enablers and barrie s oMK innproject-based organization, sustainability and development. In International Conference on Rural Innovation, Capacity Building, Knowledge Management, Entrepreneurship and Technology (ICRICKET). New Delhi. http://dx.doi.org/https://doi.org/10.13140/RG.2.1.3873.1762
42. Dahnil M, Marzuki K, Langgat J, Fabeil N (2014) Factors influencing SMEs adoption of social media marketing. Procedia Soc Behav Sci 148:119–121
43. Anuar J, Musa M, Khalid K (2014) Smartphone's application adoption benefits using mobile hotel reservation system (MHRS) among 3 to 5-star city hotels in Malaysia. Procedia Soc Behav Sci 130:552–557. https://doi.org/10.1016/j.sbspro.2014.04.064
44. Abed S, Dwivedi Y, Williams M (2015) Social media as a bridge to e-commerce adoption in SMEs: a systematic literature review. Mark Rev 15(1):39–57. https://doi.org/10.1362/146934715x14267608178686
45. Mannan B, Khurana S, Haleem A (2015) Technological Innovation challenges and opportunities in India and the developing countries. Annual IEEE India Conf (INDICON) 2015:1–6. https://doi.org/10.1109/INDICON.2015.7443854
46. Robson PJA, Bennett RJ (2000) SME growth: the relationship with business advice and external collaboration. Small Bus Econ 15(3):193–208
47. Pickernell D, Packham G, Brooksbank D, Jones P (2010) A recipe for what? UK universities, enterprise and knowledge transfer: evidence from the federation of small businesses 2008 survey. Intern J Entrep Innov 11(4):265–272
48. Khurana S, Mannan B, Haleem A (2014) Integrating innovation with sustainability: a study of practices/status for Indian manufacturing industries (SMEs). In AGBA 11th World Congress Conference. Delhi: IIT. http://dx.doi.org/https://doi.org/10.13140/RG.2.1.2824.6005
49. Kovacic A (2007) Benchmarking the Slovenian competitiveness by system of indicators. Benchmark Int J 14(5), 553–574
50. Gomes C, Yasin M (2011) A systematic benchmarking perspective on performance management of global small to medium-sized organizations. Benchmark Int J 18(4), 543–562. http://dx.doi.org/https://doi.org/10.1108/14635771111147632
51. Madsen D, Slåtten K, Johanson D (2017) The emergence and evolution of benchmarking: a management fashion perspective. Benchmark Int J 24(3), 775–805. http://dx.doi.org/https://doi.org/10.1108/bij-05-2016-0077
52. Birkinshaw J (2010) Reinventing management. John Wiley & Sons Ltd., Chichester
53. Vaccaro IG, Volberda HW, Van Den Bosch FAJ (2012) Management innovation in action: the case of self-managing teams. In: Pitsis TS, Simpson A, Dehlin E (eds) Handbook of organizational and managerial innovation. Edwar Elgar, Cheltenham, pp 138–162

54. Mihalache OR (2012) Stimulating firm innovativeness: probing the interrelations between managerial and organizational determinants. Rotterdam: Erasmus Research Institute of Management (ERIM)
55. Mannan B, Jameel S, Haleem A (2013) Knowledge management in project management. Saarbrücken: LAP LAMBERT Academic Publishing. http://dx.doi.org/https://doi.org/10.13140/RG.2.1.4921.7527
56. Cooper RG, Edgett SJ, Kleinschmidt EJ (2001) Portfolio Management for New
57. Wadhwa A, Phelps C, Kotha S (2016) Corporate venture capital portfolios and firm innovation. J Bus Ventur 31(1):95–112. https://doi.org/10.1016/j.jbusvent.2015.04.006
58. García-Teruel PJ, Martínez-Solano P (2007) Effects of working capital management on SME profitability. Int J Manag Fin 3(2):164–177
59. Sharma OP, Bambawale OM, Gopali JB, Bhaga S, Yelshett S, ingh SSK, Anan R, Singh OM (2011) Field guide Mung bean and Urd bean. Government of India, Department of Agricultural and co-operation, NCIPM, ICAR, New Delhi, India. Research- Technology Management, 47, 50–59
60. Newman A, Gunessee S, Hilton B (2012) Applicability of financial theories of capital structure to the Chinese cultural context: a study of privately owned SMEs. Int Small Bus J 30(1):65–83
61. Bhunia A (2012) Association between default behaviors of SMEs and the credit facets of SMEs owners. Eur J Busin Manag 4(1)
62. Xia L (2014) Analysis on the small and medium-sized enterprise financing bank loans availability. Appl Mech Mater 687–691:4799–4802. https://doi.org/10.4028/www.scientific.net/amm.687-691.4799
63. Mannan B, Khurana S, Haleem A (2016) Modeling of critical factors for integrating sustainability with innovation for Indian small- and medium-scale manufacturing enterprises: an ISM and MICMAC approach. Cogent Busin Manag 3(1):1–15. https://doi.org/10.1080/23311975.2016.1140318
64. Bentler P, Chou C (1987) Practical issues in structural modeling. Sociol Methods Res 16(1):78–117. https://doi.org/10.1177/0049124187016001004
65. Hair JF, Anderson RE, Tatham RL, Black WC (1995) Multivariate data analysis with readings, 4th edn. Prentice Hall Publishers, New Jersey
66. Bard JF, Balachandra R, Kaufmann PE (1988) An interactive approach to R&D project selection and termination. IEEE Trans Eng Manage 35:139–146
67. Brenner MS (1994) Practical R&D project prioritization. Res Techn Manag 37:38–43
68. Cebon P, Newton P (1999) Innovation in firms: towards a framework for indicator development. Melbourne Business School
69. Cooper R (1999) The invisible success factors in product innovation. J Prod Innov Manag 16(2):115–133. https://doi.org/10.1111/1540-5885.1620115
70. Czarnitzki D, Kraft K (2004) Firm leadership and innovative performance: evidence from seven EU countries. Small Bus Econ 22:153–173
71. Damanpour F, Schneider M (2006) Phases of the adoption of innovation in organizations: effects of environment, organization, and top managers. Br J Manag 17:215–236
72. Escribá-Esteve A, Sánchez-Peinado L, Sánchez-Peinado E (2008) Moderating influences on the firm's strategic orientation-performance relationship. Int Small Bus J 26(4):463–489
73. Kimberly JR, Evanisko MJ (1981) Organizational innovation: the influence of individual, organizational and contextual factors on hospital adoption of technological and administrative innovations. Acad Manag J 24:689–713
74. Kotler P, Trias de Bes F (2003) Lateral marketing, 1st edn. Wiley, Hoboken
75. Lawton Smith H, Bagchi-Sen S (2006) University-industry interactions: the case of the UK biotech industry. Ind Innov 13:371–392
76. OECD (2013) 'Fostering SMEs' Participation in Global Markets: Final Report', Centre for Entreprenurship, SMEs and Local Development
77. Robson PJA, Akuetteh CK, Westhead P, Wright M (2012) Innovative opportunity pursuit, human capital and business ownership experience in an emerging region: evidence from Ghana. Small Bus Econ 39:603–625

78. Temel S, Scholten V, Akdeniz R, Fortuin F, Omta O (201 3) University industry collaborationn in Turkish SMEs: investigation of a U-shaped relationship. Int J Entrep Innov 14(2), 103–115. http://dx.doi.org/https://doi.org/10.5367/ijei.2013.0109
79. Zikmund W, D'Amico M (1993) Instructor's manual to accompany Marketing (1). Minneapolis/ St. Paul, Minn.: West Pub. Co

Integrated MCDM Model for Prioritization of New Electric Vehicle Selection

Sumit Chawla, Praveen Kumar Dwivedi, Manjeet, and Lalit Batra

Abstract This study focused on finding the best electric scooter among the popular six alternatives based on the five criteria. The six alternatives taken in this study are Ola S1 STD, Ola S1 pro, Ather 450x, Simple one, TVS iQube electric, and Hero electric photon HX. The criteria taken in this study are on-road price (Delhi), driving range, battery capacity, maximum speed, motor power, and battery charging time. The specification parameter of the various electric scooters are collected and fuzzy AHP is applied for weights calculation and the TOPSIS method have been used for finding the ranking of alternatives. The best alternative identified from this research is a simple one electric scooter. This integrated model can be incorporated into any automotive industry for vehicle selection, and the results are beneficial for all new customers. The sensitivity analysis is also been performed for checking the robustness and stability of this model.

Keywords Battery charging time · Driving range · Electric scooter · Fuzzy AHP · TOPSIS

1 Introduction

Since pollution is increasing day by day in all parts of India. It is a matter of concern especially in Delhi NCR of India because Delhi is the most polluted capital of India. One of the biggest reasons for the pollution is the emissions emitted by the vehicles. Emissions released by internal combustion engines are the primary air pollutants. Nowadays there is an immense need for electric vehicles (EVs) to tackle this pollution problem. Govt. is also providing subsidies and tax relaxation for buying EVs to mitigate air pollution and reduce the dependency on crude oil [1]. So, people are switching from petrol/ diesel vehicles to EVs.

S. Chawla (✉) · P. K. Dwivedi · Manjeet · L. Batra
Department of Applied Science (Mechanical), Bharati Vidyapeeth College of Engineering, Delhi 110063, India
e-mail: chawlasumit4@gmail.com

H. Kumar et al. (eds.), *Recent Advances in Intelligent Manufacturing*, Lecture Notes in Mechanical Engineering,
https://doi.org/10.1007/978-981-99-1308-4_2

To reduce global pollution and CO_2 emissions, EV is the need of this era. 1/3rd of crude oil is being consumed by the transport sector of India [2]. CO_2 emissions are mainly generated from the transport sector which can be controlled by the adoption of EVs. With this adoption, we can convert the transport sector into a green sector [3]. Switching from internal combustion engine-based vehicles to EVs is challenging due to the high cost of EVs and limited charging infrastructure. The govt. of India is giving subsidies for buying EVs and increasing charging infrastructure for making easy adoption of EVs on road by 2030 [4]. Nowadays, every automotive manufacturer is pressurized to make EVs. Due to this, there is a variety of options available in this segment which leads to difficulty in the selection of EVs for a middle-class man. So, in this study, an integrated MCDM model is developed for the selection of the best electric scooter based on the criteria. Various criteria can be taken for identifying the best EV such as on-road price (lakh), driving range (km/charge), battery capacity (kwh), maximum speed (Kmph), motor power (W), battery charging time (hours), number of charging station, and number of service stations. The five important criteria taken for this research are on-road price (P1), driving range (P2), maximum speed (P3), motor power (P4), and battery charging time (P5). The six alternatives taken in this study are Ola S1 STD (EV_1), Ola S1 pro (EV_2), Ather 450x (EV_3), Simple one (EV_4), TVS iQube electric (EV_5), and Hero electric photon HX (EV_6). Data of specification parameters of an electric scooter is shown in Fig. 1. The data of these specification parameters have been collected from bikedekho.com as shown in Table 1.

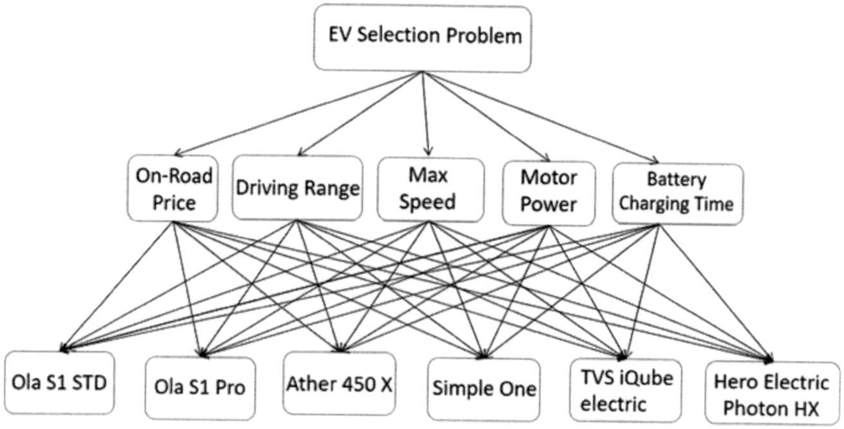

Fig. 1 Hierarchy of EV selection problem

Table 1 Specification parameters of an electric scooter

Best EVs in 2022	On-road price (Delhi) (lakh Rs)	Driving range (km/charge)	Max speed (Kmph)	Motor power (W)	Battery charging time (hours)
Ola S1 STD	0.851	121	90	8500	4.48
Ola S1 Pro	1.2	181	115	8500	4.48
Ather 450x	1.19–1.38	116	80	6000	3.58
Simple one	1.09–1.44	203	105	4500	3.44
TVS iQube electric	1.61–1.66	110–145	78	4400	4.1

2 Literature Review

EV had been already developed in the nineteenth century. But in 1912, when a large stock of crude oil developed, the price of crude oil decreased rapidly which leads to the rapid development of internal combustion engine vehicles. Because of this, EVs disappeared in 1920 [2]. But nowadays, crude oil prices have increased again due to the shortage and effect of the war between Russia and Ukraine. So EV is again in demand in this era. Generally, three types of vehicles run on Indian roads such as conventional vehicles, EVs, and hybrid vehicles. Since India is planning to achieve 30% electric vehicles out of total vehicles running in India by 2030.

S. Chawla et al. have prioritized the various new bike alternatives using the TOPSIS-MOORA model [5]. This integrated MCDM model can be applied to a new vehicle as well as for material selection [6]. S. Mall applied fuzzy AHP for various criteria related to alternate technologies finding the weights of criteria of alternate technologies, and the VIKOR method is utilized for ranking these technologies [7]. A. Ghosh et al. used the fuzzy AHP for weights calculation of criteria related to EV site selection and integrated fuzzy TOPSIS-COPRAS model is used for ranking for EV charging station [8] Rakesh Kumar et al. studied challenges faced in adopting EV in India and also discussed Govt. national electric mobility mission plan 2020 [3]. Anil Khurana et al. examined the various factors which are responsible for the adoption of EVs in India using structured equation modeling [4].

Not only India, but other countries are also forcing EV implementation. Chinese EV Manufacturers Company announced that after 2025, they will not manufacture and sell vehicles other than EVs [9]. EV adoption will enhance air quality, reduce greenhouse gases, and also reduce the country's dependency on crude oil [10].

3 Research Methodology

This paper calculates the weights of these criteria using Fuzzy AHP and identifies the best EV using the TOPSIS methodology.

3.1 Fuzzy AHP

The first step in fuzzy AHP is the development of a pairwise comparison matrix using a standard scale of relative importance. Step 2 is to convert the pairwise comparison matrix into a fuzzified form. Step 3 is to find a fuzzy geometric mean value (g_{m1}) using Eq. 1, where (u_1, v_1, w_1) and (u_2, v_2, w_2) are two fuzzy numbers and n is the number of criteria. Then, fuzzy weights $\tilde{w}_i$ are determined using Eq. 2. Finally, fuzzy weights defuzzification is done and normalized weights nw_i is determined.

$$g_{m1} = (u_1, v_1, w_1) \otimes (u_2, v_2, w_2) = \left((u_1 * u_2)^{\frac{1}{n}}, (v_1 * v_2)^{\frac{1}{n}}, \left(w_1 * w_2\right)^{\frac{1}{n}}\right) \quad (1)$$

$$\tilde{w}_i = g_{mi} \otimes (g_{m1} \oplus g_{m2} + \cdots \oplus g_{mn})^{-1} \quad (2)$$

3.2 Topsis

TOPSIS is a well-known MCDM technique. The First step of TOPSIS is to develop decision matrix X using m, the number of alternatives, and n, the number of criteria as given in Eq. 3. In step 2, decision matrix is converted into normalized decision matrix (R_{ij}) using Eq. 4, where i = 1, 2, 3 … m and j = 1, 2, 3 … n. Step 3 includes the determination of weighted decision matrix d_{ij} using Eq. 5. Then the positive ideal solution A^+ and negative ideal solution A^- is calculated using Eqs. 6 and 7, where K = beneficial based attributes and K' are non-beneficial based attributes [11]. Based on the A^+ and A^- values, euclidean distances ($S^+ \& S^-$) of each alternative is determined using Eqs. 8 and 9. Finally, the ranking of alternatives is found based on the relative closeness values C_i given by Eq. 10.

$$X = \left[x_{ij}\right]_{m \times n} \quad (3)$$

$$R_{ij} = \frac{x_{ij}}{\sqrt{\sum_{i=1}^{m} x_{ij}^2}} \quad (4)$$

$$d_{ij} = w_j \times R_{ij} \quad (5)$$

$$A^+ = \left\{d_1^+, d_2^+ \ldots, d_n^+\right\}, \text{where} : d_j^+ = \left\{\left(\text{maxi}\left(d_{ij}\right)\text{ifj} \in K\right); \left(\text{mini}\left(d_{ij}\right)\text{ifj} \in K'\right)\right\} \quad (6)$$

$$A^- = \left\{d_1^-, d_2^- \ldots, d_n^-\right\}, \text{where} : d_j^- = \left\{\left(\text{mini}\left(d_{ij}\right)\text{ifj} \in K\right); \left(\text{maxi}\left(d_{ij}\right)\text{ifj} \in K'\right)\right\} \quad (7)$$

$$S^{+} = \sqrt{\sum_{j=1}^{n} (d_j^{+} - d_{ij})^2} \quad (8)$$

$$S^{-} = \sqrt{\sum_{j=1}^{n} (d_j^{-} - d_{ij})^2} \quad (9)$$

$$C_i = \frac{S_i^{-}}{S_i^{+} + S_i^{-}}, 0 \le C_i \le 1 \quad (10)$$

4 Application of MCDM Model

The weights of the criteria are calculated by the Fuzzy AHP approach. After weights calculation, these weights are used for identifying the best alternatives using the TOPSIS method. Table 2 represents the pairwise comparison matrix of the various criteria related to the EV selection problem. Table 3 shows the fuzzified form of this pairwise comparison matrix. The normalized weights calculated for this selection problem are shown in Table 4. Table 5 determines the decision matrix for EV selection using the TOPSIS methodology. Table 6 calculates the normalized form of this decision matrix. Then, the weighted decision matrix is obtained in Table 7. Finally, the ranking of EV is obtained as shown in Table 8.

Table 2 Pairwise comparison matrix

	P_1	P_2	P_3	P_4	P_5
P_1	1	1/5	3	5	1/5
P_2	5	1	7	9	1/3
P_3	1/3	1/7	1	3	1/7
P_4	1/5	1/9	1/3	1	1/9
P_5	5	3	7	9	1

Table 3 Fuzzified Pairwise comparison matrix

	P_1	P_2	P_3	P_4	P_5
P_1	(1, 1, 1)	(1/6, 1/5, 1/4)	(2, 3, 4)	(4, 5, 6)	(1/6, 1/5, 1/4)
P_2	(4, 5, 6)	(1, 1, 1)	(6, 7, 8)	(9, 9, 9)	(1/4, 1/3, 1/2)
P_3	(1/4, 1/3, 1/2)	(1/8, 1/7, 1/6)	(1, 1, 1)	(2, 3, 4)	(1/8, 1/7, 1/6)
P_4	(1/6, 1/5, 1/4)	(1/9, 1/9, 1/9)	(1/4, 1/3, 1/2)	(1, 1, 1)	(1/9, 1/9, 1/9)
P_5	(4, 5, 6)	(2, 3, 4)	(6, 7, 8)	(9, 9, 9)	(1, 1, 1)

Table 4 Normalized weights using Fuzzy AHP

	$\tilde{r}_i$	$\tilde{w}_i$	w_i	nw_i
P_1	(0.74, 0.90, 1.08)	(0.079, 0.111, 0.156)	0.116	0.113
P_2	(2.22, 2.53, 2.93)	(0.239, 0.313, 0.423)	0.325	0.316
P_3	(0.38, 0.46, 0.56)	(0.040, 0.056, 0.081)	0.059	0.057
P_4	(0.22, 0.24, 0.27)	(0.023, 0.029, 0.039)	0.031	0.030
P_5	(3.36, 3.93, 4.44)	(0.362, 0.487, 0.641)	0.497	0.483

Table 5 Decision matrix for EV selection using TOPSIS

	P_1	P_2	P_3	P_4	P_5
EV_1	0.851	121	90	8500	4.48
EV_2	1.2	181	115	8500	4.48
EV_3	1.285	116	80	6000	3.58
EV_4	1.265	203	105	4500	3.44
EV_5	1.635	127.5	78	4400	4.1
EV_6	0.742	108	42	1200	5

Table 6 Normalized decision matrix using TOPSIS

	P_1	P_2	P_3	P_4	P_5
EV_1	0.2895	0.3356	0.4169	0.5711	0.4339
EV_2	0.4083	0.5020	0.5327	0.5711	0.4339
EV_3	0.4372	0.3217	0.3706	0.4031	0.3467
EV_4	0.4304	0.5630	0.4864	0.3023	0.3332
EV_5	0.5563	0.3536	0.3613	0.2956	0.3971
EV_6	0.2524	0.2996	0.1946	0.0806	0.4843

Table 7 Weighted normalized decision matrix using TOPSIS

	P_1	P_2	P_3	P_4	P_5
EV_1	0.0327	0.1061	0.0238	0.0171	0.2096
EV_2	0.0461	0.1586	0.0304	0.0171	0.2096
EV_3	0.0494	0.1017	0.0211	0.0121	0.1675
EV_4	0.0486	0.1779	0.0277	0.0091	0.1609
EV_5	0.0629	0.1117	0.0206	0.0089	0.1918
EV_6	0.0285	0.0947	0.0111	0.0024	0.2339

Table 8 EV ranking obtained using TOPSIS method

Alternatives	S^+	S^-	C_i	Ranking
EV_1	0.0723	0.0376	0.3423	3
EV_2	0.0261	0.0704	0.7295	2
EV_3	0.0798	0.0206	0.2053	6
EV_4	0.0218	0.0863	0.7983	1
EV_5	0.0756	0.0206	0.2140	5
EV_6	0.0867	0.0343	0.2836	4

Table 9 Sensitivity analysis

Weights criteria	Weights order/value	Ranking order
Fuzzy AHP weights	P5 > P2 > P1 > P3 > P4	EV4 > EV2 > EV1 > EV6 > EV5 > EV3
Equal weights	P1 = P2 = P3 = P4 = P5	EV2 > EV1 > EV4 > EV3 > EV5 > EV6
50% Beneficial & 50% Non-beneficial	P1 = P5 = 0.25, P2 = P3 = P4 = 0.167	EV2 > EV1 > EV4 > EV3 > EV6 > EV5
70% Beneficial & 30% Non-beneficial	P1 = P5 = 0.15, P2 = P3 = P4 = 0.233	EV2 > EV1 > EV4 > EV3 > EV5 > EV6
80% Beneficial & 20% Non-beneficial	P1 = P5 = 0.1, P2 = P3 = P4 = 0.267	EV2 > EV1 > EV4 > EV3 > EV5 > EV6

5 Results and Discussion

From Table 4, it is concluded that the battery charging time has the maximum weightage of 48.30%, and motor power has the least weightage of 3.1%. The weightage order of the criteria obtained using the fuzzy AHP is as follows: battery charging time (48.30%) > driving range (31.60%) > on road price (11.30%) > maximum speed (5.7%) > motor power (3.1%). The EV ranking obtained using the TOPSIS method is as follows: $EV_4 > EV_2 > EV_1 > EV_6 > EV_5 > EV_3$. If equal weights (60% beneficial & 40% non-beneficial) are given to all criteria, then EV ranking obtained is $EV2 > EV1 > EV_4 > EV_3 > EV_5 > EV_6$. Table 9 shows the sensitivity analysis of the given model. On varying the weights of beneficial and non-beneficial criteria, then also we got the same ranking order ($EV2 > EV1 > EV_4 > EV_3 > EV_5 > EV_6$). This analysis shows that the model is robust and stable for varying criteria.

6 Conclusion

The most important criteria identified using fuzzy AHP for new electric vehicle selection are battery charging time and driving range. The Best EV obtained using the TOPSIS methodology is a simple one (EV_4) based on the weightage obtained from fuzzy AHP. The second best EV identified is Ola S1 pro (EV_2). The least preferable EV among the alternatives is Ather 450 x (EV_3). The results can be varied based on the varying weights. If equal weightage is given to all criteria, then the best EV identified is Ola S1 pro (EV_2). The sensitivity analysis is also performed to determine the stability and robustness of the model concerning the criteria taken.

This study can be further extended by taking more alternatives and criteria. For further research, more MCDM techniques other than TOPSIS such as MOORA, ELECTRE, VIKOR, COPRAS, PROMETHEE, etc. can be applied and comparative analysis can be done.

References

1. Sun X, Li Z, Wang X, Li C (2019) Technology development of electric vehicles: a review. Energies (Basel) 13(1):1–29
2. Kumar U, Chakrabarti A (2019) Barriers in implementation of electric vehicles in India. Int J Electric Hybrid Veh 11(3):195–204
3. Rakesh Kumar MA, Padmanaban S (2019) Electric vehicles for India: overview and challenges. IEEE India Info 14(2):139–142
4. Khurana A, Kumar VVR, Sidhpuria M (2020) A study on the adoption of electric vehicles in India: the mediating role of attitude. Vision 24(1):23–34
5. Chawla S, Agrawal S, Singari RM (2019) Integrated Topsis-Moora model for prioritization of new bike selection. In: Prasad A, Gupta S, Tyagi R (eds) Advances in engineering design, LNME. Springer, Singapore, pp 755–765
6. Chawla S, Singari RM (2021) Integrated TOPSIS-PROMETHEE-MOORA model for material selection of crankcase cover. Ind J Eng Mater Sci 28:454–461
7. Mall S, Anbanandam R (2022) A: fuzzy analytic hierarchy process and VIKOR framework for evaluation and selection of electric vehicle charging technology for India. Transp Dev Econ 8:14
8. Ghosh A et al (2021) Application of hexagonal fuzzy MCDM methodology for site selection of electric vehicle charging station. Mathematics 9(4):1–28
9. Jhunjhunwala A, Kaur P, Mutagekar S (2018) Electric vehicles in India: a novel approach to scale electrification. IEEE Electr Mag 6(4):40–47
10. Bhaskar K, Pathak M, Shukla PR, Dhar S (2014) Electric vehicles scenarios and a roadmap for India. UNEP DTU Partnership, Copenhagen
11. Wang L, Chu J, Wu J (2007) Selection of optimum maintenance strategies based on a fuzzy analytic hierarchy process. Int J Prod Econ 107:151–163

LSTM Based Predictive Maintenance Approach for Zero Breakdown in Foundry Line Through Industry 4.0

T. Roosefert Mohan, J. Preetha Roselyn, and R. Annie Uthra

Abstract Achieving the customer demand as per the required delivery schedule with the available resources are always a challenging task in industries. Any unplanned catastrophic equipment failure may decrease the productivity, makes huge damage to the equipment, raises the repair cost with time, and affects delivery schedule. If these breakdowns are predicted in advance, the breakdown can be addressed before its occurrence and the demand supply chain can be met. Total Productive Maintenance (TPM) is one of the essential operational excellence tools used in industries to utilize the existing resources of a plant in an optimal way. The conventional periodic Time-Based Maintenance (TBM) and predictive Condition Based Maintenance (CBM) approach of TPM in Industry 3.0 is time consuming and not accurate enough to achieve zero down time. The Artificial Intelligence (AI) based TPM Condition Based Maintenance (CBM) approach through digital data-oriented Industry 4.0 transformation can well predict the Remaining Useful Life (RUL) of the part or assembly and based on the predictions, the breakdown is addressed in advance to eliminate breakdown by 95%. Long Short-Term Memory (LSTM) based deep learning network was developed as a regression forecasting model to predict the RUL of the part or assembly.

Keywords Long Short-Term memory network · Total productive maintenance · Predictive maintenance · Condition based maintenance · Industry 4.0

T. Roosefert Mohan (✉) · J. Preetha Roselyn
Department of Electrical and Electronics Engineering, SRM Institute of Science and Technology, Kattankulathur, India
e-mail: roosefert@gmail.com

R. Annie Uthra
Department of Computational Intelligence, SRM Institute of Science and Technology, Kattankulathur, India

H. Kumar et al. (eds.), *Recent Advances in Intelligent Manufacturing*,
Lecture Notes in Mechanical Engineering,
https://doi.org/10.1007/978-981-99-1308-4_3

1 Introduction

The rapid growth in automobile consumption, the demand in automobile manufacturing industry is growing rapidly. [1]. Any breakdown of a machine may lead to stoppage or reduction in speed of machine or production, delay in delivery schedule, decrease in production, increase in operational cost and affect the product quality. As per the studies from International Society of Automation, the global cost due to downtime per year is $ 647 Billion. Studies claim that downtime can be decreased by 30–50% in machine using effective predictive maintenance and 20–40% of machine efficiency can be increased.

Total Productive Maintenance (TPM) is one of the maintenance approaches used in industries to reduce down time [2]. Predictive maintenance can be achieved effectively using IoT [3]. As per Planned Maintenance (PM) check sheet schedule, parts are replaced over a scheduled time of interval in Time Base Maintenance (TBM) approach to eliminate breakdown. In the conventional, Autonomous Maintenance Pillar activity, Clean Lubricate Inspect and Tightening (CLIT), and PM pillar activity like TBM and Condition Based Maintenance (CBM) are carried out as per schedule to achieve zero downtime. The human error in data capturing and the poor skill of the maintenance team leads to breakdown. As the CBM activity is time based and not a continuous one, there is a chance of breakdown before the next TBM or CBM. The proactive maintenance approach is used to prevent the failure of equipment before the occurrence by monitoring the equipment healthy status and perform the maintenance activity prior to the predefined breakdown related parameter threshold limit [4]. Conventional maintenance techniques have inaccurate mathematical degradation processes and also manual intervention [3]. The causes of breakdown can be analyzed, addressed, and eliminated through TPM based approach in the existing Industry 3.0 [5]. The Predictive Maintenance activities are based on the data collected to check the healthy status of the equipment [6]. The data may misguide the maintenance personal due to the inaccurate collected data of poor skilled person. Though a 100% adherence of maintenance related checklists are achieved in industry 3.0 using systematic TPM, it is difficult to achieve zero downtime since some of the key signature parameters related to breakdown are not monitored preciously whereas intelligent, machine learning based CBM may overcome these issues [7]. Artificial Intelligence (AI) based TPM may reduce breakdown and improved the efficiency of the machine [8].

In Industry 4.0, AI or Machine Learning (ML) based predictive maintenance approach, the signature parameter related to breakdown or defect are analyzed, predicted, and trigger alert to the authenticated through servers and clouds to perform proactive approach before the breakdown [9] [10]. The vibration signal captured from a single point of an assembly can be used for predicting the RUL of equipment [11]. The benefits of IoT based automatic predictive maintenance reduces unnecessary maintenance and TBM by predicting the Remaining Useful Life (RUL) well ahead of failure using automatic alert generation to the authenticated persons for proactive maintenance activity. The collected data of signature parameter related to breakdown are processed using ML models like Multi-Layer Perceptron (MLP), Auto Regression Integrated Moving Average (ARIMA) model, and Support Vector Regression (SVR) model [12, 13]. Support Vector Machine (SVM) performs well in classification model with balanced data set [14, 15]. LSTM can be used for regression model to forecast the RUL of parts of equipment efficiently [16]. If more human dependencies are removed, the more accuracy of prediction can be achieved as mentioned in [17]. To run the machine above 20 years of old with zero downtime is very essential and hence the PM pillar needs to be addressed along with AI and Internet of Things (IoT) which enhances the system performance by eliminating the human intervention and error.

In the proposed work, all breakdown phenomenon of a foundry mould making line which restrict zero downtime are captured as per TPM approach, prioritized for prediction based on the result of Pareto analysis, multi variate signature parameters are identified and using LSTM based regression prediction model which is common to all signature parameter data, RUL is predicted for proactive activity to achieve zero breakdown. By making the LSTM model as common to all prediction model instead of multiple software, overloading the server is avoided. The proposed LSTM prediction model is compared with ARIMA as proposed in [18] to achieve the accuracy of prediction. The data captured by the sensors for the signature parameter based on the result of Pareto analysis are fed into the Programmable Logic Controller (PLC) where digitization takes place and forwarded to the SCADA for the purpose of monitoring and controlling. The captured data is sent to the GPU server/cloud to perform artificial intelligence-based LSTM prediction model and to simulate alert. Authenticated can predict the breakdown through alert to eliminate breakdown before the occurrence. The novelty and main contributions of this paper are as follows:

- Identification of critical equipment in a plant using ranking score and corresponding signature parameters using captured loss analysis for cost and time saving Industry 4.0 transformation.
- Collection and monitoring of signature parameters through PLC to the SCADA system and cloud storage.
- Development of multivariant model for predictive maintenance breakdown considering multiple parameters of the machinery till achieving world class Availability Rate.
- Development of LSTM based deep learning prediction model for time series forecasting of signature parameters to achieve zero downtime in a real time application.

2 Proposed Intelligent TPM Approach to Achieve Zero Breakdown

An IoT and AI based prediction model in industry predict the breakdown of equipment by collecting the data of related signature parameters from sensors. In the proposed work, Oil Contamination Level (OCL), hydraulic pump pressure, mixer equipment gearbox assembly vibration and hydraulic oil temperature are the identified key signature parameters in high pressure mould making line. The sensors deployed in the machineries, collect the data and transfer the information to the server. Figure 1 shows the layout of Artificial intelligent based TPM approach towards achieving zero breakdown.

The sensors or Cyber Physical System CPS captures the signals related to key signature parameters and these signals are connected to PLC to get digitized data. SCADA receives the captured data from PLC and send these data to the server then to cloud. The proposed LSTM model is implemented in the server for prediction and alert is triggered to the authorized person for proactive actions. If the breakdown related to particular phenomenon achieves zero downtime, it is standardized else based on the brain storming technique, additional sensors/CPS are deployed to capture the signature parameter to achieve zero down time for the same breakdown phenomenon. The authenticated can fetch any data related to Predictive Maintenance from the cloud for further analysis at anytime from anywhere in his mobile or PC.

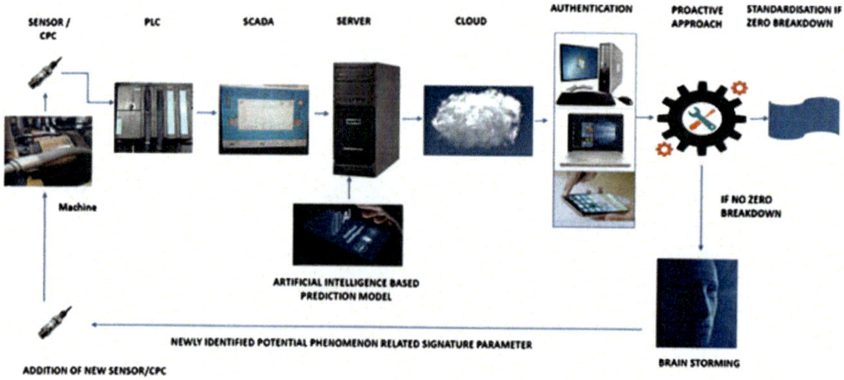

Fig. 1 Architecture of Intelligence based TPM approach

2.1 Break Down Analysis and Brain Storming Technique to Identify Signature Parameters for Industry 4.0 Transformation

From the captured breakdown phenomenon of all breakdown, five why analysis with brain storming techniques to determine the root cause are performed to identify the key signature parameters for digitalization and Industry 4.0 transformation. The identified breakdown creating signature parameter for the year 2017 is shown in Table 1. Based on Table 1, Pareto analysis is done as in Fig. 2, priority is given for the signature parameters related to more time consumed breakdown. Prediction of RUL of parts are addressed based on the priority.

Table 1 depicts the benchmark of breakdown signature parameters taken from the year 2017 in which average breakdown minutes/month and average occurrences/ months are mentioned. From the pareto, 95% of the breakdown causing parameters are initially addressed to achieve zero breakdown. This analysis will be carried out once in six months till the breakdown minutes and occurrences reaches zero. In high pressure moulding line of a foundry, around 80% of the operations are performed by hydraulic oil. The defect in oil like OCL, pressure variation and temperature change the hydraulic oil viscosity which leads to breakdown like malfunctioning of machine positioning, locking operation, and minor stoppages.

Table 1 Major loss capturing signature parameters

Sl no	Cause description	Year 2017 average Mins/ Month	Year 2017 average occurrences/ Month
1	Oil contamination	1268	11
2	Vibration	732	6
3	Oil Pressure	668	4
4	Oil temperature	438	4
5	Electrical loose Connection	128	3
6	DRIVE faults	60	2
7	Lubrication system Failure	33	1
8	PLC and control system Fault	9	0.5
9	Design issues	6	0.3
10	Poor skill	4	0.3
11	Others	2	0.7

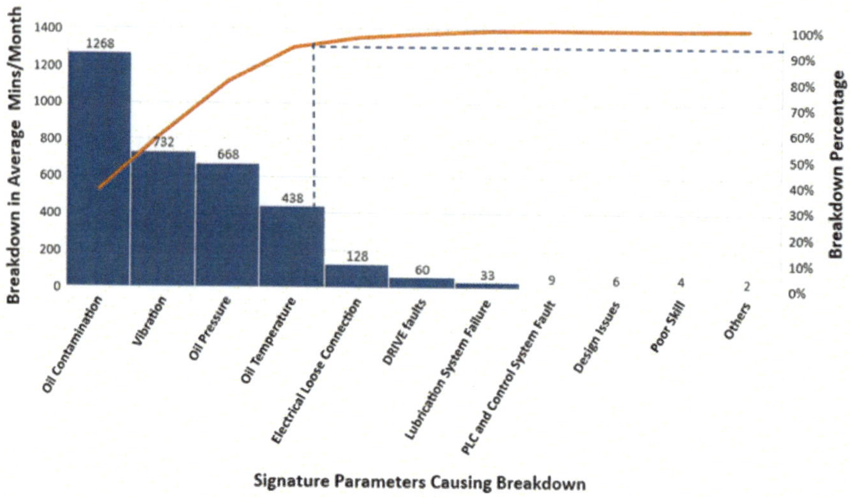

Fig. 2 Pareto analysis for captured loss

2.2 *LSTM Based Prediction*

In the proposed work, LSTM is used as time series regression forecasting model for prediction of RUL. LSTM can store the information over long period of time until it is required and activate the network corresponding to short term memory. Once the data is collected, it is analyzed with LSTM based deep learning model as shown in Fig. 3 to predict the RUL. The collected historical data are trained and tested offline in the best fit model. From the collected online data prediction is achieved using LSTM model.

Figure 4 shows the cell unit of LSTM model which consists of input gate, cell, forget gate, and output gate through which flow of information into and out of the cell takes place. The cell has a linear activation function and a self-loop before the

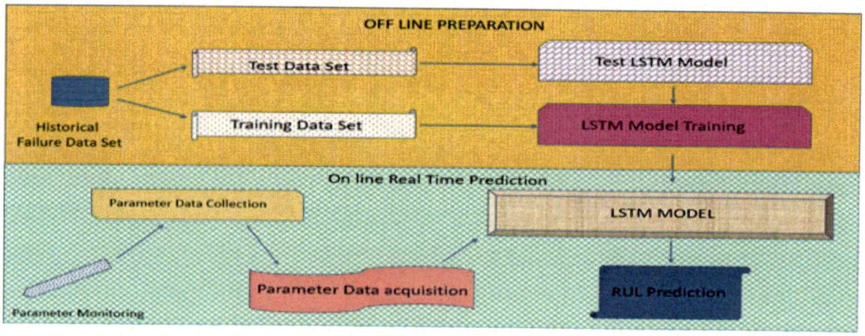

Fig. 3 Proposed prediction model architecture

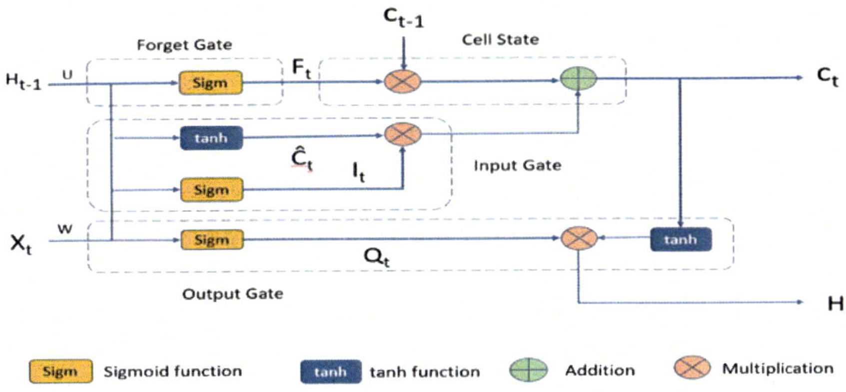

Fig. 4 LSTM cell structure

Table 2 Memory cell status of LSTM network

LSTM memory cell status		
Input gate	Forget gate	Cell state
0	0	Delete the value
1	0	Overwrite the value
0	1	keep previous Value
1	1	Add to previous value

forget gate. The value of forget gate is equal to the weight of self-loop. The forget gate computes a linear function of its inputs and using the logistic activation function gives its output either as 0 or 1. The status of the cell is based on the status of forget gate and input gate as depicted in Table 2. The LSTM cell unit also receives inputs from other cell unit in the network and are added together and sent through a tanh activation function which brings the values between −1 and 1.

The outputs of every cell in Fig. 4 is mentioned as follows:

$$\text{Ft} = \sigma\left(W_f X_t + U_f H_{t-1} + b_F\right) \tag{1}$$

$$\text{It} = \sigma(W_I X_I + U_I H_{t-1} + b_I) \tag{2}$$

$$\text{Qt} = \sigma\left(W_Q X_t + U_Q H_{t-1} + b_Q\right) \tag{3}$$

$$\widehat{\text{C}}_\text{t} = \tanh(W_C X_t + U_C H_{t-1} + b_C) \tag{4}$$

$$\text{Ct} = F_t C_{t-1} + I_t \widehat{\text{C}}_\text{t} \tag{5}$$

$$\mathrm{Ht} = Q_t(tanh(C_t)) \tag{6}$$

The data collected X, at time t is as shown in the Eq. 7.

$$X(t) = X(t-1),\ X(t-2),\ \ldots.,\ X(t-n) \tag{7}$$

where n is look back, C_t is current cell memory, C_{t-1} is previous cell memory, H_t is current cell output, Ht-1 is previous cell input, and Xt is input vector. U and W are weight vectors, and b is bias for the gates.

Stochastic Gradient Descent with Momentum (SGDM) and Adaptive Learning Rate Method (ADAM) are the two optimizers used in the proposed LSTM model for prediction and compared for prediction accuracy. The adjustable network weight used to accelerate the momentum of gradients in the right direction is shown in Eq. 8.

$$\boldsymbol{\theta}_{t+1} = \boldsymbol{\theta}_t - \alpha \Delta \mathrm{E}(\boldsymbol{\theta}_t) + (\boldsymbol{\theta}_t - \boldsymbol{\theta}_{t-1}) \tag{8}$$

$$\mathrm{g} = \Delta \mathrm{E}(\boldsymbol{\theta}_t) \tag{9}$$

where θ is a parameter vector, t is the time step, α is initial learning rate greater than zero, E(θ) is loss function, ΔE(θ) is gradient of the loss function, g is the gradient of current minibatch, and determines the contribution of the previous gradient step to the current iteration.

Adaptive Moment Estimation (ADAM) updates parameter with an added momentum term and keeps elementwise moving average using both of their squared values and parameter gradients. Adam uses the moving averages of gradient $\mathbf{m}_t$ and squared gradient $\mathbf{v}_t$ to update the network parameters as provided below:

$$\mathbf{m}_t = \beta_1 \mathbf{m}_{t-1} + (1 - \beta_1) g_t \tag{10}$$

$$\mathbf{v_t} = \beta_2 \mathbf{v_{t-1}} + (\mathbf{1} - \beta_2) g_t^2 \tag{11}$$

$$\boldsymbol{\theta}_{t+1} = \boldsymbol{\theta}_t - \frac{\alpha m_t}{\sqrt{\mathbf{v}_t} + \varepsilon} \tag{12}$$

where α is learning rate, g_t is gradient at time step t, β_1 is decay factor, and β_2 is square gradient decay factor having the default values of 0.9 and 0.999 respectively. If gradients have more noise, then the moving average of gradient is small. Hence ε is included in the equation to avoid division by zero.

2.3 Performance Metrices for Predictive Maintenance

The Key Performance Indicators (KPIs) related to breakdown is Availability Rate (AR).

$$\text{Availability Rate AR (\%)} = \frac{Machine\ Run\ Time}{planned\ \text{Pr}oduction\ Time} \times 100 \tag{13}$$

$$\text{Where-Machine Run Time} = \text{Planned Production Time} - \text{Machine Stopped Time} \tag{14}$$

To evaluate the performance of prediction model, two evaluation metrics called Root Mean Squared Error (RMSE), Root Mean Squared Percent error (RMSPE) are calculated. RMSE measures the difference between the observed value and predicted value and as follows:

$$\text{RMSE} = \sqrt{\frac{1}{n}\sum_{i=1}^{n}\left(Y_t - \hat{Y}_t\right)^2} \tag{15}$$

RMSPE measures the percent difference between the predicted and observed values as follows:

$$\text{RMSPE} = \sqrt{\frac{1}{n}\sum_{i=1}^{n}\left(\frac{Y_t - \hat{Y}_t}{Y_t}\right)^2} \tag{16}$$

where Y_t denotes the observed value at time t and $\hat{Y}_t$ denotes the predicted value at time t.

3 Results and Discussion

3.1 Description of Equipment in Industrial Machinery

Pareto analysis states that, by addressing OCL, pump oil pressure, oil temperature, and mixer assembly motor vibration 95% of breakdown can be avoided. Figure 5 depicts the installed location of oil contamination, oil pressure, and oil temperature RTD sensors in the hydraulic power pack.

Figure 6 shows that, Hydac pressure transducer type HDA3744-A-400, 0 to 400 bar, accuracy of ± 5%, two wire type is mounted hydraulic pump delivery line to measure 0 to 400 bar pressure and gives proportional output of 4 to 20 mA

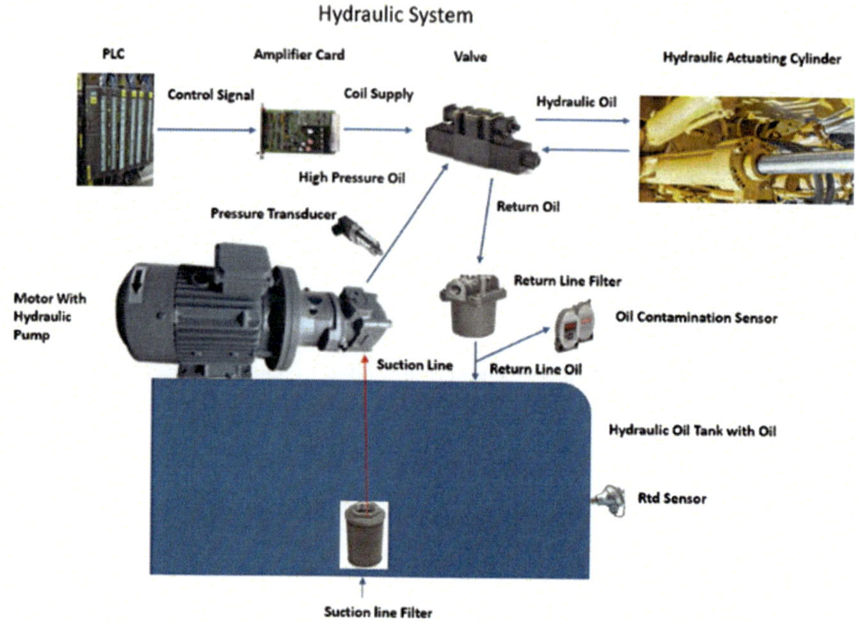

Fig. 5 Sensors position layout in hydraulic power pack

The standard set pressure is 230 bar ± 5% and hence the signal generated by the sensor is equal to 13.2 mA ± 5%.

Figure 7 shows the vibration sensor mounted on the shaft side of 160 kw motor where fluid couplings with pullies and V-belt are used to drive the gear box of mixer. The Siemens make vibration sensor type SIPLUS CMS2000, with a maximum measurement range of 10 gm, sensitivity is 500 mV/gm ± 10% is fixed at the drive end of motor. The threshold value of vertically foot mounted 160 kw, 415 V, 3 phase, 1485 rpm, 50 Hz motor with frame size 315 M, CLASS III with IM2011(IMV15) is 3.6 M/Sec2 as per the standard ISO 10816–1 is 183.542 mV.

Figure 8 shows the OCL sensor make Hydac and type CS1000 with output of 4 to 20 mA provided to measure the OCL in NAS 1638 standard. The cleanliness level

Fig. 6 Hydac pressure transducer HDA3744-A-400 in hydraulic system

Fig. 7 CMS2000 vibration sensor in shaft side of motor

of hydraulic oil from NAS 0 to NAS 14 is proportional to the sensor output 4.8 to 19.2 mA.

Figure 9 shows the PT100 RTD sensor which resistance can vary from 100 to 138.505 Ω for 0 to 1000 C. mounted on the hydraulic oil tank of oil capacity 9000 ltrs as shown in Fig. 9a and SCADA screen as shown in Fig. 9b is arranged to display the four level of oil temperature. This sensor is connected to the PLC for digitalizing and processing purpose.

Fig. 8 Oil Contamination sensor Hydac CS1000 mounted in hydraulic power pack

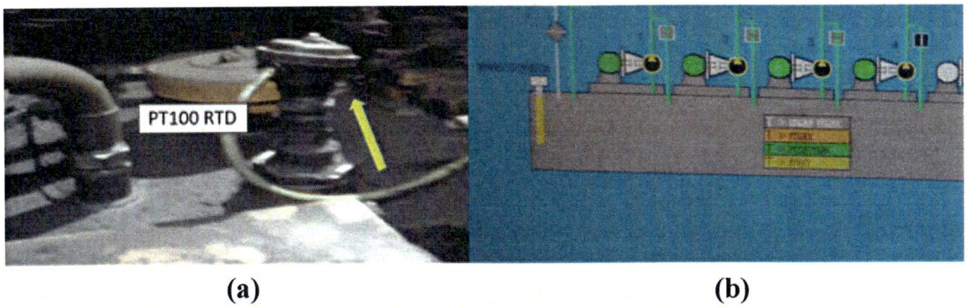

Fig. 9 **a** RTD Sensor in hydraulic Tank **b** SCADA temperature screen

3.2 Experimental Results of Prediction Model

The identified key signature parameters data of OCL, oil pressure, and mixer motor vibration are collected over a time interval of 3 min through PLC. The number of data fed to LSTM for prediction is 64 over a time interval of 192 min. Based on the recommendation of equipment supplier and based on the history of machine breakdown, the threshold values are fixed. The proposed LSTM based prediction model is developed in MATLAB 2018B version which is tested with two training optimizers namely ADAM and SGDM. The training and testing parameters are segregated as 80% and 20% respectively for all the three proposed prediction model of oil contamination, oil pressure, and motor vibration. The result of prediction model is compared with the conventional model, ARIMA model, LSTM ADAM, and SGDM model and the best to fit model is identified in the proposed work. For oil temperature IoT based system is used to predict the oil temperature using PLC, SCADA, server, and cloud to generate alert for the authenticated which induces proactive approach.

3.2.1 Prediction LSTM Results of Hydraulic Oil Pressure

In the proposed LSTM oil pressure prediction model, 80% of the collected offline hydraulic oil pressure data is used for training the model and the remaining 20% of data is used for testing.

The result of LSTM model with ADAM and SGDM optimizers are compared using the metrics RMSE and RMPSE. In the proposed application, the required oil pressure is 230 bar ± 10%. If the pressure exceeded or reduced by 10%, then the system will start malfunctioning which may affect the speed of the machine and quality of the product. Figure 10a shows the ADAM and Fig. 10b shows SGDM training for oil pressure with 80% of training data set. It uses 200 epochs to reduce RMSE. Each epoch uses an iteration of 64 data. The remaining 20% data is used for testing. Figure 11a depicts the predicted hydraulic oil pressure value using the ADAM optimizer and Fig. 11b depicts using SGDM optimizers. ADAM predicts that the oil pressure will reach the threshold value of 12.28 mA which is below 10% of the standard hydraulic pressure of 230 bar in 51,683 min which is equivalent to 35 days, 21 h and 23 min. It means the RUL of hydraulic pump is 35 days, 21 h and 23 min. From SGDM function, it is observed that the OCL will attain the 12.28 mA threshold value in 51,446 min which is equivalent to 35 days, 17 h and 26 min. Hence proactive measure can be taken to address the hydraulic pressure related breakdown before its occurrence.

The predicted values of oil pressure using the proposed LSTM using ADAM and SGDM model is compared against the ground truth and is shown in Fig. 12 from which it can be concluded that the ground truth value is almost equal to the predicted values using ADAM and SGDM.

Table.3 displays the prediction accuracy of proposed model for oil pressure in terms of RUL, prediction error, RMSE and RMSPE. From Table 3, it is clear that

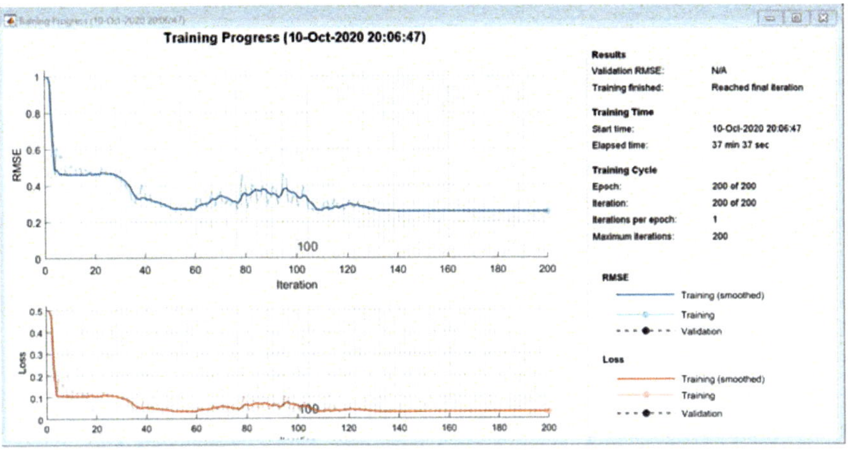

(a)

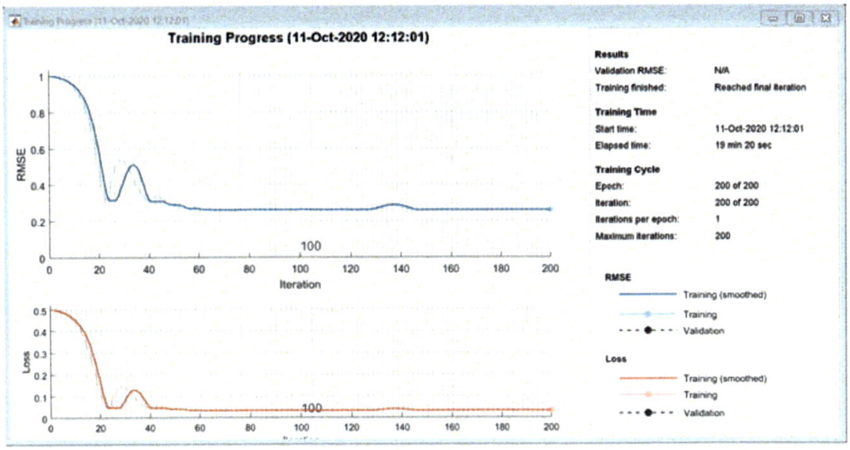

(b)

Fig. 10 **a** ADAM training **b** SGDM training

LSTM with SGDM model performs better than conventional CBM, ARIMA, and ADAM model with a lowest value of RMSE as 69.116 and RMSPE as 0.041.

The prediction accuracy is compared with conventional, ARIMA model, LSTM ADAM and LSTM SGDM model as shown in Table.4 for all three-signature parameters related to oil contamination, oil pressure and motor vibration. For all three parameters, the prediction accuracy of SGDM optimizer accuracy is reaching the requirement.

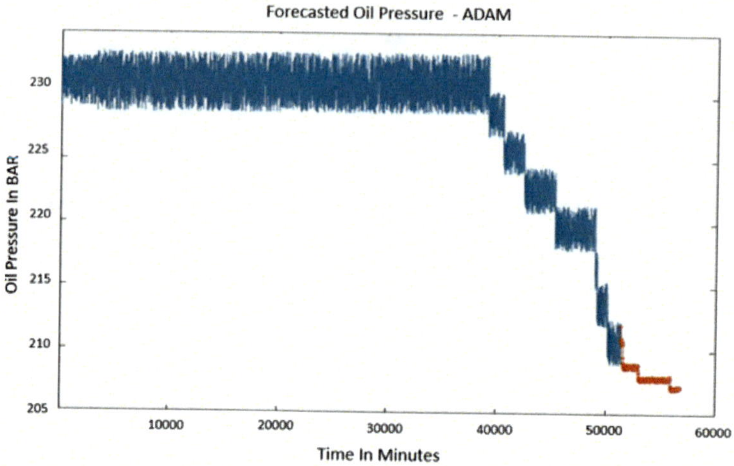

(a)

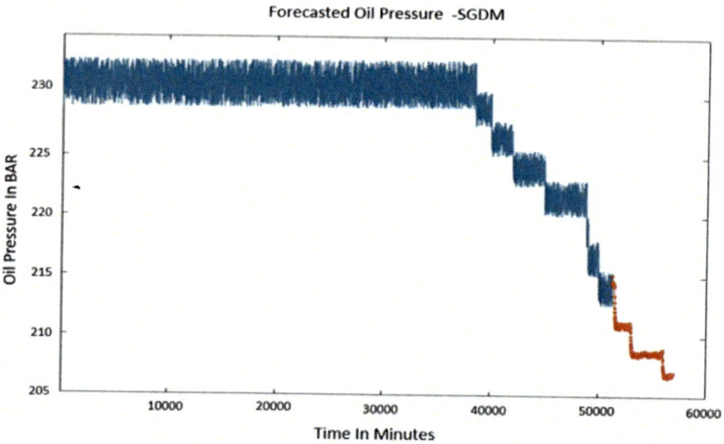

(b)

Fig. 11 **a** ADAM optimizer based prediction **b** SGDM optimize based prediction

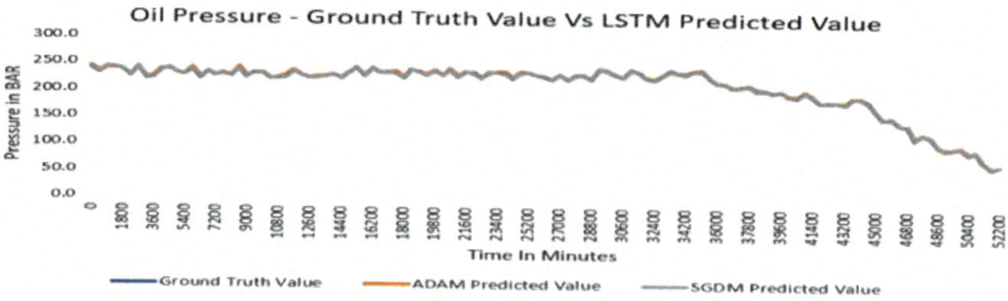

Fig. 12 LSTM predicted value versus ground truth value for oil pressure

Table 3 Prediction Accuracy of proposed model for oil pressure

Rul of hydraulic system pressure and comparision									
Sl No	Conventional TPM based CBM model RUL in mins	Actual RUL in mins	ARIMA RUL in min	LSTM RUL (SGDM) in min	LSTM RUL (ADAM) in min	Predicted error % in conventional model	Predicted error % in ARIMA model	Predicted error % in LSTM (SGDM)	Predicted error % in LSTM (ADAM)
1	150,000	182,423	182,101	182,288	182,652	17.774	−0.177	−0.074	0.126
2	150,000	99,526	99,788	99,511	99,406	−50.714	0.263	−0.015	−0.121
3	150,000	176,394	176,132	176,387	176,191	14.963	−0.149	−0.004	−0.115
4	150,000	167,832	168,096	167,901	167,648	10.625	0.157	0.041	−0.110
5	150,000	78,676	78,942	78,701	78,792	-90.655	0.338	0.032	0.147
Root Mean Square Error RMSE						44,046.821	276.197	69.116	176.274
RMSPE						47.839	0.229	0.041	0.124

Note Negative value in Error % represents premature Failure than the actual RUL

Table 4 Accuracy of proposed prediction model for oil pressure, oil contamination, and motor vibration

Parameter	Oil pressure prediction				Oil contamination prediction				Vibration prediction			
	Conventional	ARIMA	ADAM	SGDM	Conventional	ARIMA	ADAM	SGDM	Conventional	ARIMA	ADAM	SGDM
RMSE	44,046.821	276.197	176.27	69.116	33,236.453	228.784	166.428	61.231	39,891.917	312.462	201.218	104.464
RMSPE	47.839	0.229	0.124	0.041	34.234	0.168	0.101	0.039	41.453	0.273	0.146	0.071

3.2.2 Predicted Results of Hydraulic Oil Contamination Using LSTM

In the proposed work, the recommended OCL should be less than NAS 7, since 80% of the high-pressure moulding line operations are controlled by hydraulics using high precision Rexroth make servo and proportional valves with a positioning accuracy of ± 1 mm for a linear load of up to 15 Tons and with a velocity of 1800 mm/sec. The training is carried out for 200 epochs in both the optimizer cases. ADAM and SGDM training with 80% of data set and testing with 20% data set is performed for oil contamination LSTM prediction model. It uses 200 epochs to reduce RMSE. Each epoch uses an iteration of 64 data similar to all proposed LSTM prediction model and the remaining 20% data is used for testing.

Figure 13a depicts the predicted oil contamination value using ADAM optimizer and Fig. 13b shows using SGDM training optimizers. ADAM optimizer predicts that, the threshold value of 12 mA equivalent to oil contamination NAS 7 will be reached in 103,690 min which is equal to 72 days and 10 min. From SGDM optimizer function, it is clear that the oil contamination will reach the same threshold value of 12 mA in 71 days, 9 h, and 40 min which is in 102,820 min.

The predicted values of oil contamination using the proposed LSTM using ADAM and SGDM model is compared against the ground truth similar to similar to hydraulic pressure prediction model. Table 4 shows that the prediction accuracy of proposed model in terms of RUL, Prediction error, RMSE and RMSPE is clear that LSTM with SGDM model performs better than conventional CBM, ARIMA and ADAM model with MSE as 0.039 and RMSE as 61.231.

3.2.3 Prediction Results of Motor Vibration Using LSTM

ADAM and SGDM training for motor vibration is implemented in this vibration prediction model similar to oil contamination and oil pressure with 80% of data set. It uses 200 epochs to reduce RMSE. Each epoch uses an iteration of 64 data. The remaining 20% data was used for testing.

In this proposed application, the motor vibration threshold value should be below 2.3 mm/sec or 3.6 mm/Sec2 which is equivalent to 183.542 mV. The predicted motor vibration value using ADAM and SGDM optimizers are provided in Fig. 14a, b respectively. The motor vibration will reach the threshold value of 183.542 mV in 27,478 min which is equivalent to 19 days, 1 h and 58 min as predicted using ADAM function for necessary maintenance activities. From SGDM function, it is observed that the motor vibration will reach the threshold value of 183.542 mV, in 27,616 min which is equivalent to 19 days, 4 h and 16 min.

The LSTM based predicted values of motor vibration is compared against the ground truth values. From this Table 4 it is clear that LSTM with SGDM model performs better than conventional CBM, ARIMA, and ADAM model with lowest RMSE as 104.464 and RMSPE as 0.071.

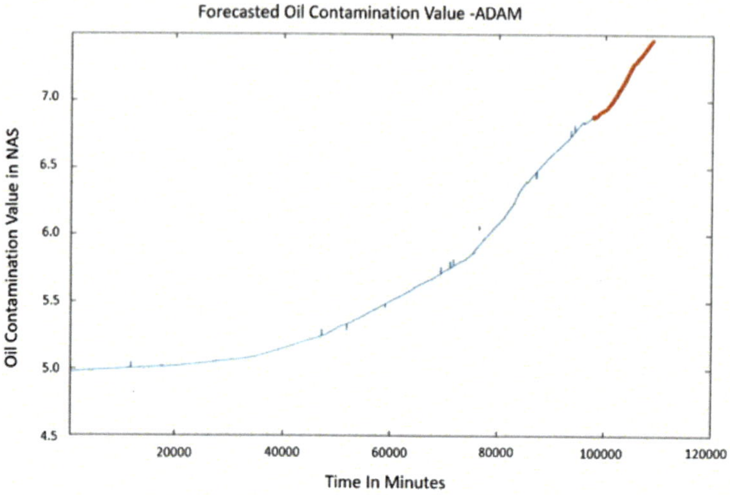

(a)

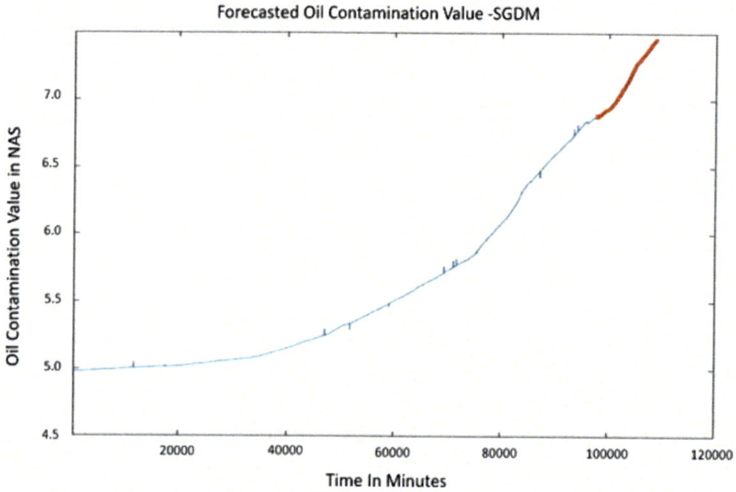

(b)

Fig. 13 **a**. ADAM Optimizer Based Prediction **b**. SGDM Optimizer Based Prediction

3.2.4 IoT Based Oil Temperature Prediction

Because of the flexibility in temperature control system, hydraulic oil temperature prediction is achieved through IoT based alarm to the authenticated for proactive approach system to achieve zero downtime related to temperature. Through PLC oil temperature is monitored and controlled by SCADA system. There are four stages oil

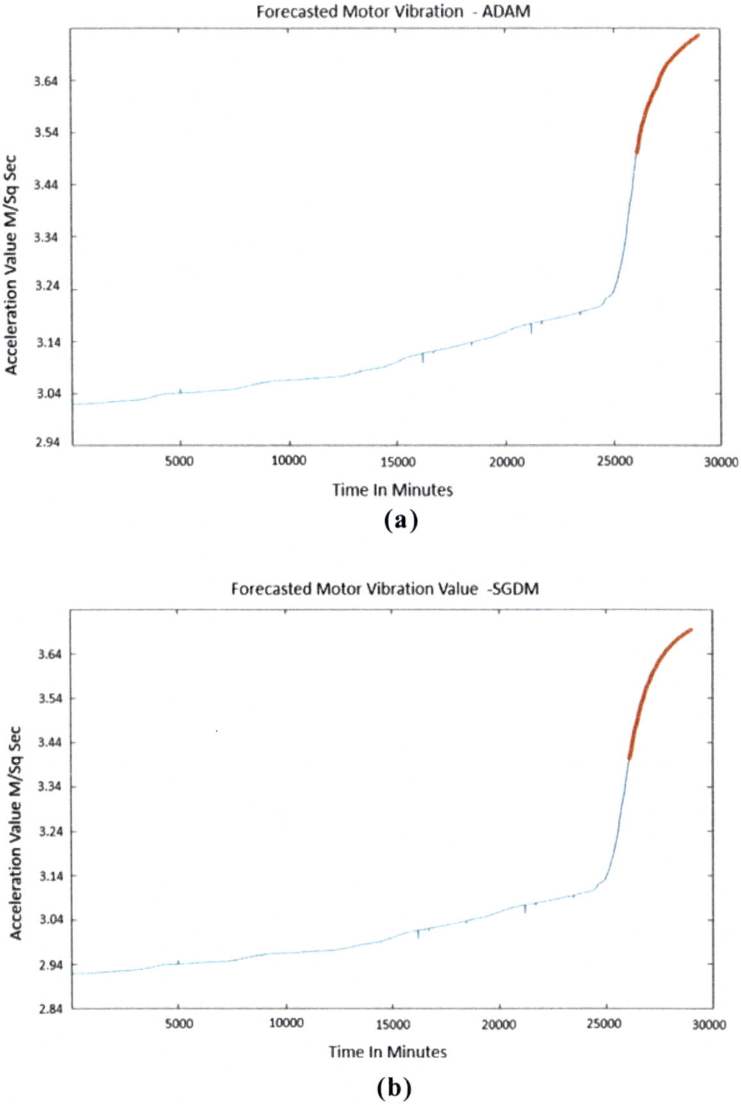

Fig. 14 **a** ADAM optimizer based prediction **b** SGDM optimizer based prediction

temperature control in SCADA namely (a) Oil temperature Very High Status when the temperature exceeds 550 C which will switch off the machine operation (b) Oil temperature High status, when the temperature is in between 450 C to 54.990 C where the alarm message is sent to the authenticated for necessary action to service the cooling system (c) Oil temperature normal status in which, the oil temperature will be 200 C to 44.990 C which is a normal temperature to run the machine (d) Oil temperature Cool status in which case the oil is in very low temperature and hence to be heated up by the PLC controlled heating system and an alert is triggered if this system is not working.

3.2.5 Discussion on Predicted Result

Based on the prediction of RUL, appropriate proactive measures have been done to eliminate the breakdown. The results of breakdown are shown in the following figures. Figure 15a shows the downtime reduction trend due to Hydraulic pressure from Q1 of year 2015 to Q4 of year 2020. The breakdown due to hydraulic pressure approached zero from Q1 of year 2020 after the implementation of LSTM prediction model for Hydraulic pressure.

Figure 15b displays the downtime due to vibration of the motors over a period of time from Q1 of year 2015 to Q4 of year 2020. The breakdown due to motor vibration approached zero from Q3 of 2019 after the implementation of LSTM prediction model. Figure 15c shows the downtime due to oil contamination which reaches zero down time from Q4 of 2018. Figure 15d depicts the downtime due to oil temperature which reaches zero down time from Q3 of year 2018.

Figure 16a displays the moulding plant downtime reduction trend from the period Q1 of year 2015 to Q4 of year 2020. Here the downtime is reduced from 5832 to 471 min which is the reduction in downtime of 91.92% after the implementation of proposed LSTM based TPM approach in the Hydraulic high pressure moulding plant. From the period Q3 of 2018 IoT based prediction model kaizen K1 was implemented. From Q4 of 2018, LSTM based oil contamination level prediction model K2 was implemented in the foundry line. From the period Q3 of 2019, LSTM based prediction model K3 for oil pressure variation was implemented and from Q1 of 2020, prediction model K4 due to motor vibration was implemented.

In the proposed work, Availability Rate (AR) is a factor of breakdown loss, tool change loss, setup and adjustment, and start-up loss. Since tool change is performed automatically within the cycle time, setup and adjustment is done online without the stoppage of machine and start-up losses is avoided by starting the line automatically in shut down time, breakdown loss is directly related to the availability rate. As shown in Fig. 16b, by implementing the conventional TPM, the AR is raised from 83.1% to 87.2% and after implementing LSTM approach, it is raised from 87.2% to 98.1%. Hence the proposed systematic TPM approach with LSTM yields 12.5% increase in AR.

4 Conclusion and Future Work

The TBM and CBM implemented in the conventional TPM have certain demerits due to human error, skill, intermittent data collection and the collected data may not be available at the time of breakdown for analysis which leads to breakdown. In the proposed work, LSTM based prediction model is used for multi signature

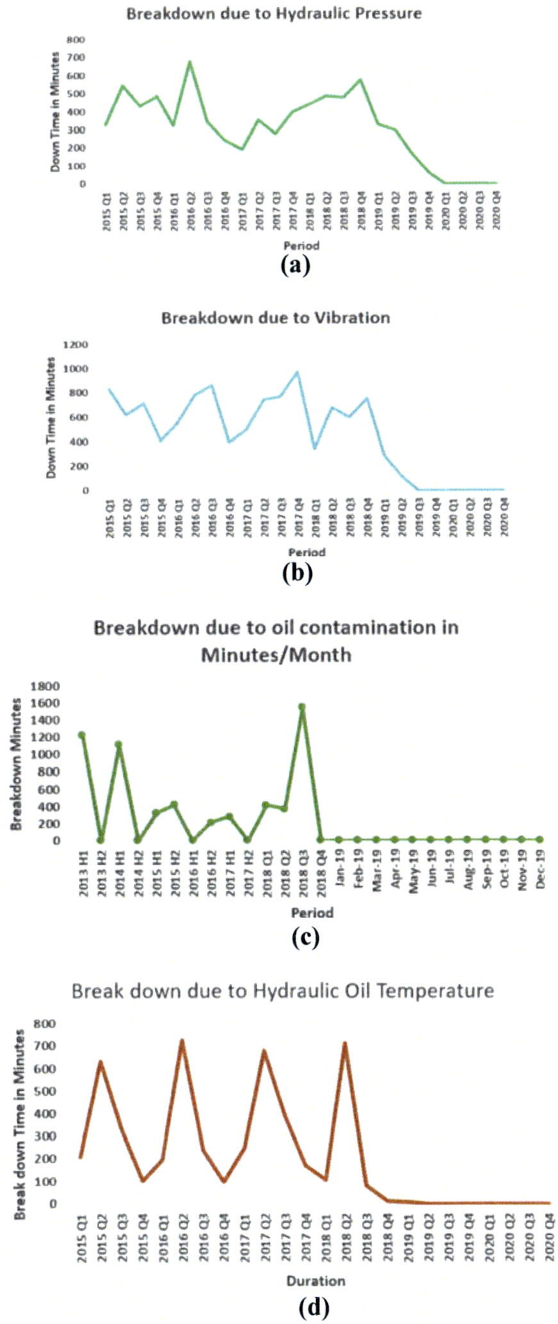

Fig. 15 **a** Breakdown trend due to pressure **b**Breakdown trend due to vibration **c** Breakdown trend due to OCL **d** Breakdown trend due to oil temperature

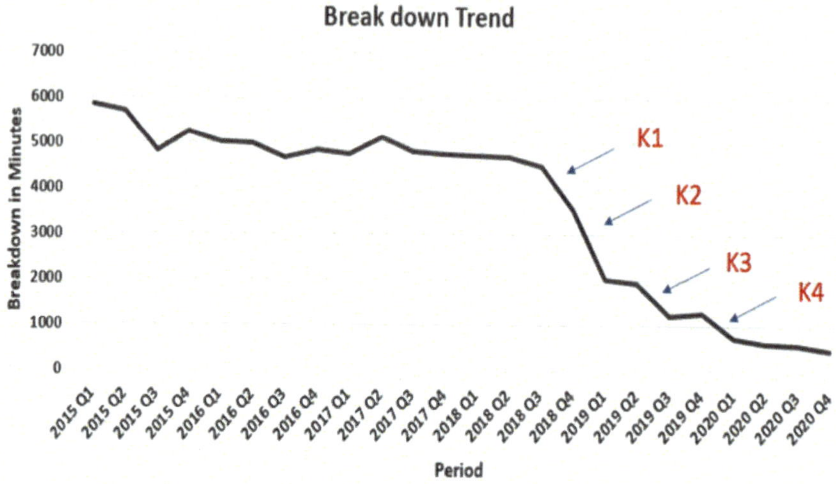

(a)

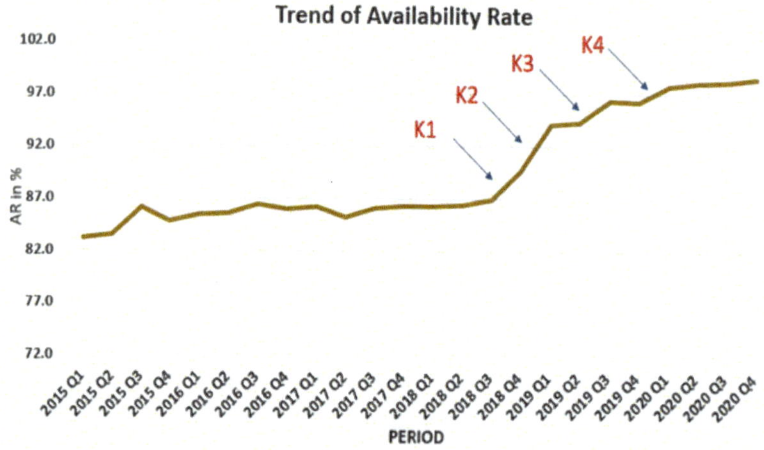

(b)

Fig. 16 **a** Plant breakdown in minutes trend **b** Plant breakdown in minutes trend

parameters as suggested by Pareto analysis which eliminates 95% of the breakdown related to all signature parameters. Using of LSTM based prediction model and IoT the Availability Rate of the equipment is increased from 87.2% to 98.1%, which means, 12.5% increase in productivity is achieved without additional manpower or equipment and with the available resources. In future, AI based prediction model with drone will be used for all condition monitoring to achieve zero breakdown.

References

1. Paulina Gackowiec (2019) General overview of maintenance strategies—concepts and approaches. Multidisciplinary Aspects of Production Engineering 2(1), 126–139
2. Mungani DS, Visser JK (2013) Maintenance approaches for different production methods. South African Journal of Industrial Engineering 24(3), 1–13
3. Yongyi Ran, Xin Zhou, Pengfeng Lin, Yonggang Wen, Ruilong Deng (2019) A survey of predictive maintenance: systems, purposes and approaches. IEEE Communications Surveys & Tutorials, XX(XX)
4. Jens Passlick, Sonja Dreyer, Daniel Olivotti, Lukas Grützner, Dennis Eilers, Michael H. Breitner (2020) Predictive maintenance as an internet of things enabled business model. A taxonomy. Electron Mark. 31(1), 67–87
5. Rimawan E, Irawan APB (2017) Analysis of calculation overall equipment effectiveness (OEE) in the implementation of total productive maintenance (TPM) PC 200–8 excavator grab and magnet type case study in cakratunggal steel mills company. Int J Sci Eng Res 8(1):1363–1368
6. Alberto Jimenez-Cortadi, Itziar Irigoien, Fernando Boto, Basilio Sierra, German Rodriguez (2020) Predictive maintenance on the machining process and machine tool. Appl Sci 10(224), 1–14
7. Dimitrios Kateris, Dimitrios Moshou, Xanthoula-Eirini Pantazi, Ioannis Gravalos, Nader Sawalhi, Spiros Loutridis (2014) A machine learning approach for the condition monitoring of rotating machinery. J Mech Sci Technol 28(1), 61–71
8. Angelo Encapera, Abhijit Gosavi, Susan L Murray (2019) Total productive maintenance of make-to-stock production-inventory systems via artificial-intelligence-based iSMART. Int J Syst Sci: Oper & Logist 8(2), 154–156
9. Thyago P Carvalhoa, Fabrízzio AAMN Soares, RobertoVita, Roberto da P Francisco, João P Basto, Symone GS (2019) Alcala systematic literature review of machine learning methods applied to predictive maintenance. Comput & Ind Eng. 127
10. Ali Reza Samanpour, André Ruegenberg, Robin Ahlers (2018) The future of machine learning and predictive analytics. Springer-Verlag GmbH Germany 2018. Digital Marketplaces Unleashed 8(30), 297–309
11. Ruben Ruiz-Gonzalez, Jaime Gomez-Gil Francisco Javier Gomez-Gil, Víctor Martínez-Martínez (2014) An SVM-Based classifier for estimating the state of various rotating components in agro-industrial machinery with a vibration signal acquired from a single point on the machine chassis. Sensors 14, 20713–20735
12. Domingos S de O Santos Júnior, João F L .de Oliveira, Paulo SG de Mattos Neto (2019) An intelligent hybridization of ARIMA with machine learning models for time series forecasting. Knowl-Based Syst 175, 2–86
13. Jeyasekar A, Kasmir Raja SV, Annie UR (2017) Congestion avoidance algorithm using ARIMA (2,1,1) model-based RTT estimation and RSS in heterogeneous wired-wireless networks. J Netw Comput Appl 93:91–109
14. Da Silva BS, Inaba FK, Salles EOT, Ciarelli PM (2020) Outlier robust extreme machine learning for multi-target regression, Expert Syst Appl 140
15. Zhao YP, Huang G, Hu QK, Tan JF, Wang JJ, Yang Z (2019) Soft extreme learning machine for fault detection of aircraft engine. Aerosp Sci Technol 91, 70–81
16. Srikanth Namuduri, Barath Narayanan Narayanan, Venkata Salini Priyamvada Davuluru, Lamar Burton, Shekhar Bhansali1 (2020) Review—Deep learning methods for sensor based predictive maintenance and future perspectives for electrochemical sensors. J Electrochem Soc 167
17. Yue Li, Yijie Zeng, Yuanyuan Qing, Guang-Bin Huang (2020) Learning local discriminative representations via extreme learning machine for machine fault diagnosis. Neurocomputing 409(7), 275–285
18. Roosefert Mohan T, Annie Uthra R, Devaraj D, Umachandran K ((2021)) Intelligent machine learning based total productive maintenance approach for achieving zero down time in Industrial machinery. Comput Ind Eng 157

Taguchi Coupled GRA Based Optimization of Shoulder Milling Process Parameters During Machining of SS-304

Vijay Singh Bhadauria, Gaurav Kumar, Husain Mehdi, and Mukesh Kumar

Abstract Milling is a flexible machine that is used to mill a wide range of industrial components, including construction and agricultural equipment, rail and mining vehicles, various types of passenger and commercial automobiles, earthmoving barges, and dams. The experimental and numerical results of shoulder milling with the VMM on SS-304 are presented in this study. This investigation aims to accomplish the effect of the milling input factor on surface roughness, material removal rate, and microhardness. The process parameters are coolant, feed, depth of cut, and speed are taken into account to optimize the result of surface roughness, material removal rate, and microhardness. Taguchi L18 technique was selected for the experimental trial with grey relational analysis. ANOVA and *F*-test were employed to search the contribution of each parameter. The result is confirmed and validated by a validation test, which illustrates that it is feasible to improve the surface roughness, material removal rate, and microhardness appreciably.

Keywords Shoulder milling · SS-304 · Taguchi method · Micro-hardness · Material removal rate · Multi-response optimization

1 Introduction

Milling is a basic machining technology commonly used in machine shops and industries today for cutting parts to exact shapes and sizes to assemble with other components in applications such as aerospace, automobile, die, and equipment design [1]. Shoulder milling is essential among the various milling operations because it can produce complicated geometric surfaces with reasonable precision and surface

V. S. Bhadauria · G. Kumar · M. Kumar
Department of Mechanical Engineering, Vidya College of Engineering, Meerut, India

H. Mehdi (✉)
Department of Mechanical Engineering, Meerut Institute of Engineering and Technology, Meerut, India
e-mail: husain.mehdi@miet.ac.in

H. Kumar et al. (eds.), *Recent Advances in Intelligent Manufacturing*,
Lecture Notes in Mechanical Engineering,
https://doi.org/10.1007/978-981-99-1308-4_4

smoothness. It can also produce several combinations when used in conjunction with the milling cutter [2]. A material's safety, dependability, and life cycle costs can be increased by machining it to suit specified service requirements [3]. The Taguchi-based grey relational analysis is utilized to establish machining parameters and optimize cutting parameters in stainless steel 304 to determine the optimal parameter combination. Ahmad Hamdan et al. [4] applied Taguchi in machining of stainless steel for performance characterization of coated carbide tool. Nayak et al. [5] have studied the performance constraints in turning of SS 304 with dry condition. An uncoated cemented carbide inserts were used. It was based on grey relational analysis. Kumar et al. [6] have worked on SS-304 to improve the surface quality by applying Taguchi. Kumar et al. [7] improved SS 321's multi performance machining properties. The grey relational analysis with Taguchi was utilized to obtain the drilling settings at optimized level. According to Karnwal et al. [8], the Taguchi technique paired with GRA can be deploy to study diesel engine performance. Sijo et al. [9] optimized the turning constraints through Taguchi. Lin and Ho [10] suggested to apply GRA with Taguchi to analyze the impacting order of parameters. Singh et al. [11, 12] improved the surface quality of AA6063T6 aluminum alloy through shoulder milling using L18 Taguchi coupled GRA. Kashyap et al. [13] improved the surface integrity of AA6082T6 aluminum alloy through shoulder milling using L18 Taguchi coupled GRA. As per above discussed literature, it was concluded that grey relational analysis along with Taguchi is the most useful techniques for optimization during shoulder milling of stainless steel. In this study process parameters under consideration are speed, coolant, feed rate and depth of cut. Surface roughness (SR), Material removal rate (MRR) and microhardness were opted as output parameters.

2 Materials and Method

Stainless Steel 304 was chosen as the workpiece material for this study because of its high application in industries. The sample is in the dimension of 75 mm × 45 mm × 10 mm, as shown in Fig. 1a. Chemical Spectrometry is used to determine the chemical composition, as shown in Table 1. A tool of solid carbide was used for machining on SS-304 pieces through a Vertical milling machine (VMM) as shown in Fig. 1b. Different machining parameters were shown in Table 2, where each parameter had three levels except coolant, which has two levels.

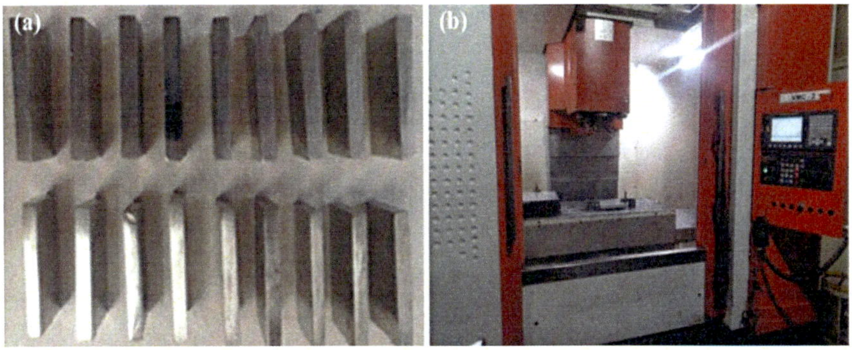

Fig. 1 **a** Base metal SS304, **b** Experimental setup

Table 1 Chemical composition of SS304

Elements	Fe	Cr	Mn	Cu	Si	Nb	V	C	Al	S
Weight %	70.3	18.3	1.7	0.36	0.34	0.04	0.06	0.04	<0.01	<0.005

Table 2 Machining range of the parameters

Parameter	Coolant	Feed (mm/min)	Depth of cut (mm)	Speed (rpm)
Notation	A	B	C	D
1	ON	2000	0.3	2000
2	OFF	2500	0.4	3000
3	–	3000	0.5	4000

2.1 Recording of Response Characteristics

Figure 2 shows the surface roughness measurement device used to measure the roughness values of each sample. After metallographic finishing, the sample was Vickers hardness tested in this study. A constant force throughout the testing was of 196.2 N, and values of micro-hardness were measured and noted. The calculations for MRR were made as per formula given below:

$$\text{Material Removal Rate} = \frac{w \times f \times d}{60} \text{mm}^3/\text{s} \tag{1}$$

where: width of block (w), feed rate (f), depth of cut (d) and each sample width is constant, i.e.,10 mm.

Fig. 2 Surface roughness tester

3 Results and Discussion

3.1 Taguchi Technique (Signal to Noise (S/N) Ratio)

Taguchi's method is the most effective research technique for refining quality while lowering costs. The Taguchi technique employs a unique OA design to examine the all assigned parameters. Grey Relational Analysis regulates the ideal degree of numerous regulated elements at the same time. Three S/N ratios are commonly utilized, and their choice is based on reaction characteristics. The criterion of "lower is better" will be used for roughness and micro-hardness and MRR, "higher is better" will be used. The following are the formulas for all three S/N ratios (Table 3) [14].

3.2 Steps of Grey Relational Analysis (GRA)

3.2.1 Normalization

The initial stage in GRA is for normalizing the trial data between 0 and 1. Normalization facts are explained in [15–17].

Table 3 Experimented responses

Experiments	A	B	C	D	Ra (μm)	MRR (mm^3/s)	Microhardness (HV)
1	ON	2000	0.3	2000	1.18	100	258.69
2	ON	2000	0.4	3000	1.4	133.3333	227.95
3	ON	2000	0.5	4000	1.74	166.6667	186.64
4	ON	2500	0.3	2000	2.02	125	169.05
5	ON	2500	0.4	3000	2.29	166.6667	185.04
6	ON	2500	0.5	4000	2.17	208.3333	193.12
7	ON	3000	0.3	3000	1.95	150	209.88
8	ON	3000	0.4	4000	1.78	200	176.57
9	ON	3000	0.5	2000	2.09	250	174.9
10	OFF	3000	0.3	4000	1.48	150	166.52
11	OFF	2000	0.4	2000	1.22	133.3333	215.58
12	OFF	2000	0.5	3000	1.98	166.6667	163.1
13	OFF	2500	0.3	3000	1.48	125	184.78
14	OFF	2500	0.4	4000	1.01	166.6667	188.54
15	OFF	25,000	0.5	2000	2.05	208.3333	213.1
16	OFF	3000	0.3	4000	2.25	150	161.21
17	OFF	3000	0.4	2000	2.3	200	239.74
18	OFF	3000	0.5	3000	1.61	250	333.05

3.2.2 Analysis of GRG (Grey Relational Grade) and GRC (Grey Relational Coefficient)

There are following formulas that are used for grey relational analysis (Table 4) [11]. The value of distinguishing coefficient (ξ) was 0.5.

$$\xi_i(l) = \frac{\nabla_{\min} + \xi \cdot \nabla_{\max}}{\nabla_{0i}(l) + \xi \cdot \nabla_{\max}} \tag{2}$$

$$\Delta_{0i}(l) = \|z_0^*(l) - z_i^*(l)\| \tag{3}$$

$$\Delta_{\max} = \max_{\forall j \in i} \min_{\forall l} \left\| z_0^*(l) - z_i^*(l) \right\| \tag{4}$$

$$\Delta_{\min} = \max_{\forall j \in i} \min_{\forall l} \left\| z_0^*(l) - z_i^*(l) \right\| \tag{5}$$

$$\gamma_i = \frac{1}{m} \sum_{l=1}^{n} \xi_i(l) \tag{6}$$

Table 4 Evaluated GRC and grade values

Experiment no.	GRC				
	SR	MRR	MH	GRG	Orders
1	0.3814	0.3333	0.5895	0.4347	13
2	0.4532	0.4216	0.4890	0.4546	12
3	0.5959	0.5305	0.3852	0.5039	11
4	0.7602	0.0000	0.3485	0.3696	18
5	0.9895	0.5305	0.3817	0.6339	6
6	0.8761	0.7153	0.3997	0.6637	4
7	0.7137	0.4728	0.4400	0.5422	9
8	0.6162	0.6725	0.3637	0.5508	8
9	0.8112	1.0000	0.3603	0.7239	3
10	0.4828	0.4728	0.3436	0.4330	14
11	0.3936	0.4216	0.4548	0.4233	15
12	0.7331	0.5305	0.3369	0.5335	10
13	0.4828	0.3979	0.3811	0.4206	16
14	0.3333	0.5305	0.3894	0.4177	17
15	0.7815	0.7153	0.4483	0.6484	5
16	0.9493	0.4728	0.3333	0.5851	7
17	1.0000	0.6725	0.5246	0.7324	2
18	0.5357	1.0000	1.0000	0.8452	1

$$\gamma_i = \frac{1}{m} \sum_{l=1}^{m} w_l \xi_i(l) \sum_{l=1}^{m} w_l \tag{7}$$

3.2.3 Best Suitable Combination of Parameters

The value of max–min in Table 5 depicts that most controlling parameter is feed rate, where GRG values of feed rate are more significant than any other process parameter. The remaining process parameters C, D, and A, as shown in Table 5, are listed in descending order of impact on multi performance. GRA selects the best shoulder milling process settings. Each shoulder milling parameter's maximum grey relational levels are A_2, B_3, C_3, and D_2, as shown in Fig. 3. As an outcome, machining parameters in shoulder milling with values $A_2B_3C_3D_2$ are a multi performance-optimized combination.

GRA selects the best shoulder milling process settings. Each shoulder milling parameter's maximum grey relational levels are A_2, B_3, C_3, and D_2, as shown in Fig. 3.

Table 5 Response table for the GRG

End milling machining parameters	Coolant	Feed rate	Depth of cut	Speed
Symbol	A	B	C	D
L-1	0.5419	0.4638	0.4642	0.5554
L-2	0.5599	0.5256	0.5354	0.5717
L-3	–	0.6633	0.6531	0.5257
Max–min	0.5599	0.6633	0.6531	0.5717

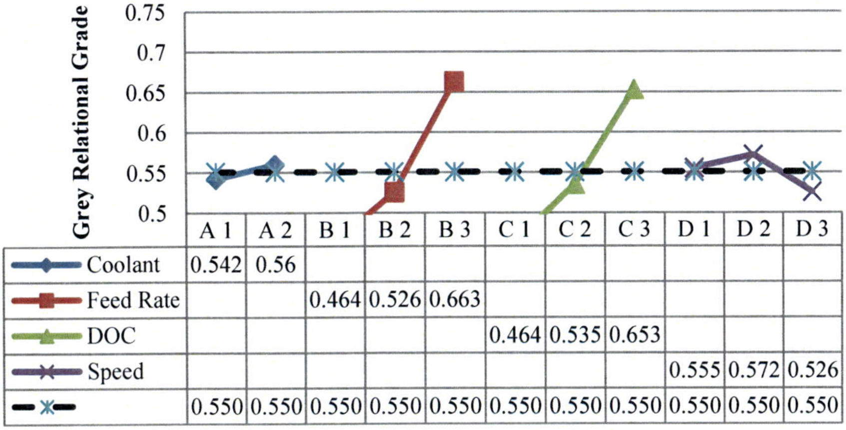

Fig. 3 Main effects plot for GRG

As an outcome, machining parameters in shoulder milling with values $A_2B_3C_3D_2$ are a multi performance-optimized combination.

3.2.4 Analysis of Variance (ANOVA)

ANOVA is practiced to identify the most important factors impacting shoulder milling. The statistical technique *F*-test was used to assess the impact of these input parameters on quality attributes. When the value exceeds four, adjustments in shoulder milling parameters have a considerable effect on quality. The estimated *F*-value for the created model is lower than the predicted *F*-value (found in the ANOVA table) at the 95% confidence interval (CI). At a 95% CI, the calculated *F*-ratio values showed that the model was adequate. According to the CI, the model is adequate in such conditions when the prob > *F* values for produced models are less than 0.05 (95% CI) [18–20]. As per actual versus predicted value, the MRR, and microhardness reveals residual fall on a straight line; errors were regularly disseminated from the regression by comparing their experimental and predicted values. In a trial version, all the coefficients were estimated and checked for significance using

Table 6 ANOVA grade value

Symbol	Source	Sum of squares	DOF	Mean square	*F* value	*P*-value	% C	
A	Coolant	0.0015	1	0.0015	0.2406	<0.0001	0.482	Significance
B	Feed rate	0.1251	2	0.0625	10.3041	0.0061	41.28	Significance
C	Doc	0.1092	2	0.0546	8.9956	0.0451	36.04	Significance
D	Speed	0.0065	2	0.0033	0.5366	0.0265	2.15	Significance
Error		0.0607	10	0.0061			20.03	
Total		0.3029	17				100.0	

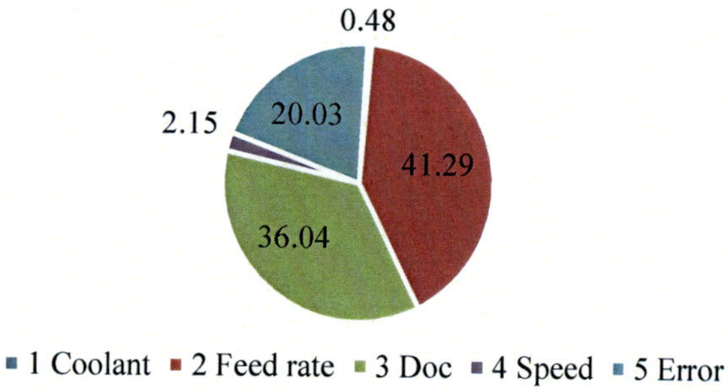

Fig. 4 Percentage contribution of machining parameters

F-test at 95% CI. Once the appropriate coefficients were established, the model was developed to determine the MRR and hardness of the developed model. As shown in Table 6, feed is the most critical parameter, accounting for 41.28% of the total, depth of cut for 36.04%, and speed and coolant for 2.15 and 0.48%, respectively. Figure 4 also shows the contributions of all parameters.

3.2.5 Validation Test

Compared to predicted grey relation grades, there is a 6.29% increase in experimental grade, which is satisfactory, as shown in Table 7.

Table 7 Validation test and its outcomes

Optimum machining parameter			
	Predicted	Experimented	Change in %
Level	$A_2B_3C_3D_2$	$A_2B_3C_3D_2$	
SR		0.5357	
MRR		1.0000	
MH		1.0000	
GRG	0.7952	0.8452	6.29

4 Conclusions

Concluded remarks are as given below:

- The ideal setup of shoulder milling process parameters is $A_2B_3C_3D_2$, which yields optimum multi performance features with coolant off, feed rate 3000 mm/min, depth of cut 0.5 mm, and machining speed 3000 rpm.
- ANOVA determined that the contribution of feed rate, depth of cut, machining speed, and coolant are 41.28, 36.04, 2.15, and 0.48%, respectively.
- This strategy yields a 6.29% improvement. As a result, Taguchi is a unique method for reducing costs and time.
- In the future, fuzzy logic, genetic algorithms (GA), and numerous nature-inspired algorithms may be utilized for optimization.

References

1. Zhang JZ, Chen JC, Kirby ED (2007) Surface roughness optimization in an end-milling operation using the Taguchi design method. J Mater Process Technol 184(1–3):233–239
2. Parmar JG, Makwana A (2012) Prediction of surface roughness for end milling process using artificial neural network. Int J Mod Eng Res (IJMER) 2(3):1006–1013
3. Jawahir IS, Brinksmeier E, M'saoubi R, Aspinwall DK, Outeiro JC, Meyer D et al (2011) Surface integrity in material removal processes: recent advances. CIRP Ann 60(2):603–626
4. Hamdan A, Sarhan AA, Hamdi M (2012) An optimization method of the machining parameters in high-speed machining of stainless steel using coated carbide tool for best surface finish. Int J Adv Manuf Technol 58(1):81–91
5. Nayak SK, Patro JK, Dewangan S, Gangopadhyay S (2014) Multi-objective optimization of machining parameters during dry turning of AISI 304 austenitic stainless steel using grey relational analysis. Procedia Mater Sci 6:701–708
6. Kumar G, Kumar M, Tomer A (2021) Optimization of end milling machining parameters of SS 304 by Taguchi technique. In: Recent advances in mechanical engineering. Springer, Singapore, pp 683–689
7. Kumar M, Kumar G, Singh OP, Tomer A (2021) Multiperformance optimization of parameters in deep drilling of SS-321 by Taguchi-based GRA. In: Recent advances in mechanical engineering. Springer, Singapore, pp 675–681
8. Karnwal A, Hasan MM, Kumar N, Siddiquee AN, Khan ZA (2011) Multi-response optimization of diesel engine performance parameters using thumba biodiesel-diesel blends by applying the Taguchi method and grey relational analysis. Int J Automot Technol 12(4):599–610
9. Sijo MT, Biju N (2011) Taguchi method for optimization of cutting parameters in Turning operations. Int J Manuf Mater Sci 1(1):44
10. Ho CY, Lin ZC (2003) Analysis and application of grey relation and ANOVA in chemical–mechanical polishing process parameters. Int J Adv Manuf Technol 21(1):10–14
11. Singh OP, Kumar G, Kumar M (2019) Multi Performance optimization of shoulder milling process parameters of AA6063 T6 aluminium alloy by Taguchi based GRA. Int J Innov Technol Explor Eng 8:420–425
12. Singh OP, Kumar G, Kumar M (2019) Role of Taguchi and grey relational method in optimization of machining parameters of different materials: a review. Acta Electron Malay (AEM) 3(1):19–22

13. Kashyap K (2021) Optimization of shoulder milling process parameters of Aa6082t6 using Taguchi coupled grey relational analysis. Turk J Comput Math Educ (TURCOMAT) 12(12):3279–3288
14. Siddiquee AN, Khan ZA, Mallick Z (2010) Grey relational analysis coupled with principal component analysis for optimisation design of the process parameters in in-feed centreless cylindrical grinding. Int J Adv Manuf Technol 46(9):983–992
15. Khan ZA, Kamaruddin S, Siddiquee AN (2010) Feasibility study of use of recycled high-density polyethylene and multi response optimization of injection moulding parameters using combined grey relational and principal component analyses. Mater Des 31(6):2925–2931
16. Jangra KK, Sharma N, Khanna R, Matta D (2016) An experimental investigation and optimization of friction stir welding process for AA6082 T6 (cryogenic treated and untreated) using an integrated approach of Taguchi, grey relational analysis and entropy method. Proc Inst Mech Eng Part L J Mater Des Appl 230(2):454–469
17. Sharma N, Khanna R, Sharma YK, Gupta RD (2019) Multi-quality characteristics optimisation on WEDM for Ti-6Al-4V using Taguchi-grey relational theory. Int J Mach Mach Mater 21(1–2):66–81
18. Mehdi H, Mehmood A, Chinchkar A, Hashmi AW, Malla C, Mohapatra P (2021) Optimization of process parameters on the mechanical properties of AA6061/Al2O3 nanocomposites fabricated by multi-pass friction stir processing. Mater Today Proc 56(4):1995–2003.https://doi.org/10.1016/j.matpr.2021.11.333
19. Nait Salah A, Mehdi H, Mehmood A, Hashmid AW, Malla C, Kumar R (2021) Optimization of process parameters of friction stir welded joints of dissimilar aluminum alloys AA3003 and AA6061 by RSM. Mater Today Proc 56(4):1675–1684. https://doi.org/10.1016/j.matpr.2021.10.288
20. Mehdi H, Mishra RS (2020) An experimental analysis and optimization of process parameters of AA6061 and AA7075 welded joint by TIG+FSP welding using RSM. Adv Mater Process Technol. https://doi.org/10.1080/2374068X.2020.1829952

Batch Reactor System (BRS): Effective Conversion of Used Cooking Oil into Biodiesel in Presence of Different Catalyst

Niraj S. Topare, Satish V. Khedkar, Sunita Raut-Jadhav, Anish Khan, and Abdullah M. Asiri

Abstract Due to the growing discrepancy between energy production and consumption, many nations are shifting their focus to biodiesel. The use of biodiesel as an appealing alternative fuel has been authorized. The production of biodiesel involves a reaction known as transesterification, in which animal fats and vegetable oils are mixed with alcohol in the presence of a catalyst. The catalyst plays a crucial role in determining the biodiesel yield and the overall sustainability of the biodiesel production process. In this study, we presented the design and development of a batch reactor system (BRS) for converting UCO (used cooking oil) into biodiesel. In order to accomplish this objective, the used sunflower cooking oil (USCO) is utilized as a feedstock. Different catalysts can be used in reactors, and their performance under different operating conditions is examined in the context of improving oil conversion. This study compared the homogeneous catalytic activity of sodium hydroxide (NaOH) to the heterogeneous catalytic activity of magnesium oxide (MgO) in the biodiesel synthesis process. The experimental tests were carried out by using NaOH and MgO catalysts with three different percentages (1.0 wt.%, 1.5 wt.%, and 2 wt.%)

N. S. Topare (✉)
School of Chemical Engineering, Dr. Vishwanath Karad MIT World Peace University, Pune-411038, India
e-mail: niraj.topare@mitwpu.edu.in; nirajtopare06@gmail.com

S. V. Khedkar
Department of Chemical Engineering, College of Engineering and Technology, Akola-444104, India

S. Raut-Jadhav
Department Chemical Engineering, Bharati Vidyapeeth (Deemed to Be University) College of Engineering, Pune 411043, India

A. Khan · A. M. Asiri
Chemistry Department, Faculty of Science, King Abdulaziz University, Jeddah 21589, Saudi Arabia

Center of Excellence for Advanced Materials Research, King Abdulaziz University, Jeddah 21589, Saudi Arabia

H. Kumar et al. (eds.), *Recent Advances in Intelligent Manufacturing*,
Lecture Notes in Mechanical Engineering,
https://doi.org/10.1007/978-981-99-1308-4_5

for the oil to alcohol molar ratio (1:5), reaction time (90 min), and reaction temperature (65°C). Experiments conducted in the designed and developed BRS have shown very promising results with 96% and 93% biodiesel yield obtained using 1.0 wt. % of NaOH and MgO respectively.

Keywords Biodiesel · Used cooking oil · Transesterification · Batch Reactor · NaOH · MgO

1 Introduction

Strengthening energy demand and global warming concerns are important factors driving renewable energy interest. The world's energy sector is already dominated by fossil fuel consumption [1]. However, future supply volatility, potentially unsustainable energy use trends, and many energy analysts have led to spending on expanding proven fossil fuel reserves and managers worldwide are searching for alternative options such as biodiesel, which can be more sustainable [2]. Fuel prices are steadily upward on the basis that cheaper alternatives to supply are being sought. Biodiesel has continued to draw rising interest from people who are improving agricultural and environmental sustainability and reducing carbon emissions to mitigate climate change [3]. Some people also consider biodiesel as a potentially important contributor to rural economic development and as a way of alleviating poverty through work generation and income generation [4, 5]. Biodiesel is thus increasingly regarded as an alternative to renewable energy which is promising and largely untapped and technical progress is increasingly being made to see its potential environmental and economic advantages [6].

Biodiesel is a replacement fuel for any sustainable biomass diesel fuel, in the most general context. It is best known as a renewable, sulfur-free, biodegradable, nontoxic, substitute for diesel oil [7]. It is also chemical and must be defined as a fuel composed of long-chain mono-alkyl fatty acid esters derived from renewable sources [8]. Strict quality standards must be met before they may be considered practical transportation fuels. Transesterification is one common process for producing biodiesel. The fuel injection system of modern diesel motors is prone to change in viscosity [9]. Reducing vegetable oil viscosity in order to increase its efficiency is one way to prevent these problems. The four methods used to overcome the problems posed by high fuel viscosity are dilution, micro emulsification, pyrolysis, and transesterification. Transesterification is the most common process and leads to vegetable oils and fats monoalkyl esters, which are now called fuel biodiesel [10–12].

The production of biodiesel often involves a process called transesterification. During this process, the triglyceride present in the feedstock undergoes a reaction with monohydride alcohol to form monoalkyl esters while a catalyst is present [13]. Since cost-effective feedstocks are necessary for the manufacturing of biodiesel, attention has been drawn to waste cooking oil as a potentially useful alternative because of its relative cheapness of it to other options than pure vegetable oil [14]. Waste cooking oil would be easier to dispose of and less likely to contaminate land and water if it were utilized to make biodiesel [15]. Also, this would be a more carbon–neutral alternative to our current combustion-based energy infrastructure [16]. There have been reports of variations in the production processes, Different molar ratios of alcohols to fatty acids, enzymatic and chemical catalysis, or noncatalytic transesterification can be used to produce biodiesel [17, 18].

WCO the term 'waste cooking oils' means vegetable oils that are used in food processing and are no longer viable for their intended use [19]. WCO originates from different sources such as domestic and industrial/commercial areas. The waste stream is potentially disturbing and needs to be properly managed [20]. The disposal of the WCO can be problematic if the oils are inappropriately disposed of through the drains in the kitchen, which can easily contribute to the obstruction of sewer pipes. Not only does petroleum diesel affect our atmosphere through hazardous material pollution, but it also physically has adverse effects on us. Petroleum diesel emissions have been caused by many illnesses and health problems. These toxins have been linked to many cases of cancer, cardiovascular and respiratory problems, asthma, and lung infection [21–23]. Repurposing it as biodiesel is one viable option for dealing with all this waste cooking oil. This approach will have positive effects on both the economy and the environment.

Due to its low cost as raw material and the fact that it does not require expensive waste removal and treatment, used cooking oil has great potential as a replacement for traditional biodiesel production [24]. Since feedstock expenses account for between 75 to 95% of the total cost of biodiesel production, making biodiesel from WCO can help keep production costs down. Plus, it lessens demands on farmland to grow biodiesel feedstock. These recycled frying oils, however, are not the same as refined or crude vegetable oils. Oil's free fatty acid (FFA) content rises when exposed to heat and water, which speeds up the hydrolysis of triglycerides. Experiments concerning the production of biodiesel from USCO were carried out as part of this study in the designed and developed oil/alcohol molar ratio batch reactor system (1:5), 90-min reaction time, reaction temperature (65°C) with the use of three different percentages of NaOH and MgO catalysts (1.0 wt.%, 1.5 wt.%, and 2 wt.%).

2 Materials and Methods

2.1 *Materials*

Raw USCO is used for the production of biodiesel and is obtained from hotels in the local area. Non-oil components of the USCO were eliminated by philter separation and moisture was extracted by heating the oil at around 120°C for 30 min. Electric heater heating is typically the best way to get the oil up to the appropriate temperature. NaOH, MgO, and 99.5% pure methanol were utilized in this study.

2.2 *Batch Reactor System (BRS)*

The batch reactor might be nothing more than a tank with some kind of agitation system installed in it. BRS was developed based on the design parameters, which included a working volume of 3.5 L at 2 bar pressure and 250 °C temperature. Using various dimensionless groups to preserve dynamic similarity and then adding empirical design correlations is a recommended technique to develop BRS. Shell dimensions for the BRS are 530 mm x 210 mm x 3 mm, and the shell jacket measures 603 mm x 165 mm x 2 mm. The actual operational volume of the reactor is 3.5 L, whereas the total volume of the reactor is 5 L. With a shaft diameter of 8 mm, a length of 210 mm, and blade dimensions of 16 mm x 12 mm x 2 mm, the agitator is used for mixing. In addition, a viewing window 220 mm x 30 mm x 3 mm with a mounting plate of 200 mm x 3 mm x 1 mm is installed on the reactor for use during operation. Air spurge provision is also offered to the BRS. For the purpose of providing heating, a special kind of heater that has a heating coil diameter of 10 mm was developed. The material used for BRS fabrication is SS 316.

Most biodiesel is made in relatively small facilities using batch reactors. Figure 1, demonstrates the designed and fabricated batch reactor system assembly. During this procedure, oil, alcohol, and a catalyst are added to a tank and stirred with an agitator for a while. The reactor's contents are drained and processed after the requisite time. Biodiesel production methods were analyzed so that a batch reactor with a framework could be designed for simple use. Due to the endothermic nature of the reaction, biodiesel production reactors incorporate heating elements to provide heat for the reactor chamber. Batch reactors start with unreacted material, react it, and end with reacted material. Figure 2 shows the designed and fabricated laboratory-scale biodiesel production experimental setup.

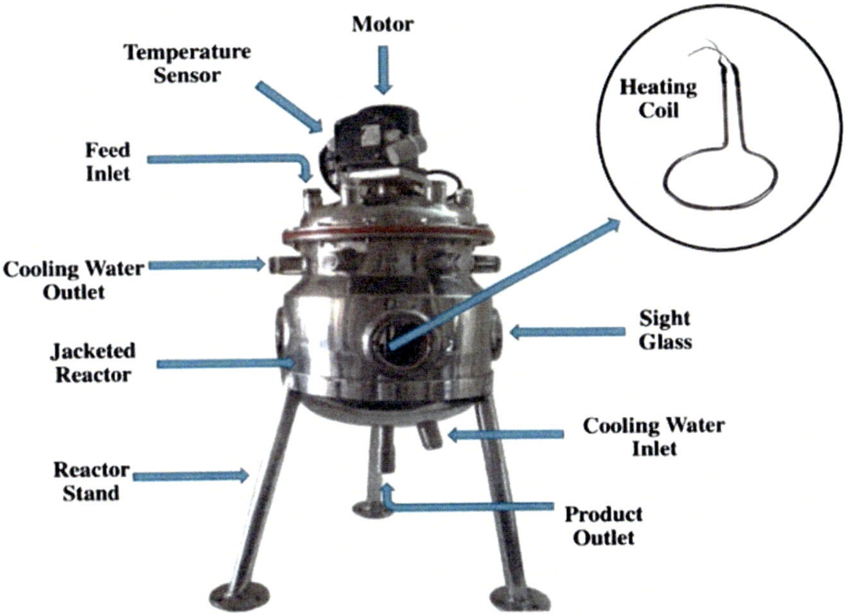

Fig. 1 Designed and fabricated batch reactor system assembly

Fig. 2 Laboratory scale biodiesel production experimental setup

2.3 *Experimental Approach (Transesterification) for Biodiesel Production*

In a designed and fabricated BRS, the transesterification reaction was carried out. Raw USCO as feed was rendered water-free, as some of the catalysts would be absorbed by any water or moisture in the system and the transesterification reaction will be slowed. For the catalyst concentration such as 1.0 wt.%, 1.5 wt.%, and 2 wt.% of NaOH and MgO at a reaction period (90 min) at 65°C and atmospheric pressure, the transesterification process was examined. The USCO sample on the weighing machine was carefully measured. The study consisted of 90 min of continuous agitation. It was feculent or turbid in the reaction mixture. The evaporative phase (methanol) is then condensed and refluxed into the reactor. The mixture was taken out from the outlet on the bottom side of the reactor until the reaction was over and placed in the separating funnel. As a result of transesterification, two phases (having different densities) are generated [25, 26]. The upper layer was composed of biodiesel, soap, and alcohol. Impurities, excess alcohol, glycerin, and catalyst were also found in the lower layer. Upper layer purification was performed, i.e. alcohol removal by maintaining the mixture at an elevated temperature of 80°C and extracting the saponified product by washing it with hot water. Figure 3 is a process flow diagram showing the production of raw USCO biodiesel.

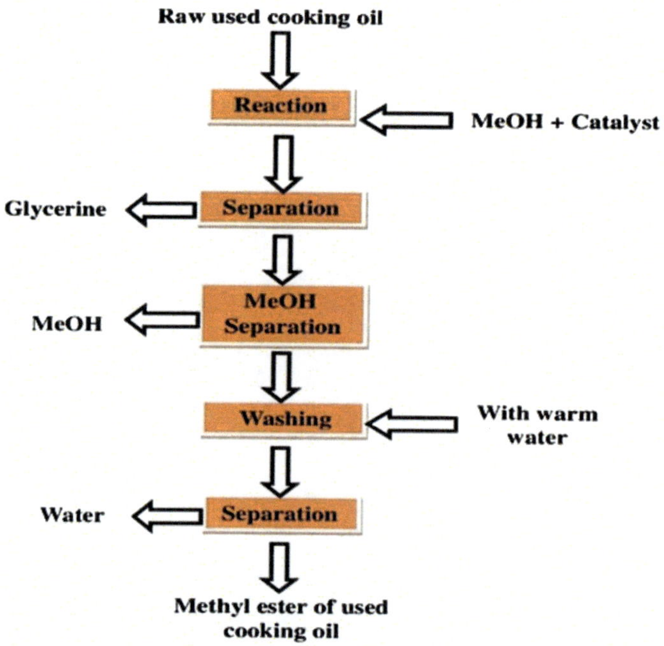

Fig. 3 Schematic representation showing the steps required to produce biodiesel from USCO

3 Results and Discussions

The reaction was conducted at 65°C for 90 min using NaOH and MgO catalysts at 1, 1.5, and 2 wt.%, respectively. The effect on the yield of USCO at 65°C of the amounts of the NaOH and MgO catalyst is shown in Figs. 4, 5 respectively. At 1.0 wt. % of NaOH catalyst, the maximum yield was obtained for USCO. It has also shown that the percentage of yield decreases as the quantity of catalyst (NaOH and MgO) increases. Results show that by using 1.0 wt. % NaOH and MgO catalysts, the yield of biodiesel derived from USCO is 96% and 93% respectively. It can be concluded that biodiesel yield by using the NaOH catalyst showed very promising results at the acceptable operating conditions of the temperature of reaction of 65°C, the reaction time of 90 min, and concentrations of 1 wt. % NaOH. Characteristics of biodiesel derived from USCO that were determined from an analysis of a sample acquired through the process with catalyst NaOH are presented in Table 1. Figure 6 shows the comparative results i.e. yield (%) for NaOH and MgO.

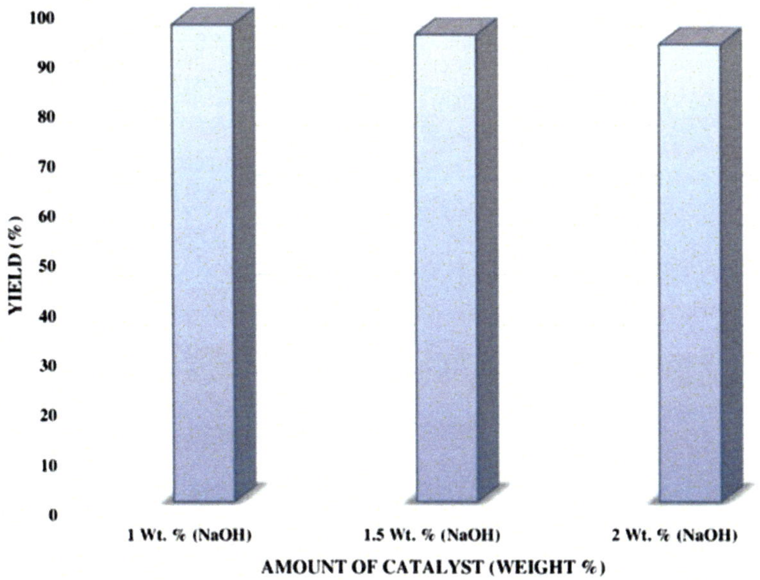

Fig. 4 The yield of USCO (%) of biodiesel production using NaOH catalyst at 65 °C, 90 min

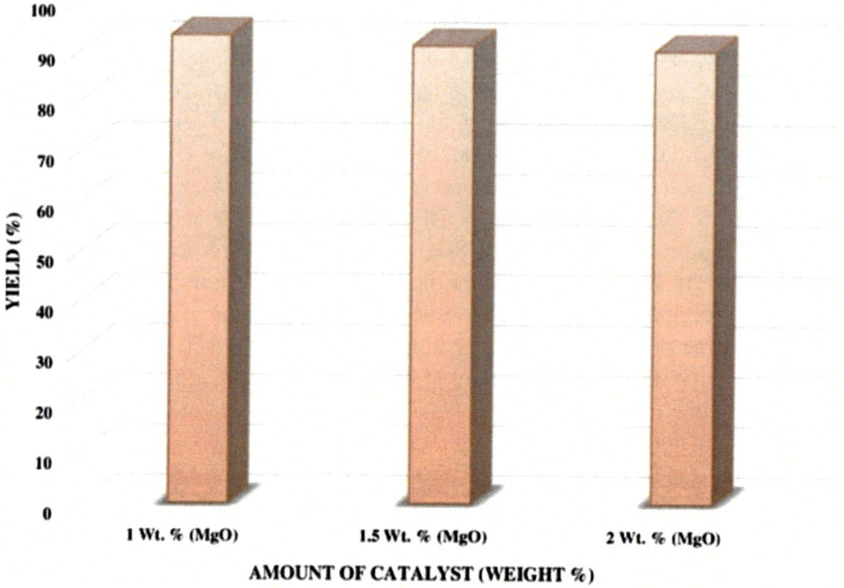

Fig. 5 The yield of USCO (%) of biodiesel production using MgO catalyst at 65 °C, 90 min

Table 1 Characteristics of biodiesel derived from USCO

Property	Biodiesel from USCO
Appearance	Clear
Color	Brownish
Density at 15°C	0.886
Flashpoint (°C)	160
Viscosity (cP)	4.92
Cloud point (°C)	−3
Sulfur contents (w/w%)	0.0135
Cetane number	55
Pour point (°C)	−8
Water content (w/w%)	0.043

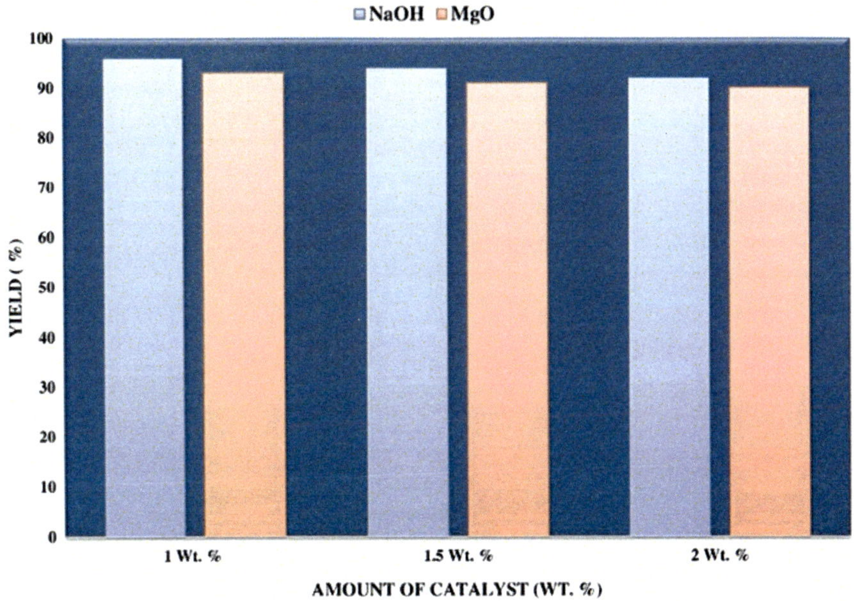

Fig. 6 Comparative results i.e. yield (%) for NaOH and MgO

4 Conclusion

Production of biodiesel in a BRS with a capacity of 3.5 L was successfully designed, fabricated, and implemented without encountering any difficulties. Transesterification of USCO and methanol in the presence of NaOH and MgO was examined at reaction conditions such as 1:5 oil to alcohol molar ratio, 90 min reaction time, and 65 °C reaction temperature. At 1.0 wt.% catalyst concentration, 65 °C reaction temperature, and 90 min reaction time, the results demonstrate that NaOH yielded around 96% biodiesel whereas MgO yielded 93% biodiesel. NaOH, a strong base, produces a greater yield of biodiesel than methoxide derived from MgO, a weaker base. Furthermore, it is observed that increasing the NaOH and MgO catalyst concentrations decreased the yield of biodiesel generated from USCO. Figure 7 represents the effective route for the conversion of USCO.

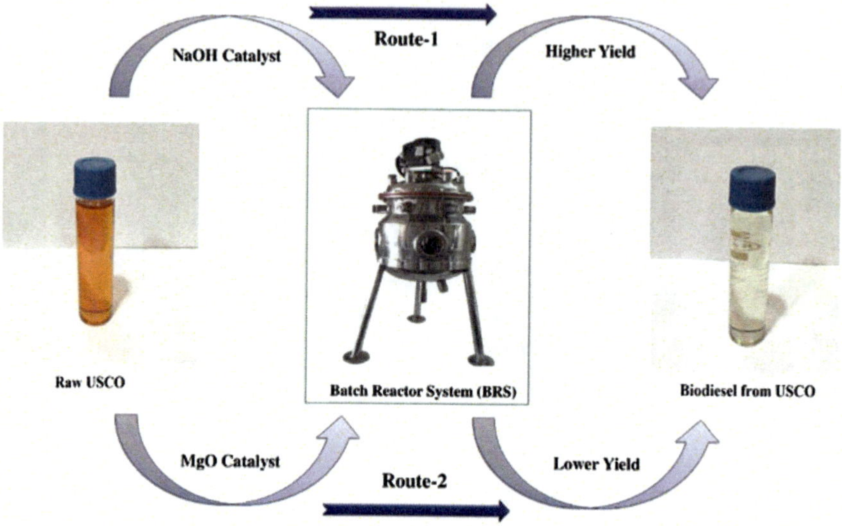

Fig. 7 Effective route for the conversion of USCO

References

1. Ibrehem AS, Salim HSA (2009) Advanced mathematical model to describe the production of biodiesel process. Bull Chem React Eng & Catal 4(2):37–42
2. Ghazi ATIM, Resul MFMG, Yunus R, Yaw TCS (2008) Preliminary design of oscillatory flow biodiesel reactor for continuous biodiesel production from jatropha triglycerides. J Eng Sci Technol 3(2):138–145
3. Kawentara WA, Budiman A (2013) Synthesis of biodiesel from second-used cooking oil. Energy Procedia 32:190–199
4. Fjerbaek L, Christensen KV, Norddahl B (2009) A Review of the current state of biodiesel production using enzymatic transesterification. Biotechnol Bioeng 102(5):1298–1315
5. Said NH, Ani FN, Said MFM (2015) Review of the production of biodiesel from waste cooking oil using solid catalysts. J Mech Eng Sci 8:1302–1311
6. Abdullah NH, Hasan SH, Yusoff NRM (2013) Biodiesel production based on waste cooking oil (WCO). Int J Mater Sci Eng 1(2):94–99
7. Valle P, Velez A, Hegel P, Mabe G, Brignole EA (2010) Biodiesel production using supercritical alcohols with a non-edible vegetable oil in a batch reactor. J Supercrit Fluids 54:61–70
8. Kumaran P, Mazlini N, Hussein I, Nazrain M, Khairul M (2011) Technical feasibility studies for langkawi WCO (Waste Cooking Oil) derived-biodiesel. Energy 36:1386–1393
9. Topare NS, Jogdand RI, Shinde HP, More RS, Khan A, Asiri AM (2022) A short review on approach for biodiesel production: Feedstock's, properties, process parameters and environmental sustainability. Materials Today: Proceedings 57:1605–1612
10. Rahadianti ES, Yerizam M (2018) Biodiesel Production from Waste Cooking Oil Indones. J. Fundam. Appl. Chem. 3(3):77–82
11. Raqeeb MA, Bhargavi R (2015) Biodiesel production from waste cooking oil. J Chem Pharm Res 7(12):670–681
12. Blinova L, Fiala J, Balog K (2014) Biodiesel production from waste cooking oil in laboratory scale. Appl Mech Mater 448–453:1656–1659

13. Hajjari M, Tabatabaei M, Aghbashlo M, Ghanavati H (2017) A review on the prospects of sustainable biodiesel production: a global scenario with an emphasis on waste-oil biodiesel utilization. Renew Sustain Energy Rev 72:445–464
14. Topare NS, Patil KD, Khedkar SV (2020) Synthesis of biodiesel from waste cooking oil and emission characteristics of Its blends. IOP Conf. Series: Materials Science and Engineering 983, 012015
15. Ajala EO, Aberuagba F, Olaniyan AM, Ajala MA, Sunmonu MO, Oduwole MM (2018) Design, construction and performance evaluation of a mini-scale batch reactor for biodiesel production: A case study of shea butter. Songklanakarin J Sci Technol 40(5):1066–1075
16. De Paola MG, Mazza I, Paletta R, Lopresto CG, Calabro V (2021) Small-Scale Biodiesel Production Plants-An Overview. Energies 14:1901
17. Lopresto CG, Naccarato S, Albo L, De Paola MG, Chakraborty S, Curcio S, Calabro V (2015) Enzymatic transesterification of waste vegetable oil to produce biodiesel. Ecotoxicol Environ Saf 121:229–235
18. Olaoye JO, Adegite JO, Salami HA (2017) Development of a laboratory scale biodiesel batch reactor. Int. Res. J. Eng. Technol. 4:1698–1704
19. Tana YH, Abdullaha MO, Hipolito CN (2016) Comparison of biodiesel production between homogeneous and heterogeneous base catalysts. Appl Mech Mater 833:71–77
20. Adokiye T, Gunorubon AJ, Kenkugile DK (2020) Modeling and control of a biodiesel transesterification reactor. Adv Chem Eng Sci 10:210–224
21. Gopi R, Thangarasu V, Vinayakaselvi MA, Ramanathan A (2022) A critical review of recent advancements in continuous flow reactors and prominent integrated microreactors for biodiesel production. Renew Sustain Energy Rev 154:111869
22. Shahare P, Harode A, Ingale A, Godbole E, Lilhare A, Deshmukh R (2018) Fabrication of Small-Scale Biodiesel Plant. Int J Innov Res Technol. 5:7–10
23. Baydir E, Aras O (2022) Increasing biodiesel production yield in narrow channel tubular reactors. Chem Eng Process 170:108719
24. Yaashikaa PR, Kumar PS, Karishma S (2022) Bio-derived catalysts for production of biodiesel: A review on feedstock, oil extraction methodologies, reactors and lifecycle assessment of biodiesel. Fuel 316:123379
25. Topare NS, Khedkar SV, Patil KD, Inamdar N (2021) Design, fabrication of multi-functional reactor system and its application for biodiesel production. AIP Conf Proc 2417(1):030002
26. Rehman MA, Seth D, Shrivastava RL (2016) Impact of green manufacturing practices on organisational performance in Indian context: an empirical study. J Cleaner Prod 137:427–448

Drivers of Industry 4.0 Operations for Logistics Management: An Analysis of Critical Performance Indicators for Last Mile Delivery

Vijay Prakash Sharma, Surya Prakash, Ranbir Singh, and Amiya Kumar Dash

Abstract The research article addresses the issues in implementing Industry 4.0 (I4.0) regarding technological advancements. This study focuses on the critical performance indicators (CPIs) regarding I4.0 adoptions and utilizes them for the analysis of efficient logistics systems. The logistics sector is growing at a rapid rate, and there is a need for the implementation of I4.0 technologies to resolve modern abnormalities in the supply chain specifically the last mile delivery (LMD). This study formulates the CPIs for I4.0 in logistics management for LMD using an Analytical Hierarchy Process (AHP). Route planning and Digitalisation are the most crucial factors in I4.0 for increasing the competency of the LMD operations. The factors are focused on logistics management in generalized form specific case studies will help to analyse the CPIs effect operationally. This study provides a quantitative approach for studying I4.0 diffusion in logistics management which can help to present a more stabilized model for logistics 4.0.

Keywords Industry 4.0 (I4.0) · Logistics Management · Critical performance indicators (CPIs) · Last mile delivery (LMD) · Analytical Hierarchy Process (AHP)

V. P. Sharma (✉) · R. Singh · A. K. Dash
School of Engineering and Technology, BML Munjal University, Gurugram, Haryana, India
e-mail: vijay.bml2017@gmail.com

R. Singh
e-mail: ranbir.singh@bmu.edu.in

A. K. Dash
e-mail: amiya.dash@bmu.edu.in

S. Prakash
Department of Operations Management, Great Lakes Institute of Management, Gurgaon, Haryana, India

H. Kumar et al. (eds.), *Recent Advances in Intelligent Manufacturing*, Lecture Notes in Mechanical Engineering,
https://doi.org/10.1007/978-981-99-1308-4_6

1 Introduction

Logistics Management is a crucial part of a modern supply chain that provides services to customers as per their needs and demand. Any organization's logistics management is key to its development and success. Logistics management facilitates the integration of several business partners or organizations to satisfy the customer's extremely variable demands [1]. Any business that experiences uncertainty may occasionally suffer significant losses. In the modern supply chain logistics management is a crucial part that needs to be studied and improved to increase the system's overall efficiency [2]. LMD is the major challenge for any logistics firm which expends almost one-third of the total transhipment cost and has the highest impact on the customer satisfaction rating. The use of technology in process management for LMD operations lowers delivery times and achieves deadlines at the lowest possible cost, the demands of all parties must be met across the whole supply chain process [3]. This can be accomplished by continuing to respond to market demands at a startling pace. To last a long time on the market in the dynamic, unstable, and competitive environment, automation, technological integration, speed, and manoeuvrability are essential. The drivers for LMD in I4.0 are needed to be investigated which will enhance the overall process. It helps to add the existing system network design to the supply chain.

I4.0 and the "fourth industrial revolution" are two words that may be used interchangeably to refer to a new phase in the organization and structure of the industrial value chain. Cyber-physical systems, such as "smart machines," are the core of I4.0 to converse and get solutions via the Internet of Things. They utilize contemporary control systems, possess embedded software, and have a unique internet address that requires permission protocol as one of the functions of blockchain [4]. Implementing I4.0 technologies lessens the supply chain's irrationality or randomness in managing logistics operations. Technological advancement in the supply chain facilitates efficient and dependable interactions between producers, suppliers, customers, and middlemen as well as a properly coordinated and integrated manufacturing and distribution system [5]. I4.0 has made its impact on each industry and the logistics sector also needed to be explored. This will help industrial engineers to develop innovative canters for supply chain and logistics.

The current trend in the logistics sector has shown a need for employing I4.0 technologies for improving customer service, cost efficiency, and technology upgradation. Government is also revising many incentives for adopting new advanced sustainable technologies in the service sector. This study is focused on investigating the drivers of I4.0 for logistics management about the fundamental features driving the next industrial revolution. This research article will address the following research questions (RSQ):

RSQ1: What is the current technological advancement in the logistics sector?

RSQ2: What are the I4.0 CPIs driving the logistics management phenomenon?

RSQ3: What is the impact of the I4.0 CPIs on the efficiency parameters of the supply chain specifically for LMD?

The research questions mentioned above help to analyse the drivers of I4.0 operations for logistics management by investigating the CPIs, especially for LMD. The most important factors are identified by literature and a survey for their establishment. These factors are analysed by the popular Analytic Hierarchy Process (AHP) methodology with regard to various crucial parameters. This study is a conscious effort to investigate the I4.0 technologies adoption in Logistic management for LMD operations. This study provides the fundamental knowledge about the driving factors, and it helps to provide initial priority order for adoption of I4.0 in LMD. This study helps to improve the existing supply chain network and present the various criterion for sustainability.

2 Literature Background

Industrial development had to move through a number of stages, including new strategies, new techniques, new forms, and their application, before we arrived at the most sophisticated version of the industry. Along with this industrial innovation, different social ideals were also being developed. The following figure shows how logistics are passed down through generations (Fig. 1).

At the beginning of the industrial phase, the experimentalists produced the development of an engine that moved man from manual labour to machines that made it more comfortable for the transportation of services of all kinds. This was important for logistics as well since it made the transit and storage processes simpler and more accessible. The second stage covers almost half of the preceding century and lasts for around 150 years. The absorption process of the massive production of commodities

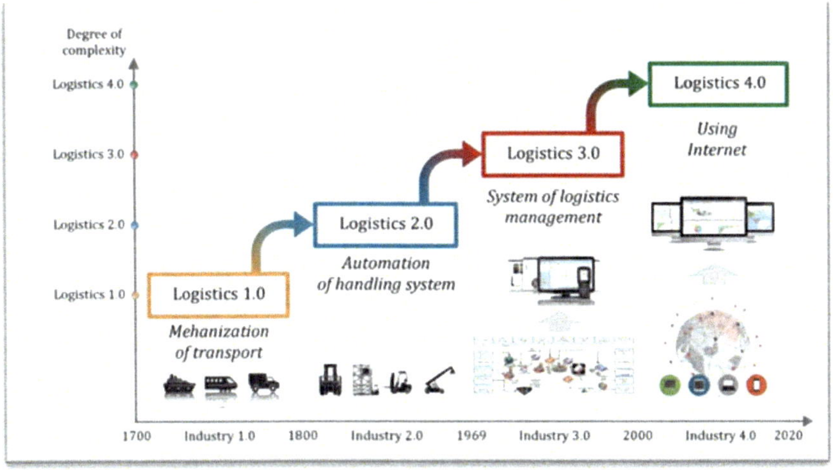

Fig. 1 History of technological advancement in the logistics sector (adapted from [6, 7])

and raw resources is what this stage depends on the most. Because logistics has reached a new peak in all of its related disciplines, this has a significant influence on the technological society as well [7, 8]. The automation industry has also expanded. Information and Communication Technology (ICT) sector and its applications are mostly covered in the following stage.

In the logistics and supply chain process, the computer is extremely important. Companies are increasingly relying on ICT fields for communication with new members and in new models. ICT is employed for innovative applications, fresh perspectives, novel data-exchange techniques, and their exterior connections with celebrations. The last or last stage started throughout this century, which had a minimal impact on ICT in activities connected to manufacturing. Digital and intelligent actions have largely covered a sizable region of market-related data, whether it be the one who is requesting or providing the thing. ICT had aided logistics in more ways than only. In 2011, the term "I4.0" was originally used in Germany. Various names for this technology are in use worldwide, including Smart Factories, Industrial IoT, and Smart Industry [5, 9].

2.1 Diffusion in Logistics Management

I4.0 creates a network of connections between people, technology, and the environment in which they live. This resulted in the virtual or imaginary world, which is entirely dependent on computers, fully replacing the manual handling procedure. ICT has made it possible to digitalize and connect things as well as their fresh ideas. I4.0 introduces new phases for a product's life cycle, including the creation of new products based on consumer demand and all associated tasks up until delivery [8]. Cyber-Physical Systems (CPS), Internet of Things (IoT), Digital Twin, Internet of Services (IoS), Big Data, Autonomous Drones & Robotics, Cloud Computing (CC), Augmented Reality (AR), Blockchain and horizontal-vertical combination of the system are the main elements that make up the concept of I4.0 [9, 10]. On the basis of intelligent machines, robots, and other automated equipment, smart factories are created. The study in the domain of I4.0 produced a spectacular result by demonstrating that investment in the field will rise over the course of the next year since it affects the majority of the industrial sector. This explains about the recent advancements in the logistics sector regarding technology upgradation [8, 11].

2.2 Prominent Characteristics of Logistics 4.0

Logistics 4.0, which debuted in 2011, enhanced every stage in the sector, including marketing, inventory management, supply chain management, etc. I4.0 is not partially diffused by logistics management for creating an auto-sustainable and responsive supply chain. Logistics 4.0 has the following features with regard to critical performance indicators (CPIs):

Digitalisation (AA1) sourced at [12–14]- It provides easiness to the data exchange process between the company and a complete combination of all the values of the supply chain. The logistics sector has undergone a major upheaval because of digital technologies [15]. Businesses may now use digitization to adjust their current policies and procedures, take on new projects, and even reinvent consumer experiences in order to meet evolving market needs and business conditions [11].

Technology integration (AA2) sourced at [16–18]- IoT had provided new forms of smart technologies where everything is mostly connected with automation products. Cloud computing had not only helped the supplier but also for the buyers to make their clumsy tasks easier and more convenient to handle. This integration process had also played a vital role in handling vast data in a proper and sequenced manner [13]. I4.0 had also worked on dismantling the process so that the burden will not be only on a single step. The improvement in the field of robotics and automation had also covered a great gap for moving to the zenith of the industry functioning [19].

Mass Customization (AA3) sourced at [20–22]- Mass customization refers to the capacity to produce products for customers financially and without sacrificing efficiency. The goal is to increase the production process' focus on the consumer. I4.0 absolutely flips this on its head. Due to sensors and communication technologies, machines on the production floor may now communicate, gather data, and transmit directives on their own [23]. These measures may be associated with other business units and the supply and delivery networks, increasing efficiency savings even further. Real-time activities include product traceability, OEE optimization, supply chain coordination, and more [15].

Route Planning (AA4) sourced at [24–26]- Companies in the supply chain are hungry for an automated solution to help them realize their potential, especially the ones that are dynamic, such as those with dynamic truck routing. I4.0 must be implemented in order to get organizations ready for dynamic supply chain systems. Although the Dynamic Vehicle Routing Problem (DVRP) has been the topic of several studies, little research has been done on how I4.0 may affect DVRP. An urgent need to develop and comprehend the concept of I4.0 has indeed been identified, and it has been proposed to apply it to dynamic vehicle routing problems time [3]. This acknowledgment has come from extensive reviews of recent literature on supply chain networks, dynamic vehicle routing issues, and I4.0.

Omni-channel retailing (AA5) sourced at [27–29]- The capability of a logistic planning system to rationally select amongst potential supply sources. This is true for choosing suppliers as well as choosing a source of supply and a network of providers for internal locations. The contemporary trend of the Omni-channel is the result of the quick evolution of sales and distribution channels in the previous age, which was primarily focused on the customer-cantered shopping experience and the increased significance of more tailored items and quicker delivery [30]. The enhancement of production and service systems is being driven by the rising demand for individualized products and services as well as disruptive information and production technologies. The RSQ2 is very well explained in this section.

3 Methodology

This research article initially determines the CPIs from the published recent literature and a questionnaire survey was prepared through multiple brainstorming sessions among the six experts. The survey was conducted in ten different logistics firms. 50 respondents replied to the questionnaire with a Likert scale of 1–5. The responses regarding the various CPIs are collected regarding I4.0 factors (AA1-AA5) and analysed about four important criterions, that is cost efficiency (CC1), technological advancement (CC2), customer satisfaction and urbanisation (CC3), government initiatives and policies (CC4) using AHP methodology. The flow chart of the research design is presented in Fig. 2.

3.1 AHP Method and Utilisation

AHP is a mathematical modelling tool that helps us to determine the quantitative nature of the observed data and their relevance with regard to various criteria and has the ability to convert the empirical surveys into quantitative data for analysis [20, 31]. In this paper AHP presents a hierarchical approach at four levels. This article compares the factors of I4.0 which are Digitalisation (AA1), Technology

Fig. 2 Steps adopted for conducting research design

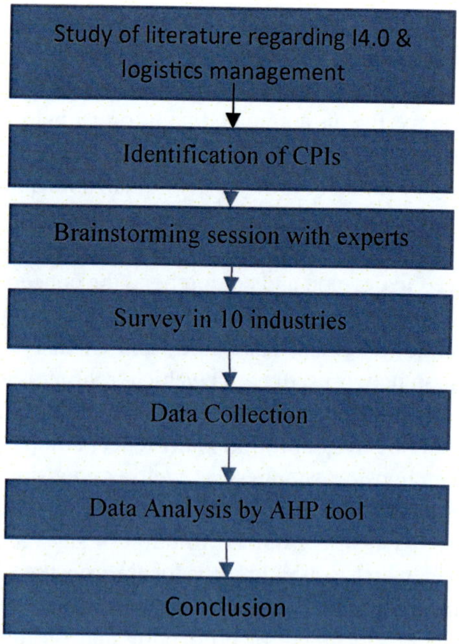

Integration (AA2), Mass Customization (AA3), Route Planning (AA4), and Omni-channel retailing (AA5) with respect to the criteria's cost efficiency (CC1), technological advancement (CC2), customer satisfaction and urbanization (CC3), government initiatives and policies (CC4).

3.2 AHP Implementation

AHP is implemented to get the results regarding various CPIs for all the major factors involved in the diffusion of I4.0 technologies in logistics management. Table 1 presents Project I4.0 crucial factors nomenclature, and Table 2 presents the evaluation criteria table for calculating the efficiency of the supply chain. In step 1 role of I4.0 technologies for logistics management is analysed in Table 3. It is very clear from Table 3 that customer satisfaction and urbanization lead to the implementation of I4.0 factors in logistics management at a higher impact level while cost efficiency decreases at the initial stage of introducing I4.0 technologies in the supply chain. In step-2 to step-6 each factor (AA1-AA5) is evaluated concerning criteria (CC1-CC5) respectively from Tables 4 and 5.

Step-1

See Table 3.

Step-2

See Table 4.

Table 1 Project I4.0 crucial factors nomenclature

S.NO	Options	Abbreviation/Code
1	Digitalization	AA1
2	Technology integration	AA2
3	Customization	AA3
4	Route planning	AA4
5	Omni-channel retailing	AA5

Table 2 Evaluation criteria table

S.NO	Criteria	Abbreviation/Code
1	Cost efficiency	CC1
2	Technological advancement	CC2
3	Customer satisfaction and urbanisation	CC3
4	Govt. Incentives and policies	CC4

Table 3 Criteria concerning I4.0 in logistics management for achieving efficiency

	CC1	CC2	CC3	CC4	Eigen Vector (EV)	Weight (WT)	Comp Eigen Vector (CEV)
CC1	1	1/3	1/3	3	0.7598	0.183	0.9434
CC2	3	1	1/5	5	1.3161	0.317	1.8445
CC3	3	5	1	1/5	1.3161	0.317	2.4876
CC4	1/3	1/5	5	1	0.7598	0.183	1.8924

$\lambda_{max} = 7.1678$, Consistency Index = 1.0559, Consistency Ratio = 1.1733

Table 4 Factors concerning cost efficiency

	AA1	AA2	AA3	AA4	AA5	EV	WT	CEV
AA1	1	3	1	3	3	1.9332	0.3497	2.0385
AA2	1/3	1	3	3	1	1.2457	0.2253	1.4298
AA3	1	1/3	1	1/5	3	0.7248	0.1311	0.8766
AA4	1/3	1/3	5	1	3	1.1076	0.2003	1.3282
AA5	1/3	1	1/3	1/3	1	0.5173	0.0936	0.5459

$\lambda_{max} = 6.219$, Consistency Index = 0.3047, Consistency Ratio = 0.2721

Step-3

See Table 5.

Step-4

See Table 6.

Step-5

See Table 7.

Step-6

See Table 8.

Table 5 Factors concerning technological advancement

	AA1	AA2	AA3	AA4	AA5	EV	WT	CEV
AA1	1	5	3	5	5	3.2719	0.5071	2.8555
AA2	1/5	1	5	1/5	3	0.9029	0.1399	1.0248
AA3	1/3	1/5	1	1/3	1/3	0.3749	0.0581	0.3534
AA4	1/5	5	3	1	1/5	0.9029	0.1399	1.1463
AA5	1/5	1/3	3	5	1	1	0.155	1.177

$\lambda_{max} = 6.5569$, Consistency Index = 0.3892, Consistency Ratio = 0.3475

Table 6 Factors concerning customer satisfaction and urbanization

	AA1	AA2	AA3	AA4	AA5	EV	WT	CEV
AA1	1	1/5	1/3	1/5	1/5	0.3056	0.0522	0.2646
AA2	5	1	3	1/3	1/3	1.1076	0.1893	1.1591
AA3	3	1/3	1	3	1/3	1	0.1709	1.37
AA4	5	3	1/3	1	3	1.7188	0.2938	2.0612
AA5	5	3	3	1/3	1	1.7188	0.2938	1.7335

$\lambda_{max} = 6.5883$, Consistency Index = 0.3971, Consistency Ratio = 0.3545

Table 7 Factors concerning govt. incentives and policies

	AA1	AA2	AA3	AA4	AA5	EV	WT	CEV
AA1	1	1/3	1/3	1/5	1/3	0.3749	0.0574	0.3113
AA2	3	1	1/5	1/3	1/3	0.5818	0.0891	0.5273
AA3	3	5	1	1/5	1/5	0.9029	0.1382	0.8988
AA4	5	3	5	1	3	2.9542	0.4522	2.4868
AA5	3	3	5	1/3	1	1.7188	0.2631	1.5443

$\lambda_{max} = 5.7684$,Consistency Index = 0.1921 ,Consistency Ratio = 0.1715

Table 8 Final priority order for I4.0 Factors for improving performance efficiency

Digitalisation	0.2518
I4.0 technology integration	0.1619
Customization	0.1219
Route planning	0.2569
Omni-channel retailing	0.2075

4 Results and Analysis

It is very prominent from Tables 4 and 5 that digitalization (AA1) plays the most important role in controlling the expenditures of I4.0 technology implementation and diffusion in logistics. Customization at the mass level will be crucial but not much helpful in cost reduction. A lot of costs are involved in adding customization features in product delivery in the logistics sector. Route planning emerged as the crucial factor for customer satisfaction as it affects variables like the right product delivery at right time to the right person. Urban cities have a lot of traffic conjunctions leading to delays in delivery and dissatisfaction among customers. It also creates a loss for the company due to the cancellation of orders which are already dispatched from the source supplier. Proper route planning and omnichannel retailing with multiple suppliers also help government employment initiatives and policies entrepreneurship at small scale enterprise level as found in Tables 6 and 7.

The final combined study considering all four criteria CC1, CC2, CC3 and CC4 for I4.0 drivers' diffusion in logistics presents the Final priority order for I4.0 Factors for improving performance efficiency following priority as shown in Table 8.

$$AA4 < AA1 < AA5 < AA2 < AA3$$

Route planning and digitalization are the two most crucial factors that indicate the overall performance of the supply chain can be increased in I4.0 era if they are taken care of well. These drivers of I4.0 have the maximum impact on logistics sector efficiency. Route planning (AA4) and Digitalisation (AA1) are the most important CPIs that are needed to be planned for effective LMD operations in Logistics 4.0. Effective route planning will increase the time and cost efficiency simultaneously in LMD operations. Similarly, digitalisation remains the fundamental and most important feature of Logistics 4.0 for transparency and effectiveness of any organisation. Omni-Channel retailing and I4.0 technology integration stand at third and fourth positions showing that multiple retailers and suppliers are needed to be established at primary stage before the implementation of I4.0 technologies. This will make a proper utilisation of I4.0 technologies for LMD operations. Customization is the last parameter in the scenario as it really depends on customer wish to opt for this specific feature. It could be a crucial factor at maturity stage of Logistics 4.0 leading to higher customer satisfaction and delivery efficiency. The RSQ3 is explained in this section.

5 Conclusion

Logistic 4.0 had provided a lifeline or a new life to most of the companies that were on the verge of extension or need to flourish in this competitive environment. As there is no exact data for the number of changes that this logistic 4.0 had made in the field of the industry but according to the data collected it is shown properly that this had made a remarkable upliftment of the companies. Digitalization and route planning can really help the logistics sector to improve its efficiency at LMD operations. It is very important to create a sensor-based tracking system for real-time data reporting and efficient route planning. I4.0 provided a complete correlation between the real and imagined product specification and supply chain. It has helped in providing an instant solution to problems and their related feedback. I4.0 technology has provided advancement in all the processes of the industry. I4.0 has increased business efficiency and provides easily available data and their changing flexibility in the modern supply chain. This is very evident from the above study that the I4.0 system implementation in logistics sector will definitely enhance the working capability of firms. There is a need for complete digitalisation and proper route planning for efficient LMD operations which can lead to economic and organisational growth. This study provides the major driving factors that are needed to be involved while planning an efficient smart supply chain. The priority level will help engineers and supply chain managers to design an I4.0 system for modern warehouses and logistics networks. It is also very evident from the study that drivers of I4.0 are needed to be co-existing together at different supply chain stages where customer satisfaction is the final goal.

6 Future Scope

I4.0 requires is a huge amount of wealth required during the time of implementation. A company had to completely change its ways of governance of product formation which required plenty of time and money. I4.0 requires highly skilled and efficiently equipped labour for running cyber-physical system. It requires complete information regarding the new methods, machines, and strategies for adoption of I4.0 at large scale. New and advanced machines which may be automated or may be robotized are required for this which generally increased the inventory cost of the company which is a big cause of concern for small companies. There is high flexibility in data about usage I4.0 systems for improvement of supply chains and logistics networks for LMD leading to requirement of dynamic routing methodologies. Most reputed and advanced companies are required to adopt these smart technologies presenting a guideline to SMEs and MSME sector. A standardized system for all industry's performance measurements is difficult to make and customized industry-specific study for Industry 4.0 can be done on firm-to-firm, company-to-company, and industry-to-industry bases for determining crucial performance indicators. These indicators will various industries to improve their efficiency in operational stages for future studies. These CPIs can act as a fundamental block for design a complex I4.0 model/ framework for efficient LMD operations in modern supply chains.

References

1. Wang M, Wood LC, Wang B (2020) Logistics innovation capability and its impacts on the supply chain risks in the Industry 4. 0 era. Mod Supply Chain Res Appl 2(2), 83–98. https://doi.org/10.1108/MSCRA-07-2019-0015
2. Bank R, Murphy R (2013) Warehousing sustainability standards development. IFIP Adv Inf Commun Technol. IFIPAICT 414, 294–301, Springer (2013), doi: https://doi.org/10.1007/978-3-642-41266-0_36
3. Bachofner M, Lemardelé C, Estrada M, Pagès L (2022) City logistics: challenges and opportunities for technology providers. J Urban Mobil 2(1), 100020. https://doi.org/10.1016/J.URBMOB.2022.100020
4. Fonseca LM (2018) Industry 4 . 0 and the digital society : concepts, dimensions and envisioned benefits. In Proceedings of the international conference on business excellence vol. 12 no. 1, pp. 386–397. https://doi.org/10.2478/picbe-2018-0034
5. Ola et al. (2017) Logistics 4 . 0 and emerging sustainable business models. Adv Manuf 5(4), 359–369. https://doi.org/10.1007/s40436-017-0198-1

6. Radivojević G, Milosavljević L (2019) The concept of logistics 4.0. 4th Logist. Int. Conf., LOGIC vol. 1, pp. 283–293. https://logic.sf.bg.ac.rs/wp-content/uploads/LOGIC_2019_ID_32.pdf
7. Wang K (2016) Logistics 4.0 solution-new challenges and opportunities. In 6th international workshop of advanced manufacturing and automation, vol. 6, no. 11, pp. 68–74. Atlantis Press. https://doi.org/10.2991/iwama-16.2016.13
8. Hu H, Wu J, Wang Z, Cheng G (2018) Mimic defense: a designed-in cybersecurity defense framework. IET Inf Secur 12(3):226–237. https://doi.org/10.1049/iet-ifs.2017.0086
9. Liboni LB et al. (2019) Smart industry and the pathways to HRM 4 . 0 : implications for SCM. Supply Chain Manag 24(1), 124–146. https://doi.org/10.1108/SCM-03-2018-0150
10. Fink AA, Klöckner M, Räder T, Wagner SM (2022) Supply chain management accelerators: types, objectives, and key design features. Transp Res Part E Logist Transp Rev 164(8), 102728. https://doi.org/10.1016/j.tre.2022.102728
11. Liao Y, Deschamps F, De Freitas E, Loures R (2017) Past, present and future of Industry 4 . 0 - a systematic literature review and research agenda proposal. Int J Prod Res 55(12), 3609–3629. https://doi.org/10.1080/00207543.2017.1308576
12. Agostini L, Nosella A (2020) The adoption of Industry 4 . 0 technologies in SMEs : results of an international study. Manag Dec 58(4), 625–643. https://doi.org/10.1108/MD-09-2018-0973
13. Wang Y, Sarkis J (2021) Emerging digitalisation technologies in freight transport and logistics: current trends and future directions. Transp Res Part E Logist Transp Rev 148(4), 102291. https://doi.org/10.1016/J.TRE.2021.102291
14. Pan S, Trentesaux D, McFarlane D, Montreuil B, Ballot E, Huang GQ (2021) Digital interoperability in logistics and supply chain management: state-of-the-art and research avenues towards. Phys Int Comput Ind 128(6), 103435. https://doi.org/10.1016/J.COMPIND.2021.103435
15. Castelo-Branco I, Oliveira T, Simões-Coelho P, Portugal J, Filipe I (2022) Measuring the fourth industrial revolution through the Industry 4.0 lens: The relevance of resources, capabilities and the value chain. Comput Ind 138(6), 103639. https://doi.org/10.1016/j.compind.2022.103639
16. Pessot E, Zangiacomi A, Rocchi V (2020) What matters in implementing the factory of the future manufacturing regions. J Manuf Technol Manag 32(3), 795–819. https://doi.org/10.1108/JMTM-05-2019-0169
17. Wagire AA, Rathore APS, Jain R (2020) Analysis and synthesis of Industry 4 . 0 research landscape using latent semantic analysis approach 31(1), 31–51. https://doi.org/10.1108/JMTM-10-2018-0349
18. Lu Y (2017) Cyber physical system (CPS)-based industry 4.0: a survey. J Ind Integ Manag 2(3), 1750014. https://doi.org/10.1142/S2424862217500142
19. Kerin M, Pham DT (2019) Smart remanufacturing : a review and research framework. J Manuf Technol Manag 31(6), 1205–1235. https://doi.org/10.1108/JMTM-06-2019-0205
20. Bakhtari AR, Waris MM, Sanin C, Szczerbicki E (2021) Evaluating industry 4.0 implementation challenges using interpretive structural modeling and fuzzy analytic hierarchy process. Cybern Syst 52(5), 350–378. https://doi.org/10.1080/01969722.2020.1871226
21. Zhong RY, Dai QY, Qu T, Hu GJ, Huang GQ (2013) Robotics and computer-integrated manufacturing RFID-enabled real-time manufacturing execution system for mass-customization production 29(1), 283–292. https://doi.org/10.1016/j.rcim.2012.08.001
22. Ploos Van Amstel R, D'hert G (1996) Performance indicators in distribution. Int J Logist Manag 7(1), 73–82. https://doi.org/10.1108/09574099610805458
23. Hietanen A, Pieters R, Lanz M, Latokartano J, Kämäräinen JK (2020) AR-based interaction for human-robot collaborative manufacturing. Robot Comput Integr Manuf 63(6), 101891. https://doi.org/10.1016/j.rcim.2019.101891
24. Bai X, Lam JSL (2021) Freight rate co-movement and risk spillovers in the product tanker shipping market: a copula analysis. Transp Res Part E Logist Transp Rev 149(5), 102315. https://doi.org/10.1016/j.tre.2021.102315
25. Chackelson C, Errasti A, Ciprés D, Lahoz F (2013) Evaluating order picking performance tradeoffs by configuring main operating strategies in a retail distributor: A Design of Experiments approach. Int J Prod Res 51(20), 6097–6109. https://doi.org/10.1080/00207543.2013.796421

26. Gentili M, Mirchandani PB, Agnetis A, Ghelichi Z (2022) Locating platforms and scheduling a fleet of drones for emergency delivery of perishable items. Comput Ind Eng 168(6), 108057. https://doi.org/10.1016/j.cie.2022.108057
27. Wu J, Wang H, Shang J (2019) Multi-sourcing and information sharing under competition and supply uncertainty. Eur J Oper Res 278(2), 658–671. https://doi.org/10.1016/J.EJOR.2019.04.039
28. Perera S, Dawande M, Janakiraman G, Mookerjee V (2019) Retail Deliveries by Drones: How Will Logistics Networks Change? SSRN Electron J 29(9), 2019–2034. https://doi.org/10.2139/SSRN.3476824
29. Osakwe CN, Hudik M, Říha D, Stros M, Ramayah T (2022) Critical factors characterizing consumers' intentions to use drones for last-mile delivery: does delivery risk matter? J Retail Consum Serv 65, 102865. https://doi.org/10.1016/J.JRETCONSER.2021.102865
30. Kumar M, Basu P, Avittathur B (2018) Pricing and sourcing strategies for competing retailers in supply chains under disruption risk. Eur J Oper Res 265(2):533–543. https://doi.org/10.1016/j.ejor.2017.08.019
31. Sureshchandar GS (2022) Quality 4.0 – understanding the criticality of the dimensions using the analytic hierarchy process (AHP) technique. Int J Qual Reliab Manag 39(6), 1336–1367. https://doi.org/10.1108/IJQRM-06-2021-0159

Stable Zone Identification During Machining on CNC Lathe Using ANFIS

Pankaj Gupta, Sachin Gupta, and Bhagat Singh

Abstract Wavelet Denoising and Local Mean Decomposition (WDLMD) decompose the vibration signals obtained experimentally into Product Functions (PFs). The most important PFs that reveal chatter information are chosen from among these. These well-known PFs are combined to create a signal that depicts real-time discourse. Following that prediction model for the Chatter Index (CI) was established utilising the reconstructed chatter signal. According to the findings, increasing spindle speed improves machining stability. However, there is a significant rise in chatter as the depth of cut and feed rate are raised. Additionally, for the best stable zone prediction, the modeling of CI w.r.t. cutting parameters has been applied using ANFIS.

Keywords Stability analysis · Local mean decomposition · Wavelet denoising · CNC lathe · ANFIS

1 Introduction

Chatter is a serious impediment to manufacturing industry automation in the machining process [1–3]. As a result, monitoring the machining process is critical for detecting and controlling regenerative chatter [4, 5]. Once the chatter has been recognized, suppressing it is a difficult operation. Chatter can be reduced by adjusting cutting parameters to their optimal values, according to some researchers. The SLD can be employed to anticipate the best combination of spindle speed and cut depth [6]. The most extensively used method for predicting a stable cutting zone is the SLD [7, 8]. It explains the stable and unstable boundary in terms of cutting parameters depth of cut and spindle speed [9]. But feed rate is lacking, which should not be neglected. Because of insufficient engineering support, it is sometimes impossible to forecast the optimal cutting range using SLD. As a result, a technique to detect

P. Gupta (✉) · S. Gupta · B. Singh
Jaypee University of Engineering and Technology, Guna, MP, India
e-mail: pankajguptamech88@gmail.com

H. Kumar et al. (eds.), *Recent Advances in Intelligent Manufacturing*,
Lecture Notes in Mechanical Engineering,
https://doi.org/10.1007/978-981-99-1308-4_7

the commencement of chatter, as well as processing and exploration of recorded experimental signals, is required.

The microphone is one of the sensors that plays a significant part in the capture of chatter signals [10]. During milling, researchers employed a microphone to collect acoustic signals [11]. The audio signal is used for non-destructive analysis which is vibration signals [12]. The obtained chatter signals are also said to be affected by ambient noise, masking the true nature of chatter features [13]. The most extensively utilised chatter detection approaches are time domain analysis [14], time–frequency domain analysis [15, 16], and frequency domain analysis [17] are the most widely used chatter detection techniques. Huang [18] used cutting force variation to define cutting stability. Chatter identification has also been investigated in face milling operations using FFT spectrum analysis of prominent frequencies and energy ratio [19]. The peaks found in these tests are related to chatter frequencies as well as undesired contaminations and ambient sound. As a result, removing these noise contents before processing the signal for feature extraction is critical.

Various signal processing techniques has been used in the past by researchers. In this paper authors propose a novel technique called WDLMD. Wavelet denoising is a technique for cleaning up chatter signals of unwanted environmental noise and impurities [20, 21] and then LMD is used to demodulate the signals into a number of PFs, separating the chatter frequencies. The CI statistical parameter was examined for chatter feature extraction when the chatter was discovered.

2 Methodology

For better illustration, methodology is shown in Fig. 1.

2.1 *Theoretical Analysis*

Understanding the chatter phenomenon requires mathematical modelling of the rotating mechanism. This created model has been used to simulate the turning process. The mathematical model was created for orthogonal turning operation in (sDoF) is represented as

$$m\frac{d^2x(t)}{dt^2} + c\frac{dx(t)}{dt} + kx(t) = \Delta F(x, t) \tag{1}$$

where m is the mass of the system, k is the spring constant and c is the damping coefficient.

Since, natural frequency $w_n = \sqrt{k/m}$ and damping ratio $\zeta = c/2\sqrt{\mathrm{km}}$, Hence, the equation of motion yields

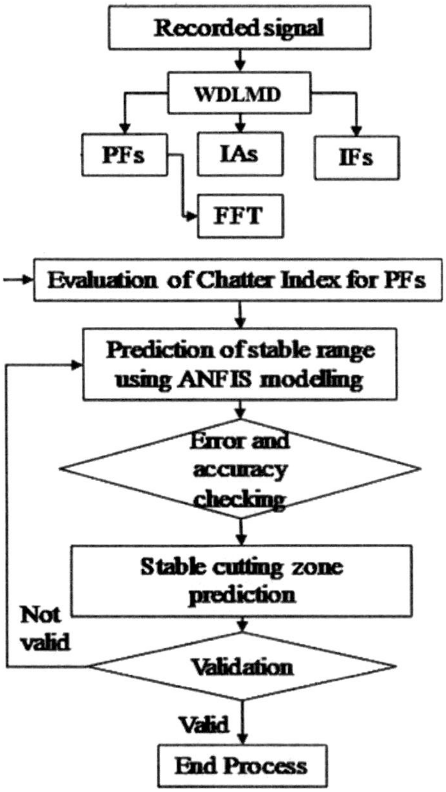

Fig. 1 Methodology

$$\frac{d^2x(t)}{dt^2} + 2\zeta w_n \frac{dx(t)}{dt} + w_n^2 x(t) = \frac{\Delta F(x,t)}{m} \quad (2)$$

Wavelet Denoising Technique (WDT)

The most popular analytical method for frequency domain analysis is spectral analysis based on the Fourier transform. The Fourier transform, however, is unable to offer any data on how the spectrum varies over time. In contrast to the assumption made by the Fourier transform, machining signals are never stationary. WDT can clean up pollutants while protecting sensitive data. The two steps of this method are wavelet decomposition and thresholding. The noisy signal is decomposed in the first stage. The signal length determines the level of decomposition (signal size must be divisible by two levels). After passing the noisy signal through low and high-pass filters, coefficients are calculated.

WDT steps: Apply a wavelet transform to a noisy signal to obtain noisy wavelet coefficients at a level that allows for adequate differentiation. Depending on the level and the threshold technique, choose the appropriate threshold limit. The threshold wavelet coefficients are subjected to an inverse wavelet transform to produce a denoised signal.

Soft Thresholding

$$W_{\lambda\_S} = \begin{cases} W_\lambda - \lambda & |W_\lambda| \geq \lambda \\ 0 & |W_\lambda| < \lambda \\ W_\lambda + \lambda & |W_\lambda| \leq -\lambda \end{cases} \quad (3)$$

where W_λ is the noisy wavelet coefficient and λ is the threshold.

Hard Threshold

$$W_{\lambda\_H} = \begin{cases} W_\lambda & |W_\lambda| \geq \lambda \\ 0 & |W_\lambda| < \lambda \end{cases} \quad (4)$$

Coefficient values are either kept or removed via hard thresholding.

Local Mean Decomposition (LMD)

As shown in Fig. 2, the LMD technique involves the following steps.

3 Real-Time Turning Experiment Results and Analyses

3.1 Experimentation

The CNC lathe: MTAB XL-TURN, workpiece: Al-6061-T6, AHUJA AGN-480 dynamic unidirectional microphone and carbide insert: TTS04 is utilised to validate the suggested WDLMD process. The full factorial design for CI as formulated in Eq. (5) is stated in Table 1.

$$\text{Chatter Index (CI)} = \frac{1}{N}\sum_{n=1}^{N} |x(t)| \quad (5)$$

3.2 WDLMD on Signal

In one of the research, Matlab software was used to filter out the noise from the machining signal using db5 wavelet coefficients as shown in Fig. 3, at a level that allows for adequate differentiation. Depending on the level and the threshold technique, choose the appropriate threshold limit. The thresholded wavelet coefficients

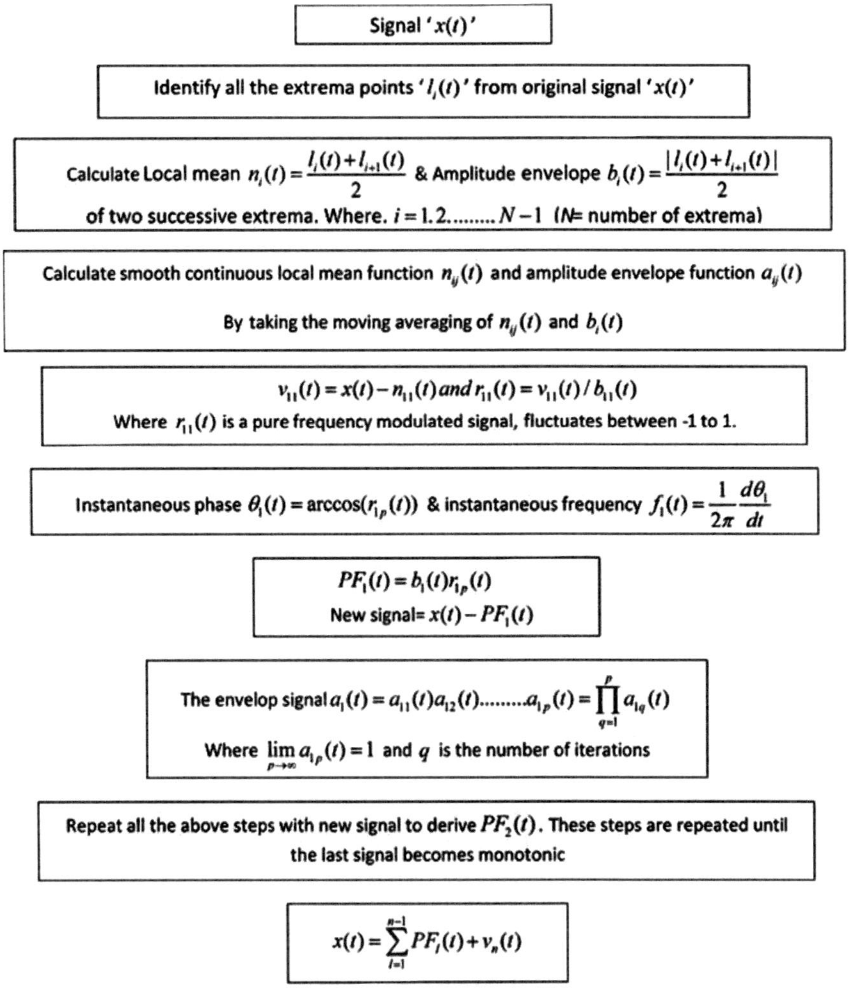

Fig. 2 Flow chart of LMD

are subjected to an inverse wavelet transform to produce a denoised signal. LMD has also been used to WDT-derived denoised signals. The reconstructed signal is shown in Fig. 4 after the FFT analysis of all PFs was done for visual comprehension.

3.3 Adaptive Neuro-fuzzy Inference System (ANFIS)

Fuzzy logic conceptualized with neural network are called ANFIS [22, 23]. The benefits of both methodologies are kept in this strategy, resulting in enhanced forecast

Table 1 Turning parameters and chatter index (CI)

(d) mm	(n) rpm	(f) mm/min	(CI)	(d) mm	(n) rpm	(f) mm/min	(CI)
0.2	1000	30	0.86	0.3	1500	40	1.7
0.2	1000	35	0.91	0.3	2000	30	3.42
0.2	1000	40	1.16	0.3	2000	35	3.45
0.2	1500	30	1.13	0.3	2000	40	3.73
0.2	1500	35	1.8	0.4	1000	30	1.02
0.2	1500	40	1.8	0.4	1000	35	0.93
0.2	2000	30	2.9	0.4	1000	40	1.26
0.2	2000	35	3	0.4	1500	30	2.34
0.2	2000	40	3.23	0.4	1500	35	2.43
0.3	1000	30	0.99	0.4	1500	40	2.64
0.3	1000	35	0.89	0.4	2000	30	3.07
0.3	1000	40	1.13	0.4	2000	35	3.14
0.3	1500	30	1.4	0.4	2000	40	3.47
0.3	1500	35	1.47				

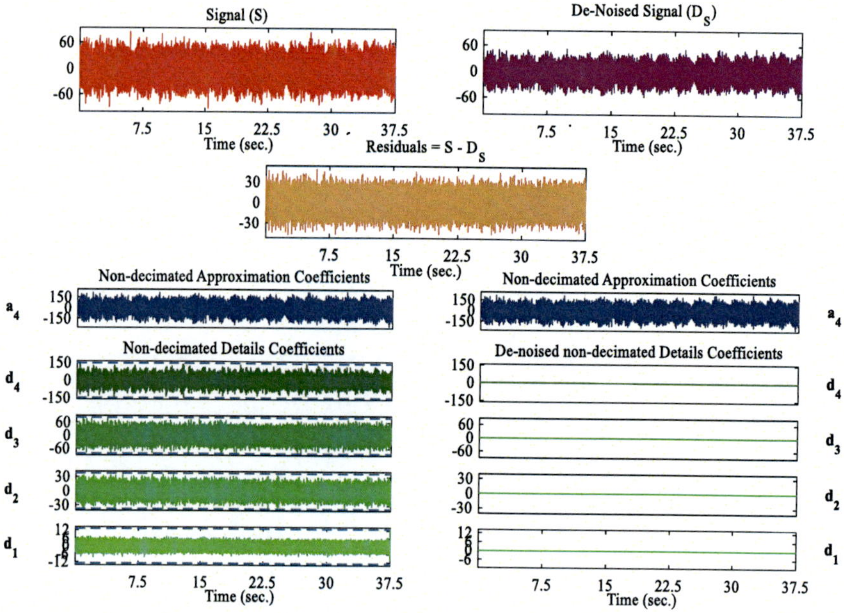

Fig. 3 Wavelet denoising of signal

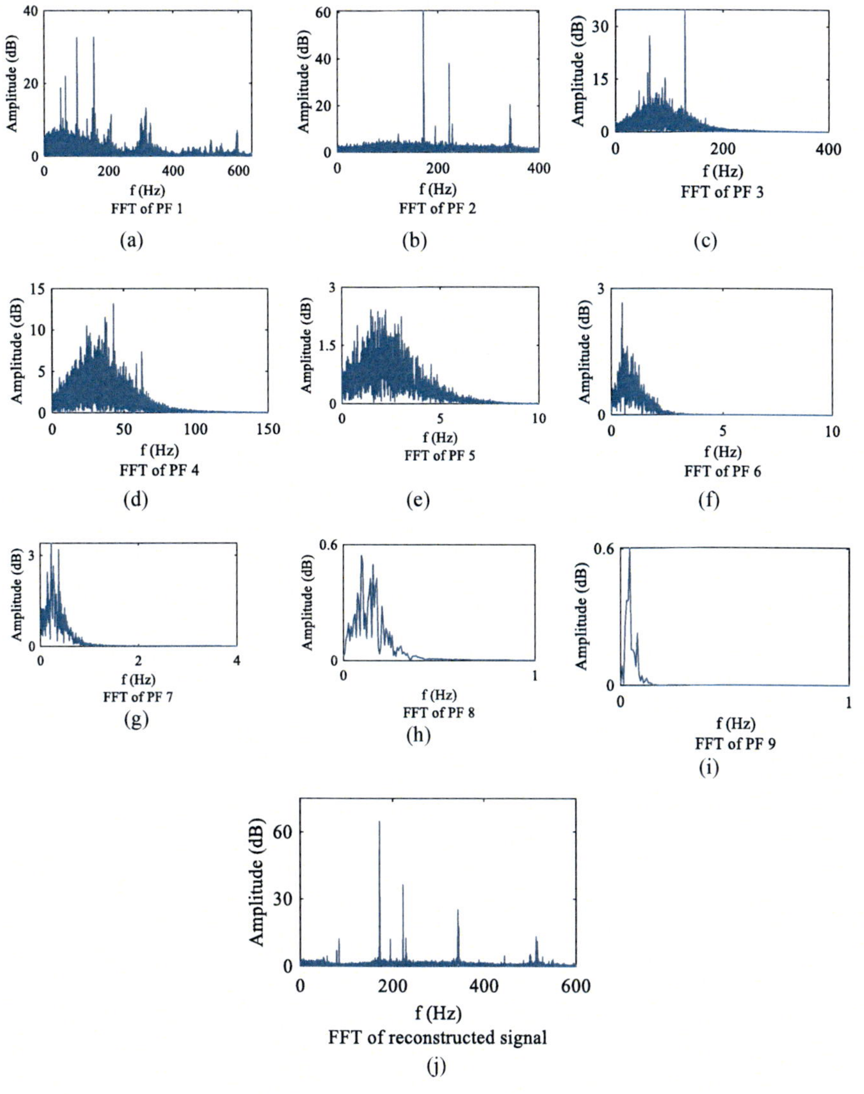

Fig. 4 FFT of PFs and reconstructed signal

accuracy. ANFIS model can be trained and if–then fuzzy rules built to calculate the membership function for input and output variables. Figure 5 depicts Sugeno's fuzzy logic paradigm and ANFIS architecture, which is a multilayer feed-forward network made up of nodes and directed links that connect them.

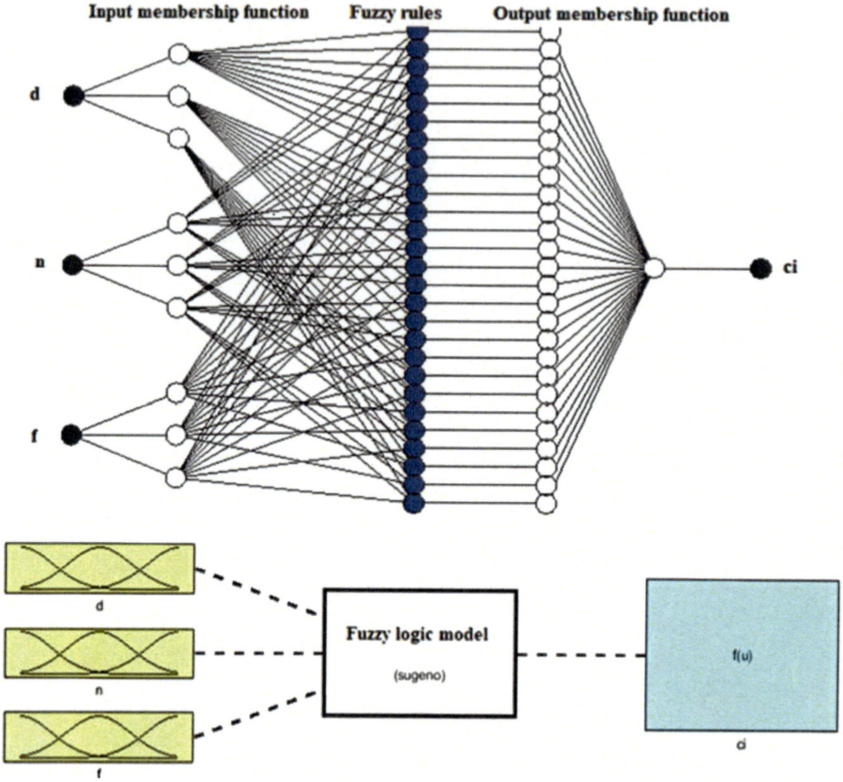

Fig. 5 Sugeno's fuzzy logic paradigm and the ANFIS architecture

The fuzzy inference diagram for a FIS is displayed by the Rule Viewer, which is invoked with the MATLAB command rule- view. The ANFIS model's Rule Viewer is shown in Fig. 6. This figure also depict the training of input data w.r.t CI.

Validation of the Created ANFIS model's Predictive Power

Predicted values are contrasted with experimental values as given in Table 2, to evaluate the ANFIS model's predictive skills, which has been determined that the average percent divergence is only 3.68 percent. The model that has been created is important as a result.

4 Result and Discussion

Determine safe machining ranges based on the minimum value of CI and the variance in CI as a function of the input parameters. 3D graphs of CI w.r.t. d/n/f are shown in Fig. 7. When these figures are analyzed using two parameters at a time but the effect

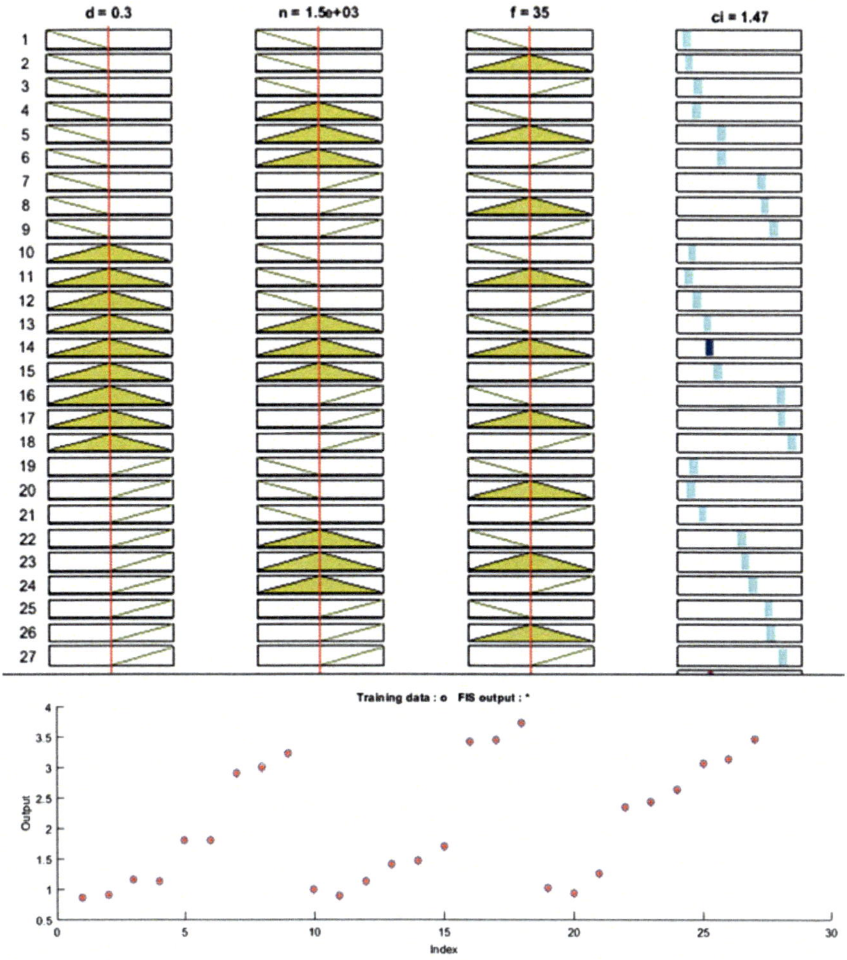

Fig. 6 ANFIS rule viewer and data training

of feed rate has also been studied in this study. It has been deduced that feed rate is another crucial element that must be appropriately adjusted in order to reduce noise. The developed stable range are; viz. d (0.25–0.36) mm, n (1000–1500) rpm, and f (33–37.2) mm/min,

Table 2 Experimental versus predicted values of CI

d (mm)	n (r.p.m)	f (mm/rev)	CI (predicted)	CI (experimental)	%Deviation
0.2	1000	40	1.16	1.12	3.44
0.2	2000	30	2.9	2.67	7.93
0.2	2000	35	3	3.07	2.33
0.3	1000	35	0.89	1.03	15.7
0.3	1000	40	1.13	1.22	7.96
0.3	2000	30	3.42	3.48	1.75
0.3	2000	35	3.45	3.48	0.86
0.3	2000	40	3.73	3.72	0.26
0.4	1000	35	0.93	1.07	15.05
0.4	1000	40	1.26	1.32	3.68
			Average % deviation		3.68

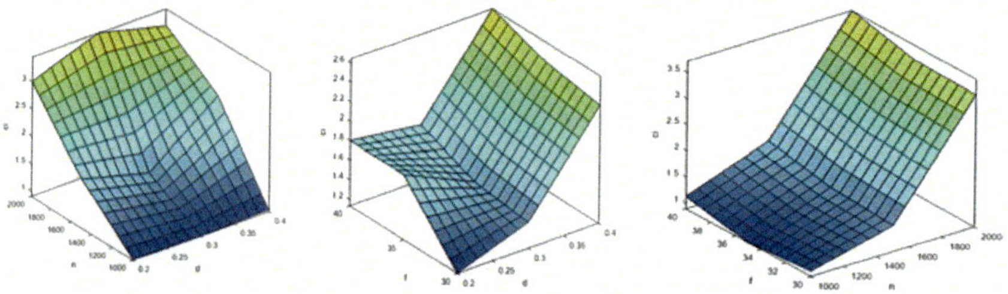

Fig. 7 Variation of CI w.r.t. d/n/f

5 Conclusion

In this paper, a unique methodology for determining the appropriate range of cutting parameters that will result in stable machining is proposed. The ANFIS model's average deviation is 3.68% indicating that the generated model is significant. The results show that for the examined range of input parameters, satisfactory surface quality may be achieved by keeping the input parameters within n (1000–1500) rpm, d (0.25–0.36) mm, and f (33–37.2) mm/min. The proposed methodology may aid researchers and machinists in achieving steady machining.

Funding There has been no funding for this project from any organization.

References

1. Munoa J, Beudaert X, Dombovari Z, Altintas Y, Budak E, Brecher C, Stepan G (2016) Chatter suppression techniques in metal cutting. CIRP Ann 65:785–808
2. Mancisidor I, Pena-Sevillano A, Dombovari Z, Barcena R, Munoa J (2019) Delayed feedback control for chatter suppression in turning machines. Mechatronics 63:102276
3. En X, Zhang Y, Huang X, Li Y, Yang GJMBDoS (2020) Time-varying reliability and global sensitivity analysis of regenerative chatter stability in turning considering tool wear. Machines 1–21
4. Wan M, Feng J, Ma Y-C, Zhang W-H (2017) Identification of milling process damping using operational modal analysis. Int J Mach Tools Manuf 122:120–131
5. Yang Y, Zhang W-H, Ma Y-C, Wan M (2016) Chatter prediction for the peripheral milling of thin-walled workpieces with curved surfaces. Int J Mach Tools Manuf 109:36–48
6. Maamar A, Bouzgarrou BC, Gagnol V, Fathallah R (2017) Time domain stability analysis for machining processes. In: Advances in acoustics and vibration. Springer, pp 77–88
7. Qiu J, Ge RJIJoMS (2018) An improved stability lobe and turning chatter characteristic investigation 149:338–348
8. Copenhaver R, Schmitz T, Smith SJCA (2018) Stability analysis of modulated tool path turning 67:49–52
9. Lu K, Lian Z, Gu F, Liu HJMS (2018) Model-based chatter stability prediction and detection for the turning of a flexible workpiece. Signal Process 100:814–826
10. Delio T, Tlusty J, Smith S (1992) Use of audio signals for chatter detection and control
11. Cao H, Yue Y, Chen X, Zhang X (2018) Chatter detection based on synchrosqueezing transform and statistical indicators in milling process. Int J Adv Manuf Technol 95:961–972
12. Filippov A, Nikonov AY, Rubtsov V, Dmitriev A, Tarasov SYJJoMPT (2017) Vibration and acoustic emission monitoring the stability of peakless tool turning. Exp Model 246:224–234
13. Shrivastava Y, Singh B (2019) A comparative study of EMD and EEMD approaches for identifying chatter frequency in CNC turning. Eur J Mech-A/Solids 73:381–393
14. Rafal R, Pawel L, Krzysztof K, Bogdan K, Jerzy W (2015) Chatter identification methods on the basis of time series measured during titanium superalloy milling. Int J Mech Sci 99:196–207
15. Liu C, Zhu L, Ni C (2017) The chatter identification in end milling based on combining EMD and WPD. Int J Adv Manuf Technol 91:3339–3348
16. Uekita M, Takaya Y (2017) Tool condition monitoring technique for deep-hole drilling of large components based on chatter identification in time–frequency domain. Measurement 103:199–207
17. Fu Y, Zhang Y, Zhou H, Li D, Liu H, Qiao H, Wang XJMS (2016) Timely online chatter detection in end milling process. Signal Process 75:668–688
18. Huang P, Li J, Sun J, Zhou J (2013) Vibration analysis in milling titanium alloy based on signal processing of cutting force. Int J Adv Manuf Technol 64:613–621
19. Wei C-C, Liu M-K, Huang G-H (2016) Chatter identification of face milling operation via time-frequency and Fourier analysis. Int J Automat Smart Technol 6:25–36
20. Bonda AGY, Nanda BK, Jonnalagadda S (2020) Vibration signature based stability studies in internal turning with a wavelet denoising preprocessor. Measurement 154:107520
21. Zaida H, Bouchelaghem AM, Chehaidia SE (2021) Experimental study of tool wear evolution during turning operation based on DWT and RMS. In: Defect and diffusion forum. Trans Tech Publications, pp 392–405
22. Jang J-S (1993) ANFIS: adaptive-network-based fuzzy inference system. IEEE Trans Syst Man Cybern 23:665–685
23. Kumar S, Singh B (2019) Chatter prediction using merged wavelet denoising and ANFIS. Soft Comput 23:4439–4458

A Compherehive Study on Supply Chain Management Using Artificial Intelligence: An Indian Railway Perspective

Manoj Kumar, Vipin, and Ashish Agarwal

Abstract If this trend is not reversed, railroads' growth and existence will be in jeopardy because freight service is their main source of revenue. Railways are fast losing market share in the case of freight service. This essay examines how passengers and other stakeholders feel about the current state of IR. Using the example of the Indian Railway as a case study, this essay analyses the supply chain management framework in the context of public procurement. The article discusses supplier relationship management's impact on performance indicators. It manages public procurement. The study discusses Indian railway supply chain management. The findings could help policymakers incorporate commercial management into public administration.

Keywords Indian railway · Key performance indicator · Public procurement

1 Introduction

The Ministry of Railways oversees Indian Railways (IR). Material Management Department of the Indian Railway follows public procurement rules and procedures to ensure ongoing material supply for rolling stock manufacture and maintenance, repair, and overhaul of railway assets [1]. An efficient and effective supply chain management is vital to the Indian Railways for production and maintenance of railway assets, inventory management with budgetary constraints. Strict rules

M. Kumar (✉)
Research Scholar, Delhi Technological University, Delhi—110042, India
e-mail: manojkumar@abes.ac.in

Vipin
Department of Mechanical Engineering, Delhi Technological University, Delhi—110042, India

A. Agarwal
School of Engineering and Technology, IGNOU, New Delhi—110068, India

M. Kumar
Assistant Professor, ABES Engineering College, Ghaziabad, India

H. Kumar et al. (eds.), *Recent Advances in Intelligent Manufacturing*,
Lecture Notes in Mechanical Engineering,
https://doi.org/10.1007/978-981-99-1308-4_8

and procedures and transaction-based approaches hinder Supply Chain Management (SCM). A tiny percentage of bill-of-materials savings translates to enormous financial benefits and improved Indian Railways efficiency [2]. This report examines Indian Railways' upstream procurement concerns (SCM). Indian Railways is one of the largest government organisations, employing 14 lakh people and spending 1.05 lakh crore rupees annually. Railways must maintain a significant number of assets to run trains efficiently and boost capacity by constructing 500 locomotives and 3000 carriages annually. Nearly 20% of the Railways' budget and assets are used for material purchase and administration [2]. Indian Railways formed a separate service, "Indian Railway Stores Services," to highlight the importance of purchase. UPSC recruits engineers to manage the procurement system and assure maintenance and production material availability. With decreasing budgetary support and rising fuel prices, and social obligations limiting fare increases, railways must think outside the box to make their systems more efficient, right-size their organisations, and critically examine procurement, manufacturing, and maintenance practises to eliminate redundant procedures [3]. Considering the volume and value of purchases, it's vital to relook at the Railways' supply chain system and employ the latest techniques in supply chain management to decrease wasteful spending. Supply chain management is an efficient technique to right-size a company. Supply chains can be redesigned scientifically to manage inventory, utilise backward integration, satisfy end-user needs, and maintain a tight balance between material availability and utilisation [4]. Unnecessary operations can be eliminated to free up labour for more productive uses.

2 Supply Chain Management

Supply chain is a network of organizations that link raw material supply to ultimate delivery. Raw materials are purchased, stored, loaded, and shipped to end users. Transportation networks are among the most critical building blocks in a country's economic development [5]. The performance of a country's transportation network affects travel and goods transfer. Freight transport uses rail, road, air, water, and pipelines. Material management has progressed greatly in the last 50–60 years. In 1940–50, material management was called Chief Storekeeper. His priorities were material, warehousing, and logistics. Stock outs were more costly than excess inventories. In 1970–1980, as competition increased, businesses realized the relevance of MM as a tool for boosting competitive advantage and profitability. Material Manager is added. The focus was on decreasing material costs, value engineering, vendor development, and inventory management. In the 1990s, inbound and outbound logistic functions were combined to use common resources and skills. Now Supply Chain management is a full-fledged operation [6]. It encompasses multiple disciplines and provides a framework for study. The service provider's downstream supply chain influences customers' goods purchases.

3 Relation, Supply Chain Management, and Indian Railways

Supply specifications, contract terms, and conditions govern supplier interactions in public procurement and IR. Product design and value engineering suffer from poor supplier incentives. Supplier R&D and value engineering are limited. Suppliers are hesitant to invest in process capabilities, productivity, and tools due to uncertain future orders. In lump sum competitive bidding, bidders give minimum quality. Bill of materials savings require supplier engagement [7]. Long-term contracts offer cheaper costs, supplier responsiveness, and smaller inventory, etc. Long-term contracts do fall under public procurement, despite popular belief. UNCITRAL allows framework agreements as a mechanism of public procurement. The last four years of Indian railways stock procurement statistics were reviewed. 95% of purchases are from approved suppliers. Indian railway contributes significantly to these vendors' operations. Even though the contracts are lump sum, IR procures from the same set of suppliers, which means the supplier has a long-term commercial relationship with IR. Railway business has been recession-free and offers development potential. Vendor acceptance without engineering cost estimates causes cartelization, mistrust, and corruption. Small improvements can boost Indian Railway's efficiency and effectiveness. Each thing needs a unique code in a codification system. IR assigns item codes based on end use [8]. If different types of equipment use the same spare parts, the rule that each item has its own unique item code could be broken. There are ongoing efforts to unify the item code, but they don't always work. From the point of view of procurement, it may be better to classify item codes by industry category. The end use of items is used to divide purchases into subgroups. This led to different subgroups buying the same type of item. So, from a procurement point of view, grouping by industry category is better. Purchases are made for each item separately and on a lump-sum basis every year [9]. This increases cycle times and repeats work during annual contract completion. System of lump sum procurement of all items year to year has certain demerits, such as high cycle time, poor responsiveness, high inventory and high stock out situation, repetition of contracting effort, high cost of procurement due to uncertainty of future business and fixed set up, cost, tooling cost, arm's length supplier relation management, poor incentive for innovation and value engineering, etc. Ongoing vendor approval [10]. Technical and financial supplier competency is considered when approving, although engineering cost estimation is not discussed or bargained. This often causes cartels. Development and source approval lack openness. Pre-shipment inspections ensure incoming raw material quality. This increases system inventory, cycle time, poor responsiveness, and costs. In Fig. 1, we have initialized the factors of supply chain management to Indian railway perspective.

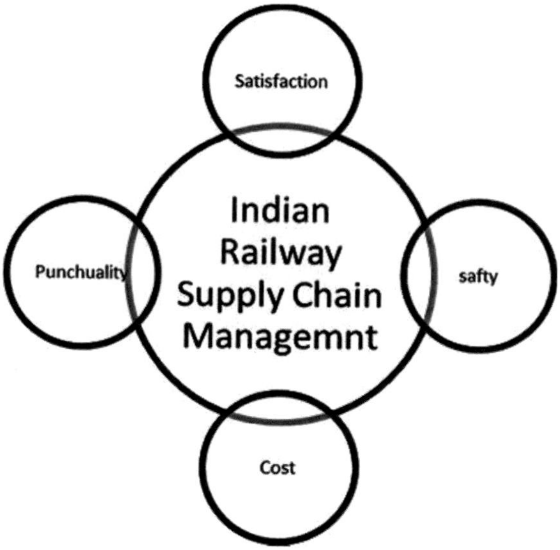

Fig. 1 Indian railway supply chain management factors

4 Machine Learning in Supply Chain Management

Retailers, wholesalers, shippers, manufacturers, and end users are all integral nodes in the supply chain. Because consumers are the driving force behind every supply chain, it's crucial to cater to their needs and preferences [11]. The availability chain facilities have learned to work together to meet demand. The entities collaborate to reduce supply chain costs. In the absence of such interactions, important and ideal supply Chain networks mismatch. Known and unknown factors cause the disparity. Alignment of company interests, long-term relationship management, unwillingness to share information, complexity of large-scale supply chain management, competency of supply chain management staff, performance management and incentive systems to assist supply chain management cause gaps [12]. With e-business and technology becoming more fluid and dynamic, corporations are reluctant to have long-term connections. Another factor could be not knowing buyer demand and creating more in advance. Machine Learning Models have replaced the traditional approach of knowing and forecasting demand. Supply chains are complicated and challenging, involving functions like purchasing, contracts, procurement, warehousing, production, packaging, shipping, and distribution. Each function is complex, thus combining them takes time, effort, and money. It's important to include quick-response tactics. Previously, basic decisions like product delivery to the client took a long time, but with AI and machine learning, this is now easier and the goods may be supplied within 24 h. This paper examines current and potential uses of Machine Learning in Supply Chain Management (Fig. 2).

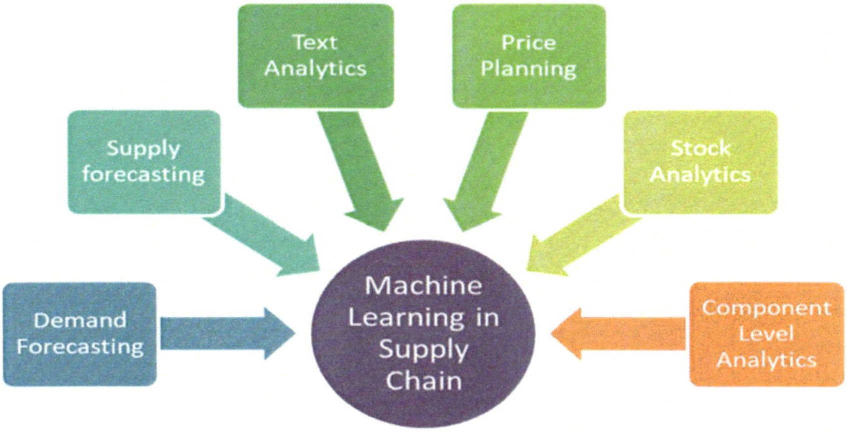

Fig. 2 Cases for machine learning in supply chain management

4.1 Predictions Approaches

ML has several types. Supervised, unsupervised, and reinforcement learning are well-known. Additional types [9].

4.1.1 Supervised Learning

Most people use Supervised Learning. Supervised Learning trains a computer program using example data. As the output is also known, this learning process seeks to uncover rules that relate input and output data and apply them to fresh data. The program is being trained.

4.1.2 Unsupervised Learning

Unsupervised learning is a knowledge-discovering mechanism. This method of learning lacks correct answers and pre-labeled target values [13]. "Learning without an instructor" describes this method. Unsupervised learning clustering is well-known. The approach categorizes inputs by common patterns by identifying similarities. Association Rules, Self-Organizing Maps, Multidimensional Scaling, Nonlinear Dimension Reduction.

Fig. 3 Process

4.1.3 Reinforcement Learning

Reinforcement learning requires iterative determination of the ideal solution, which is unknown at the start of the learning phase. This mechanism rewards logical approaches and punishes mistakes. This allows the system to respond to complicated external forces. Using rewards and punishments, the system seeks its own solutions. Machine Learning tasks can be categorized. Supervised learning includes classification and regression. Clustering and ARM are unsupervised learning problems. Reinforcement learning is another type. Figure 3 shows selected methods of ML, grouped by the task they commonly solve. This list of approaches is far from complete, and several can solve multiple tasks. Each technique has multiple algorithms. ML approaches and algorithms can provide complications. Error sources develop for two reasons. One problem is the data. Insufficiently representative training data and a lack of sufficient data for model training are still common issues. Bad data is another issue. First, patterns must be recognized in data sets. Identifying trends is difficult if the data contains errors or outliers. Unneeded features are another problem. Selecting suitable training characteristics is critical for ML performance.

4.2 Machine Learning Process

Real-world machine learning apps are frequently part of a larger project. Such initiatives have process model frameworks.

Figure 3's approach is used for ML projects and is a common standard. The six-phase method includes tasks. Starting with "Business Understanding" and finishing with "Deployment" The framework improves results while lowering project costs and time (Fig. 4).

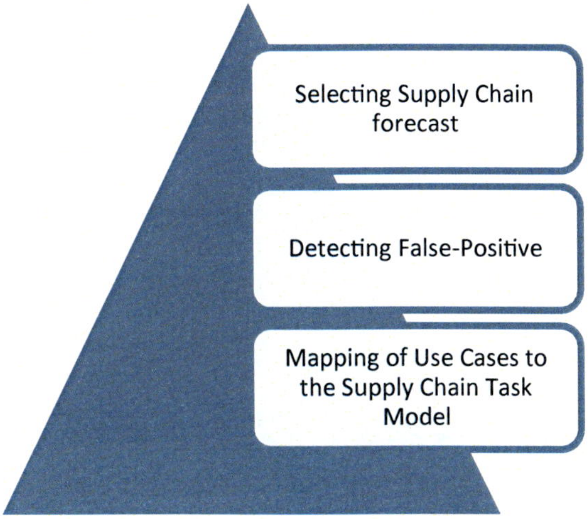

Fig. 4 Machine learning application areas in supply chain management

5 Outcomes and Key Findings

AI helps firms optimize processes, schedule production, manage supply risk, minimize supply chain interruption, decrease language barriers, improve customer service and quality control, and forecast demand. With these benefits, firms will confront several hurdles to maximize the new technology. Shortage of processing power, high cost, long implementation process, linguistic and social cue complexity, lack of data scientists and analysts, sharing data, legacy systems, and security concerns are the major difficulties today. As this technology advances, additional AI uses and difficulties will arise. AI and Machine Learning technologies allow computers to "learn" and mimic human activities in a continuous and autonomous process based on information and interactive processes. This lets AI generate more confident assumptions about new Indian Railway System supply chain management scenarios. AI has improved railway logistics efficiency and SCM parameters. Our research will focus on SCM and logistics outsourcing AI applications (3PL). Improved monitoring and visibility of supply chain processes allows dynamic decision making and optimization.

6 Conclusions and Implications

The paper explains how machine learning may improve supply chain management; especially lead time prediction and minimization. Lead time is vital for supply chain planning and customer satisfaction, thus it's been studied. Compared to standard programming, it rearranges tasks. Classical programming uses rules to calculate the desired result. In machine learning programming, known data and the desired output are represented to find unknown rules. This unique approach to issue solving is useful in business, especially supply chain management. Due to the high number of hidden and unpredictable components in a complex ecosystem like the supply chain, modeling it using normal methodologies can be difficult or impossible. Responding to demand unpredictability and reducing prices is crucial for Indian customers, who have several low-cost options. Thus, customers are crucial to value delivery. The Indian railway system should integrate its supply chain with customers to build customer trust by meeting their needs. Supply-chain methods enhance manufacturing to make items more competitive. Collaborative planning, forecasting, and restocking can help retailers manage product assortment. This will also aid supply issues.

References

1. Ellram LM (1991) Supply chain management: the industrial organization perspective, Int J Phys Distrib Logist Manag, 21(1), 13–22
2. Harris NG (1992) Punctuality and performance in planning passenger railways, Handbook. Transport Publishing Co., Ltd., Derbyshire England, pp 130–142
3. Olsson N, Haugland H (2004) Influencing factors on train punctuality—results from some Norwegian studies. Transp Policy 2(4):387–397
4. Faber N, de Koster M, Smidts A (2013) Organizing warehouse management. Int J Oper Prod Manag 33(9):1230–1256
5. Fleisch E, Tellkamp C (2005) Inventory inaccuracy and supply chain performance:a simulation study of a retail supply chain. Int. J. Production Economics 95:373–385
6. Ganeshan R, Boone T, Stenger AJ (2001) The impact of inventory and flow planningparameters on supply chain performance: An exploratory study. Int J Prod Econ 71(1–3):111–118. https://doi.org/10.1016/s0925-5273(00)00109-2
7. Gattorna J, Day A, Hargreaves J (1991) Effective logistics management. Logist Inf Manag 4(2):2–86. https://doi.org/10.1108/09576059110143603
8. Kumar SK, Tiwari MK, Babiceanu RF (2010) Minimization of supply chain cost with embedded risk using computational intelligence approaches. Int J Prod Res 48(13):3717–3739
9. Liu H, Yao Z, Zeng L, Luan J (2019) An RFID and sensor technology-based warehouse center: assessment of new model on a superstore in China. Assem Autom 39(1):86–100
10. Krishnaswamy, KN, AppaIyer Sivakumar, Mathirajan M (2006) Management research methodology: Integration of principles, Methods Tech
11. Ülgen V, Forslund H (2015) Logistics performance management in textiles supply chains: best-practice and barriers. Int J Product Perform Manag 64(1):52–75
12. Taylan O, Darrab IA (2012) Fuzzy control charts for process quality improvement and product assessment in tip shear carpet industry. J Manuf Technol Manag 23:402–420
13. Stalidis G, Karapistolis D, Vafeiadis A (2015) Marketing Decision Support Using Artificial Intelligence and Knowledge Modeling: Application to Tourist Destination Management. Procedia—Soc Behav Sci 175:106–113

A Quantification of Supply Chain Management Factors Using Artificial Intelligence

Manoj Kumar, Vipin, and Ashish Agarwal

Abstract

Purpose A supply chain manages the flow of goods and services, money, or information from the product's beginning to its end. As new technology is produced, the way organizations move information and things changes. AI is altering the supply chain for every industry, including car production. Supply chain management (SCM) involves corporate partners and end consumers. For this purpose, we have quantified the parameters of supply chain management respect to Indian railway system.

Objective This study proposes a novel way for analyzing and implementing collaborative AI-inspired SCM processes. AI in this setting is driven by developing supply networks and information systems. Design-science approaches have become more popular in information systems research, whereas systems thinking are effective for analyzing complicated challenges.

Methodology The paper proposes an AI based method for analyzing and optimizing collaborative supply networks. The topicality and possible benefits of integrating AI for supply chain collaboration are highlighted, as are the shared traits of systems thinking and design-science research.

Outcomes Prediction is an important for effective parameters in supply chain management.

Keywords Supply chain management · Artificial intelligence · Services · Reliability

M. Kumar (✉)
Delhi Technological University, Delhi 110042, India
e-mail: manojkumar@abes.ac.in

ABES Engineering College Ghaziabad, Ghaziabad 201009, India

Vipin
Department of Mechanical Engineering, Delhi Technological University, Delhi 110042, India

A. Agarwal
School of Engineering and Technology, IGNOU, Delhi, New Delhi 110068, India

H. Kumar et al. (eds.), *Recent Advances in Intelligent Manufacturing*,
Lecture Notes in Mechanical Engineering,
https://doi.org/10.1007/978-981-99-1308-4_9

1 Introduction

Supply Chain Management (SCM) assures physical, financial, and information flows across all supply chain components (suppliers, subcontractors, wholesalers, retailers, customers, etc.). This is a goal for any company seeking financial competitiveness, delivery, and service excellence [1]. Globalized trade, complicated trade flows, speedier competition, and sustainable development obligations make this more difficult. Using IT to communicate with supply chain partners is crucial in today's volatile environment. Systems include EDIS, ERP, AI, and RFID. AI in supply chain management can improve strategic and tactical functions such as business planning, supply chain development, demand prediction, and risk analysis [2, 3]. AI will optimize the supply chain framework and its complicated tracked transit patterns in both input and output material flows. It improves product/process quality management and stock management using machine learning algorithms, such as contemporary object detection/recognition technologies. It also improves product/process quality management, enabling strategic decision making and optimization. AI enables more complicated, wide-ranging, and effective shop-floor management. As well as improving inventory management, maximizing utilization, and preventing supply chain disruptions from machinery and equipment shutdowns. In the 1980s, supply chain management was designed to include end-users and original suppliers in business processes [4]. Initial providers deliver valuable goods, services, and data to customers and other stakeholders. A company's project requires procurement of new goods, services, and resources. From start to finish, the term refers to project services and supplies. Material procurement includes the supply chain. Procurement gets an organization's products, whereas Supply Chain Management completes them. Procurement departments must handle more than SCM. SCM and procurement help companies maximize profits. Reduce balancing costs, supplier quality, supply assurance, and supplier innovation. Suppliers, manufacturers, logistics providers, and others form the supply chain.

2 Smart and Digital Supply Chain Management

Supply Chain Management (SCM) maximizes customer value and generates long-term competitive advantage. Supply chain organizations design and run efficient supply chains. SCM is a supplier network's address. Companies desire tighter supplier connections for competitive advantage. Competitive advantage reduces customer expenses and delivery times. Another is providing the correct product at the proper time, place, and cost. Modern supply chain revolution integrates IoT sensors and data analytics for AI in mobility. These disruptive breakthroughs produce brand-new, intelligent SCM capabilities in warehousing, manufacturing, quality, maintenance, transportation, and logistics. Mahamuni provides examples of some of the measurable advantages that businesses employing SCM have documented in their publications. Thanks to IoT technology, unexpected downtime was reduced by 48%, overall equipment efficiency increased by 16%, the defect rate was decreased from 4.9 to 2.5%, and the cycle time for introducing new products was cut by 23% (from 15 to 11 months). In order to facilitate communication, define digital supply chain as the exchange of information between providers (commercial production, design, research, or competition) [5]. All the data stored and handled across the chain will be used in the supply chain's forecasting model. Supply Chain is more than departmental inventory distribution. Digitalization is helping companies shorten product launch times. Pharmaceutical businesses want to digitalize for better product commercialization. First-wave digital revolution focuses on employing technology and data analytics to better understand customers. The next digital wave will change operations.

3 Recent Evaluation of SCM and AI

AI and other digital technologies are revolutionizing Supply Chains, allowing them to make autonomous decisions based on real-time data analysis. Thus, enabling hitherto inaccessible software solutions and automation. Information once collected by humans will be gradually machine-generated, allowing more precise judgments and faster reactions to disturbances, transforming the supply chain into a strong interconnected system. Future supply chains will be able to self-steer, monitor the environment, react to changes, learn from prior circumstances, and simulate conceivable scenarios, gaining advanced flexibility and agility. AI is altering the automobile supply chain, specifically. AI helps firms optimize processes, schedule manufacturing, manage supply risk, minimize supply chain interruption, remove language barriers, improve customer service, and forecast demand [6]. With these benefits, firms will confront several hurdles to maximize the new technology. Shortage of processing power, high cost, long implementation process, linguistic and social cue complications, lack of data scientists and analysts, sharing data, legacy systems, and security concerns are today's main challenges. As AI evolves, more applications will be created [7]. When automotive businesses apply AI, they can improve procedures, manufacture more efficiently, and save money [8]. IR will improve customer service, strengthening consumer-SCM connections and encouraging repeat consumers. SCM and its services can build stronger relationships, encouraging growth and development. This new technology will alter every part of a railway system, so grab some popcorn and prepare [5].

4 Issues in Supply Chain Management

Supply chain management may minimize costs, increase revenues, expedite turnover, and improve core competency [9]. On every node of the supply chain, vast amounts of data have been collected and stored, and it's growing like a snowball. With so much data, it's hard for a company to determine the rules between suppliers and customers using its own business data, then analyses and make decisions [10]. Nodal firms in the supply chain cannot analyses or use information without a sophisticated data analysis and processing tool. If nodal firms don't develop core competencies, the supply chain will suffer [11]. AI, sophisticated data analysis, and processing functions will help supply chain management. Supply chain management is an array of management principles and methods that views the supply chain as a whole with the same strategic goal. It stresses information sharing and strategic partnership among supply chain members [12]. Supply chain management has various issues.

5 Research Methodology

This project aims to construct a model by integrating different operations, such as SCM issue identification, AI data analysis, and result interpretation to identify intelligent risk reduction techniques. This system required multidisciplinary expertise and multiple data gathering and assessment methods [13]. First, we researched SCM and AI literature to construct a conceptual model. Developing a sophisticated model requires a risk management team and AI. SC and IT specialists were interviewed to improve and validate the conceptual model. Risks vary with industry, business, and organization structure [14]. Each company's risk profile and attitude are unique. The proposed paradigm was tested to collect practitioner ideas, comments, and criticism. Case company used to understand implementation challenges [15]. Kaggle open source data was used. Any firm can employ its supply chain management services. IR used a single platform to optimize process management and gain competitive advantage. This Indian railway was chosen to use the recommended data-driven risk management strategy because they had risk management concerns.

6 Quantification

Linear relationships show a consistent ratio of rise or drop, indicating a tendency. Linear regression is the link between the dependent variable's value and the primary function's linear function. Statistics are utilised often. Multiple linear regression assumes a linear relationship between independent variables (X_1, X_2, … X_n) and dependent variables (Y_1, Y_2, Y_n) (Y). Input variables with irrelevant or noisy data limit multivariate linear regression's predictive ability. This research evaluates the

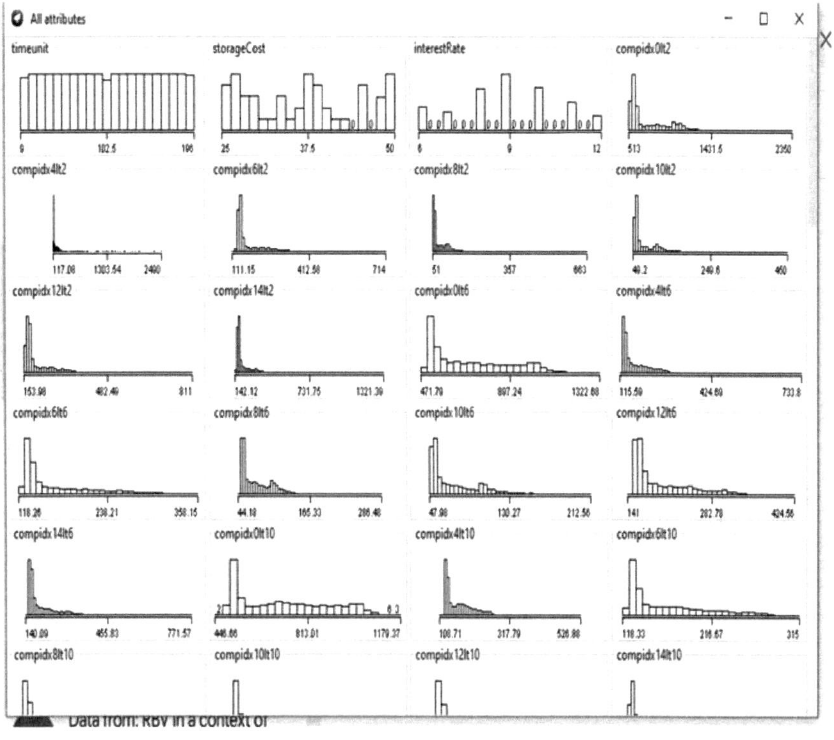

Fig. 1 Visualization

wave information prediction approach to improve navigation and maritime traffic control due to frequent marine accidents. Algorithm uses linear regression. Waves are found initially. Using linear regression, this paper predicts destructive waves. Linear regression is seen in Fig. 2.

IR (Indian railway) may employ prediction tools to determine how risk factors affect their cash value, revenue, and growth rate. Regression analysis can predict continuous risk levels. SCM factors can measure risk exposures. Estimating IR services using variables. AI often detects anomalies. Unusual events and trends can be detected. Service variables and their sub-factors should be analyzed for their effects on events. To establish efficient service plans, analyses such patterns. Classification strategies give objects to security categories. Depending on the data format and decision-preference, maker's SVM, decision tree, and kNN can be employed. Cluster analysis can categories risky goods, areas, and stakeholders. Clustering is an unsupervised learning method with no pre-defined categories. Various studies may be undertaken to measure the association between IR services, such as the relationship between performance and customer dissatisfaction ratings or services and behaviour characteristics. Figure 1 is a set visualization that helps decision-makers translates SCM difficulties into problems. The framework comprises management problem categories, services that help select the most beneficial activity and AI tasks that may supply answers. Before specifying AI, decision-makers should decide how to use AI results for risk management (Fig. 2).

Figure 3 illustrates the linear regression result. The solid line shows actual input data and the dotted line predicts it. You can compare the target to the actual net gain. R-square = 31.4439, which suggests causative factors vary by 60%. Coefficient of determination measures how well a linear model fits a data set. It's the proportion of the response variable's variance that the applied model can explain. Coefficient of determination is usually positive.

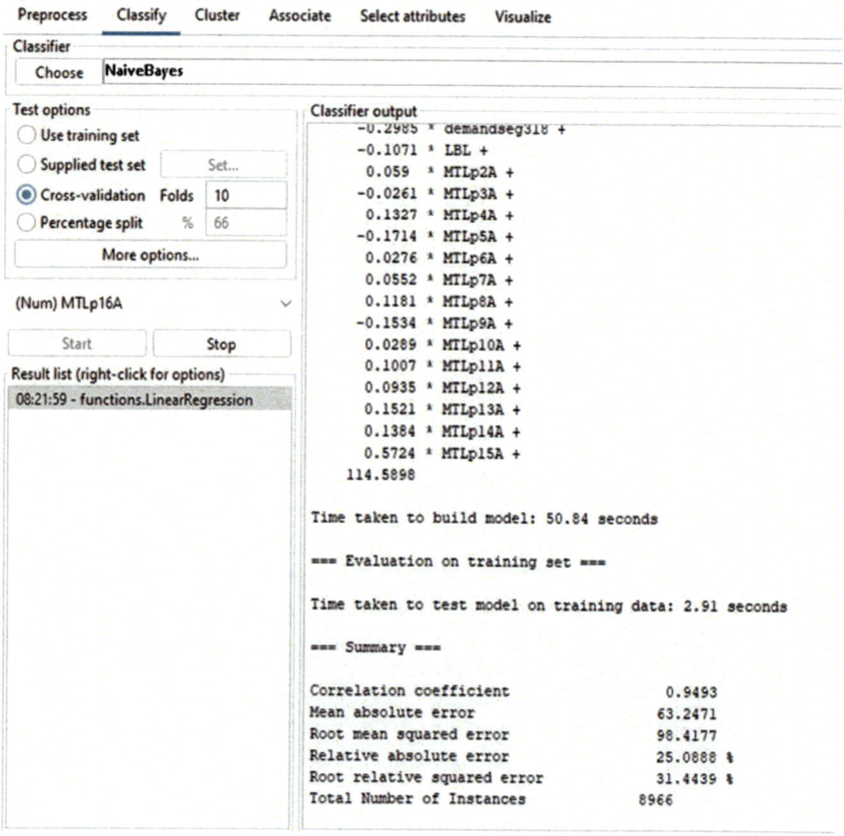

Fig. 2 Linear regression

Fig. 3 Error

7 Summary

AI uses machine learning, neural networks, statistics, pattern repository systems, knowledge acquisition, information extraction, HPC, digital and visual technologies, etc. The continual growth of supply chain management and the discovery of hidden circumstances, trends, and linkages from vast data sets will help enterprises enhance decision quality and supply chain management efficiency, leading to competitive supply chain strengths. We forecast profits using SCM industry data and linear regression. To fix the Over Fitting problem, increase the volume of training data or minimize the number of services. SCM influences Indian railways. AI tools include SVM, decision tree, and kNN. AI can transform SCM and address several Indian Railway System difficulties. AI has also reduced human labour.

References

1. Jung YG, Choi JA, Cha BH (2014) Analysis of radioactive contamination normal level of numerical isotope using clustering methods. J Inst Internet, Broadcast Commun (JIIBC) 14(6):41–46
2. Min H, Heo JY (2014) Document clustering scheme for large-scale smart phone sensing. J Inst Internet, Broadcast Commun (JIIBC) 14(1):253–258

3. Hong YS, Park CK, Cho SS, Suck-Joo Hong (2014) Intelligence transportation safety information system. Int J Internet, Broadcast Commun (IJIBC) 6(2):20–24
4. Matsuo Y (2012) Artificial intelligence and Deep Learning. ISBN 979-11-86008-23-2. In: Seber GAF, Lee AJ (eds) Linear regression analysis. John Wiley & Sons Publishers, USA
5. Lim DH, Kim JS, Lee BG (2016) Wave information estimation and revision using linear regression model. J Korea Multimed Soc 19(8):1377–1385
6. Cho YH, Kim IH (2010) Predicting the performance of recommender systems through social network analysis and artificial neural network. J Intell Inf Syst 16(4):159–172
7. Ryoo EC, Ahn HC, Kim JK (2013) The audience behavior-based emotion prediction model for personalized service. J Korea Intell Inf Syst Soc 19(2):73–85
8. Chunxi W, Jianjun Z (2005) Supply chain information integration based on data warehouse. Comput Eng Appl 36:P183-186
9. Dongrong X, Dayong W (2007) Research on supplier selection in e-business supply chain environment. Micro-Comput Inf F(7):P168–169
10. Lei Y, Changqing C (2001) Information supply chain model of data warehouse. Comput Eng Appl 22:P121-123
11. Zefang X, Zhongneng Z (2004) decision support system based on data warehouse. Comput Eng 30(12):P360-362
12. Thun JH, Hoenig D (2011) An empirical analysis of supply chain risk management in the German automotive industry. Int J Prod Econ 131(1):242–249
13. Huang X, Tan BL, Ding X (2015) An exploratory survey of green supply chain management in Chinese manufacturing small and medium-sized enterprises: pressures and drivers. J Manuf Technol Manage 26(1)
14. Soosay CA, Hyland P (2015) A decade of supply chain collaboration and directions for future research. Supply Chain Manage Int J 20(6):613–630
15. Kersten W et al (2017) Trends and strategies in logistics and supply chain management - digital transformation opportunities. BVL International

Preparation and Characterization of CuO-Au Hybrid Nanofluid with Different Mixing Ratio

Mohd Aidil Iqhwan, Ooi Jen Wai, and Prem Gunnasegaran

Abstract Nanofluids have gained a lot of attention from researchers in recent decades due to the growing need in industrial applications as a promising heat transfer fluid. In the quest to further enhance the heat transfer capability of nanofluids, researchers have delved into the mixture of two or more type of nanoparticles with base fluids which is termed as hybrid nanofluids. This novel high-class working fluids are often tailored to meet specific applications as it manages to bring about the benefits of both type of nanoparticles. However, the agglomeration process still remains a challenge to date in the field of nanofluids. Hence, preparation stage of the hybrid nanofluids is significant to mitigate the impact of agglomeration rate. In this paper, the preparation process of CuO-Au hybrid nanofluids and the thermophysical properties of the prepared hybrid nanofluids is studied. The thermophysical properties of the prepared hybrid nanofluids is characterized through thermal conductivity, viscosity, and density. At 40 °C, the highest thermal conductivity for gold (Au) mono nanofluid is 2.65 W/mK, while the highest thermal conductivity for CuO-Au hybrid nanofluid with the mixing ratio of (3:7) is 1.53 W/mK. The viscosity for hybrid nanofluids is higher compared to mono nanofluids. Set with mixing of a higher ratio of Au nanoparticles yields higher density.

Keywords Nanofluid · Hybrid nanofluid · Thermal conductivity · Viscosity · Density

1 Introduction

Heat transfer fluid (HTF) is one of the most important materials for transferring heat from a heat source to other heat demands, which is commonly used in industrial and residential applications. Nanofluid is a widely utilized HTF due to its advantageous

M. A. Iqhwan · O. J. Wai · P. Gunnasegaran (✉)
Institute of Power Engineering, Universiti Tenaga Nasional, Jalan IKRAM-UNITEN, 43000 Selangor, Malaysia
e-mail: prem@uniten.edu.my

H. Kumar et al. (eds.), *Recent Advances in Intelligent Manufacturing*, Lecture Notes in Mechanical Engineering,
https://doi.org/10.1007/978-981-99-1308-4_10

characteristics and features. Mono nanofluid and hybrid nanofluid are different type of nanofluids. Mono nanofluid consists of a single nanoparticle dispersed in a solvent, whereas hybrid nanofluid comprises of two different nanoparticle types. For hybrid nanofluid application, the electric double layer repulsion between nanoparticles can be increased in order to improve the stability of the suspension in this application (one layer is formed on the surface of nanoparticles by the corresponding ions within the crystal lattice and the second diffusion layer is formed on the base fluid) [1]. There are two method that are commonly used to prepare the hybrid nanofluid which are one-step method and two-step method [2]. For the one-step method, the nanoparticle will undergo the synthesis with base fluid in one step while for the two-step method will proceed with the two stages which are stirring and ultrasonication process. Preparing hybrid nanofluid using the two-step method is more economical and easier compared to one-step method. Asokan et al. [3] using two-step method to prepare the Al_2O_3-CuO hybrid nanofluid. Numerous researchers have examined the properties of many nanofluid types. CuO-Au hybrid nanofluids at a low mass concentration of 0.1% with varying mixture ratios and the addition of surfactant are the subject of less investigation. To develop enhanced HTFs with greater thermal conductivity efficiency, researchers have concentrated on producing materials with greater thermal conductivity [4]. In heat transfer applications, numerous forms of HTF are used. For examples, the utilization of nanofluids. Nanofluid is a nanoscale suspension that can be created from a single nanoparticle. This nanofluid falls into two categories, which are mono nanofluid and hybrid nanofluid. Mono nanofluids are the most common form of nanofluids. A suspension of two different types of nanoparticles in a base fluid such as water, ethylene glycol, or petroleum oil is an example of a hybrid nanofluid. Numerous investigations have been spurred by hybrid nanofluids, as has been demonstrated. Theoretically, base fluid has a lower heat transfer efficiency than a nanofluid composed of a single type of nanoparticle [5]. Furthermore, hybrid nanofluid has superior performance compared to mono nanofluid, which may be categorized based on its thermophysical features [6]. By combining a proper of two types of nanoparticles, it can utilize the synergistic energy which can improve the characteristic of the hybrid nanofluid [7]. In the numerical section, the single-phase model was compared to two-phase numerical techniques that are more commonly used [8]. As proven, the heat transfer coefficients for particles with higher thermal conductivities are larger than those for particles with lower thermal conductivities, which are also increased at higher Reynolds numbers [9, 10]. According to experimental research [11, 12], the ratio of 60:40, Al_2O_3-CuO hybrid nanoparticle has the greatest performance compared to other sample ratios of 20: 80, 40:60, and 50:50, respectively. This shown that the different ratio of nanoparticle can influence the thermophysical of the hybrid nanofluid. Several factors will be focused in order to fulfill all of the objectives established in current work. Copper (II) oxide (CuO) and gold (Au) nanoparticles will be used in the present work. By mixing these nanoparticles, a CuO-Au water-based hybrid nanofluid will be prepared. This study will then analyze the characterization and thermophysical properties of hybrid nanofluids with different mixing ratios. In addition, examine the optimum stability time of hybrid nanofluid in the ultra sonification bath.

2 Materials and Methods

In this research, CuO and Au nanoparticles with an average particle size of less than 50 nm and 100 nm, respectively were used in the preparation of the mono and hybrid nanofluid. In addition, SDBS surfactant with technical grade was used mixed with several sets of mono and hybrid nanofluid. All of these materials were purchased from an authorized chemical supplier, Sigma-Aldrich. The nanofluids were prepared with different mixture ratio of nanoparticles at a constant mass concentration percentage. All of these nanofluids (CuO: Au) were prepared at mixture ratio of 10:0 (Set 1), 7:3 (Set 2), 5:5 (Set 3), 3:7 (Set 4), and 0:10 (Set 5), respectively. Distilled water was used as the base fluid for all experimental work in this investigation. The nanofluid was weighted using the analytical balance following the prepared compounding formulation. The nanofluid samples were stirred using the magnetic stirrer for 2 h at a constant temperature of 50 °C. Aster the stirrer process, the nanofluid undergoes the ultrasonication process for 6 h at the same 50 °C constant temperature value. The solid mass concentration for these nanofluid was fixed at 0.1%. Referring to Table 1, illustrate the properties of various nanoparticles and surfactant used in this current study.

In order to explore the potential and the advantages of nanofluid in engineering thermal application, it is essential to characterize the dispersion stability of nanofluid. There are many methods that can be used to evaluate the stability of the nanofluid. One of the common methods is analyze using the zeta potential. Zeta potential is the most effective method to quantify the dispersion stability of the prepared nanofluids based on the potential at the slipping or shear plane of the colloid partial which moves under an electric field. The stability of all prepared nanofluid was evaluated using the Litesizer 500 instrument at several different temperature range of 20 to 60 °C. Furthermore, this research also evaluated the stability of the nanofluid using visual sedimentation inspection method. The visual properties of the hybrid nanofluid were to be observed during this process. The sedimentation inspection can be undertaken with the naked eye according to the established parameter grading scale as tabulated in Table 2.

Table 1 Properties of copper (II) oxide and gold nanoparticles

Types of particles	Copper (II) oxide	Gold
Molecular formula	CuO	Au
Average particle size	<50 nm	<100 nm
Molecular weight (g/mol)	79.55	196.97
Color	Black	Black

Table 2 Sedimentation parameter scale for the stability evaluation of hybrid nanofluid

Rating	0	1	2	3	4	5
Sedimentation	No	Very little	Little	Intermediate	High	Very high

The thermal conductivity of the prepared nanofluids were evaluated using the TEMPOS thermal analyzer. This TEMPOS device uses the transient hot-wire method where a pulse of heat was applied to the device needle. The temperature response of the nanofluid was measured on the adjacent needle during and after the application of the heat pulse. For the measurement, approximately 50 ml of nanofluid samples was poured into a small glass container used in the test. The reading of the thermal conductivity was taken at several different temperature value range between 20 and 40 °C. The thermal conductivity reading was measured repeated for 3 times for each nanofluid at each different temperature.

The viscosity of the various mono and hybrid nanofluids was evaluated by using the rheometer device. The reading of the viscosity was recorded at several different temperature in range between 20 and 60 °C. Approximately 30 ml of prepared nanofluids was used during the measurement process.

The density of the nanofluids was evaluated using the DMA™ 1001 density analyzer. The measurement was conducted with the correlation of 10 °C increments of temperature from 20 to 60 °C. Before the started to measure the density of the nanofluids, the density device was calibrated with air and deionized water following the procedure stated in the user manual. As stated in the user manual, the device has an accuracy of ± 0.0001 g/cm^3. The nanofluids sample was loaded in a 10 ml syringe and injected into the instrument's U-shaped tube glass.

3 Results and Discussion

3.1 *Dispersion Stability of Mono and Hybrid Nanofluids*

By conducting the zeta potential analysis, the dispersion stability of the various nanofluid can be evaluated. Figure 1 illustrates zeta potential value versus different temperature of CuO-Au hybrid nanofluids. CuO-Au mono and hybrid nanofluid show a lower value of zeta potential with ranged ±0 to ±22.10 mV. Based on the zeta potential value, it indicated that these sets of nanofluid are not stable and have a higher possibility of agglomeration. Among these sets, set 1 with CuO mono nanofluid shows to have the highest value of zeta potential at all different temperature. However, CuO-Au hybrid nanofluid with a different mixture ratio has a higher value of zeta potential compared to Au mono nanofluid. For the visual sedimentation inspection method, the results were tabulate in Table 3. All of the rating scale was given based on the evaluated for 7 days of continuously observation.

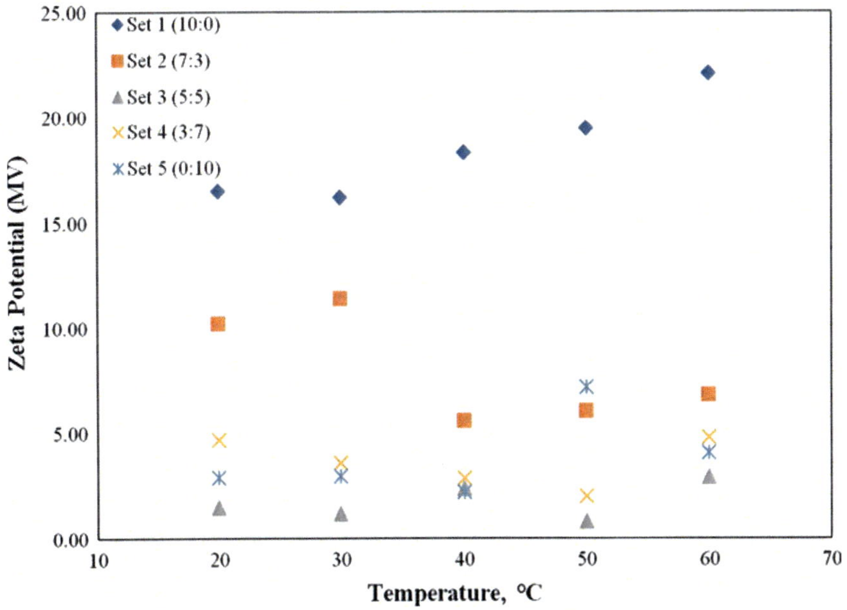

Fig. 1 Zeta potential of CuO-Au hybrid nanofluid versus temperature

Table 3 Sedimentation rating for each set from day 1 to day 7 of observation day

Sample	Mixed ratio	Sedimentation rating						
		Day 1	Day 2	Day 3	Day 4	Day 5	Day 6	Day 7
Set 1	10:0	0	0	0	0	0	0	1
Set 2	7:3	0	1	1	2	2	3	3
Set 3	5:5	0	1	2	2	3	3	3
Set 4	3:7	0	1	2	2	3	3	3
Set 5	0:10	0	0	0	1	1	2	2

3.2 *Thermal Conductivity of Mono and Hybrid Nanofluids*

Figure 2 shows the pattern of the thermal conductivity of CuO-Au hybrid nanofluid at different point of temperature in range 20 to 40 °C. Based on Fig. 2, at the temperature of 20 °C, CuO and Au mono nanofluid show higher values of thermal conductivity 0.6758 W/mK and 0.5900 W/mK, respectively, compared to CuO-Au hybrid nanofluid. As the value of the temperature increase, the reading of thermal conductivity for all Sets are increasing too. Focusing on Fig. 2, the set 5 which consists of 100% of gold nanoparticle water-based, the value of thermal conductivity shows a drastic increase with higher temperature. This is due to the physical properties of the

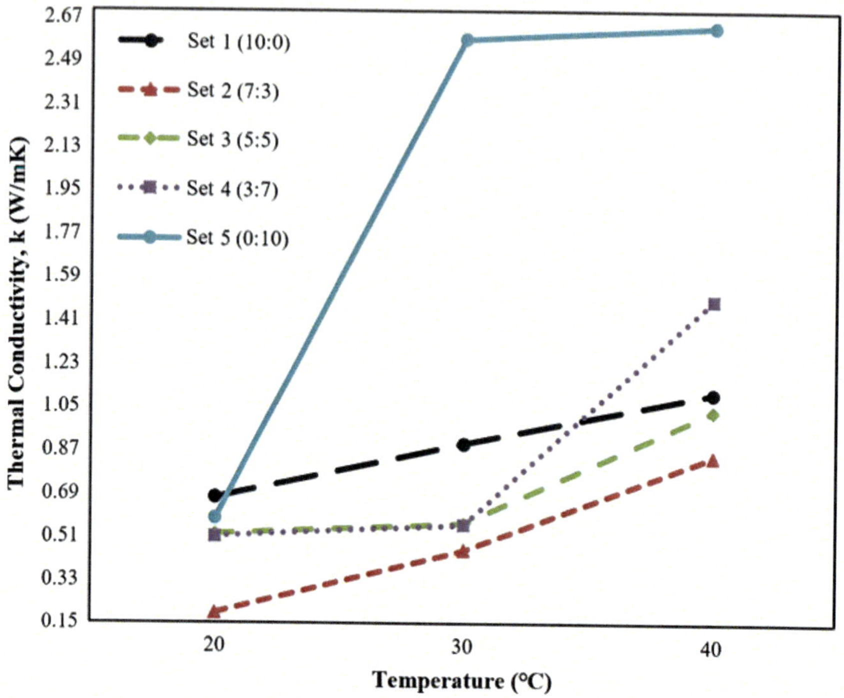

Fig. 2 Thermal conductivity of CuO-Au hybrid nanofluid versus temperature

metal nanoparticle thermal conductivity which is a higher value than metal oxide as reported by Suganthi and Rajan [13].

3.3 *Viscosity of Mono and Hybrid Nanofluids*

Figure 3 shows the viscosity results for the CuO-Au nanofluid. Based on the results, all sets for CuO-Au nanofluid have the same trend. As the temperature increased, most of the viscosity value decreased and started to fluctuate as the temperature more than 50 °C. Set 1 and set 5 are mono nanofluids which consist of CuO nanoparticles and Au nanoparticles, respectively. At the temperature of 20 °C, the viscosity value of the Au mono nanofluid is higher compared to the CuO mono nanofluid. However, as the temperature continuously increased, the data shows a comparable result for mono nanofluid. It is also depicted from Fig. 3 that hybrid nanofluid has higher value of viscosity compare that with mono nanofluids at all temperatures.

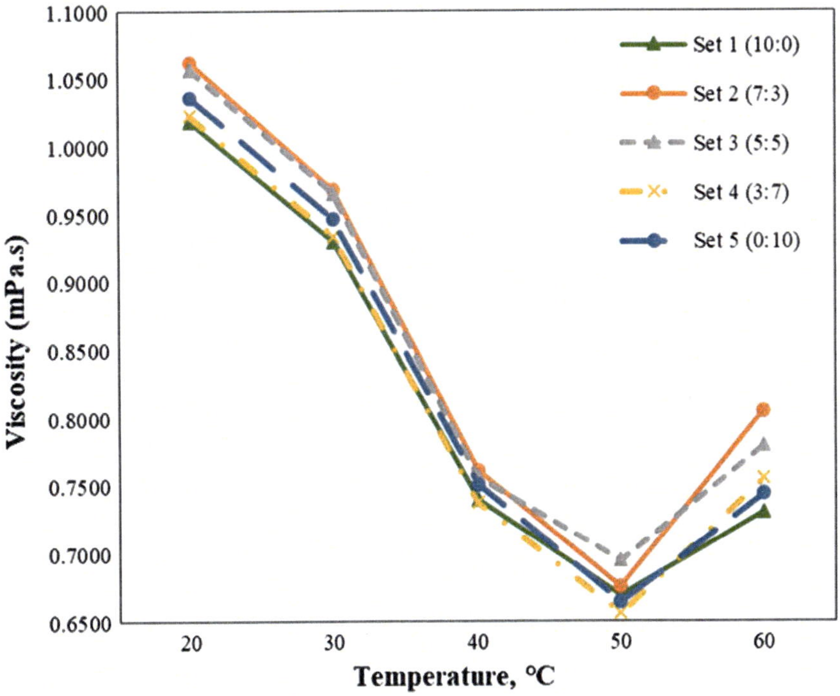

Fig. 3 Viscosity of CuO-Au hybrid nanofluid versus temperature

3.4 Density of Mono and Hybrid Nanofluids

Figure 4 shows the data recorded for CuO-Au hybrid nanofluid density for all the sets. Based on the results, all sets show a reduction value for the density of the nanofluid. At the temperature of 20 °C, set 1 and set 4 have the same highest value of density compared to other sets. When the temperature of the nanofluid is increasing, the value of the density decrease. Sets with a higher amount of CuO nanoparticles have a higher reduction value compared to sets filled with Au. This is because gold is one of the densest metals on the earth and denser compared to CuO.

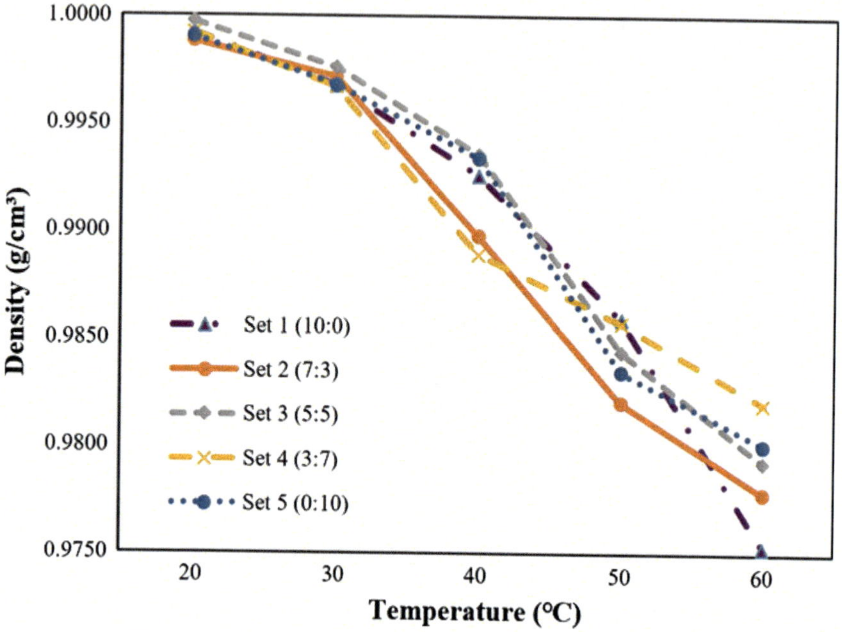

Fig. 4 Density of CuO-Au hybrid nanofluid versus temperature

4 Conclusion

To evaluate the stability of hybrid nanofluids, visual sedimentation inspection, ultrasonication, and zeta potential tests have been carried out. Visual sedimentation inspection has been conducted for 7 days straight for all the samples. CuO-Au mono hybrid nanofluid shows intermediate sedimentation on day 2 and very high sedimentation started from day 3 of the observation day. For ultrasonication method, it has proceeded at the initial phase of the nanofluid preparation. It is discovered that the nanofluid should undergo 6 h of ultrasonication process in order to produce a homogenous nanofluid with no sedimentation at the bottom of the beaker. Based on the zeta potential results, CuO mono nanofluid is more stable compared to other sets of nanofluids. The thermophysical properties of CuO-Au hybrid nanofluid are evaluated throughout the present study, which are thermal conductivity, viscosity, and density. For the thermal conductivity, the results are taken with an increment of 10 °C starting from 20 °C until 40 °C while for the density and the viscosity, the results are recorded until 60 °C. The thermal conductivity for all 5 sets of nanofluid is increased directly proportional to the increment of 10 °C of the temperature up to 40 °C. At the temperature of 20 °C, CuO and Au mono nanofluid have a higher value of thermal conductivity compared to hybrid nanofluid. As the temperature increases, the trend of the thermal conductivity remains the same where the mono nanofluid has a higher value compared to hybrid nanofluids except for set 4 (3:7) at temperature of

40 °C. The viscosity results are recorded to have same pattern of value towards the increment of 10 °C temperature for all sets. As the temperature increased, most of the viscosity value decreased and started to fluctuate as the temperature more than 50 °C. Sets with a higher ratio of gold (Au) demonstrate to have denser density.

5 Recommendation

Mono nanofluid and hybrid nanofluid are very promising heat transfer fluids that can be totally utilized and practiced in heat transfer industrial and domestic. As the usage of these nanofluids is very beneficial, thus more studies and research should be conducted on this kind of heat transfer fluid. Therefore, there are some suggestions and recommendations based on the outcomes of this current research. Conduct a nanofluid preparation with different types of mono and hybrid nanofluid using the same method but different time duration of magnetic stirrer and ultrasonication process. Prepare several mono and hybrid nanofluids with different mass concentration percentages and different mixture ratios of nanoparticles. In addition, use different types of surfactants with different amount ratios to evaluate the enhancement and effect on the mono and hybrid nanofluid. Perform heat capacity testing, FESEM, and TEM analysis on the prepared mono and hybrid nanofluid. Lastly, prepared the nanofluid using different types of based fluid such as ethylene glycol and oil.

Funding The authors would like to thank the Ministry of Higher Education Malaysia (MOHE) for funding this study under the Fundamental Research Grant Scheme (FRGS) of Grant No. FRGS/1/2019/TK09/UNITEN/03/1.

References

1. Che Sidik NA, Jamil M, Aziz Japar WMA, Muhammad Adamu I (2017) A review on preparation methods, stability and applications of hybrid nanofluids. Renew Sustain Energy Rev 80:1112–1122
2. Wei Xian H (2018) Review on preparation techniques, properties and performance of hybrid nanofluid in recent engineering applications. J Adv Res Fluid Mech Thermal Sci 45:1–13
3. Asokan N, Gunnasegaran P, Wanatasanappan VV (2020) Experimental investigation on the thermal performance of compact heat exchanger and the rheological properties of low concentration mono and hybrid nanofluids containing Al_2O_3 and CuO nanoparticles. Thermal Sci Eng Progress 20:17–27
4. Mercan H (2020) Thermophysical and rheological properties of hybrid nanofluids. Hybrid Nanofluids Convection Heat Transf 101–142
5. Bubbico R, Celata GP, Annibale F, Mazzarotta B, Menale C (2015) Comparison of the heat transfer efficiency of nanofluids. Chem Eng Trans 43:703–708
6. Kshirsagar DP, Venkatesh MA (2021) A review on hybrid nanofluids for engineering applications. Mater Today: Proc 44:744–755

7. Hayat T, Nadeem S (2017) Heat transfer enhancement with Ag-CuO/Water hybrid nanofluid. Results Phys 7:2317–2324
8. Bajestan E, Charjouei M, Niazmand H, Daungthongsuk W, Wongwises S (2016) Experimental and numerical investigation of nanofluids heat transfer characteristics for application in solar heat exchangers. Int J Heat Mass Transf 92:1041–1052
9. Mikkola V, Puupponen S, Granbohm H, Saari K, Ala-Nissila T, Seppala A (2018) Influence of particle properties on convective heat transfer of nanofluids. Int J Thermal Sci 124:187–195
10. Hamid KA, Azmi WH, Nabil MF, Mamat R, Sharma KV (2018) Experimental investigation of thermal conductivity and dynamic viscosity on nanoparticle mixture ratios of TiO_2-SiO_2 nanofluids. Int J Heat Mass Transf 116:1143–1152
11. Wanatasanappan VV, Abdullah MZ, Gunnasegaran P (2020) Thermophysical properties of Al_2O_3-CuO hybrid nanofluid at different nanoparticle mixture ratio: An experimental approach. J Mol Liq 313:34–58
12. Ali N, Teixeira JA, Addali A (2018) A review on nanofluids: fabrication, stability, and thermophysical properties. Hindawi J Nanomater 1–33
13. Suganthi KS, Rajan KS (2017) Metal oxide nanofluids: review of formulation, thermo-physical properties, mechanisms, and heat transfer performance. Renew Sustain Energy Rev 76:226–255

Scenario Analysis and New Future Business Models for the Indian Automotive Industry

K. Ravi and P. Naveenchanran

Abstract The world's fastest expanding sector is the automotive industry, which is one of India's significant businesses because of its robust democracy and international partnerships. According to this circumstance, there is a growing need in India to expand the compact vehicle industry. India is now building the fifth-largest economy in the world, with a nominal GDP of $2.94 trillion dollars and a real GDP of $10.51 trillion dollars (PPP). The goal of the current analytical investigation was to produce inventory based on Porter's five forces model by analyzing the most recent updates and patterns. Due to economic and industrial reforms, the low energy consumption of the automotive sector has made it an appealing location for global automobile manufacturing, which is coupled with a rise in per capita and disposable income of the employed population. Nevertheless, the study finds that the Indian automotive industry's overall competitiveness in the present might be characterized as modest. As a result, with the backing of government measures, the anticipated growth rate in the next years and prospects in the patchy Indian market would decline significantly.

Keywords Automotive industry · Democracy · Economic · Indian market · Manufacturing

K. Ravi (✉)
Research Scholar, Department of Automobile Engineering, Bharath Institute of Higher Education and Research, Chennai, Tamil Nadu, India
e-mail: kraviphd2020@gmail.com

P. Naveenchanran
Department of Automobile Engineering, Bharath Institute of Higher Education and Research, Chennai, Tamil Nadu, India

H. Kumar et al. (eds.), *Recent Advances in Intelligent Manufacturing*,
Lecture Notes in Mechanical Engineering,
https://doi.org/10.1007/978-981-99-1308-4_11

1 Introduction

The export sector, agriculture, greater savings rates, and middle-class families are the key drivers of the economy's success. The "Make in India" manufacturing initiative, as well as the digital India and startup India programmes, are the recent government's main priorities. In 2019–2020, the industry sector is anticipated to increase by 2.5%, compared to 6.9% growth and a 29.6% proportion of GVA in 2018–2019. It contributes to 26% of the total GDP as shown Fig. 1. The principal source of energy is the world's third-largest coal producer. The chemical industry is the world's second largest industry. Indicative of industrial production (IIP), the industry sector is anticipated to increase by 0.6% in 2019–2020, following a 5% increase in 2018–2019. In 2019–2020, the major steel industry expanded by 5.2%, compared to 3.6% in 2018–2019. In 2019, 1.1943 billion phone connections were made in India [1–3].

By 2020, it should have reached its goal of being the fifth-largest manufacturing country in the world. It wants to create 100 million jobs with a 25% share of the GDP by 2022. India will have the chance to grow its economy through the Make in India initiative since it has numerous businesses set up for high-tech manufacturing, including GE, HTC, Siemens, Toshiba, and Boeing. Customers' purchasing power will increase as a result of these configurations. The manufacturing sector is anticipated to expand by 2% between 2019 and 2020. Italian business Tecnocap has announced a joint venture with Tecnocap Oriental to enter India in 2020 with a 75%

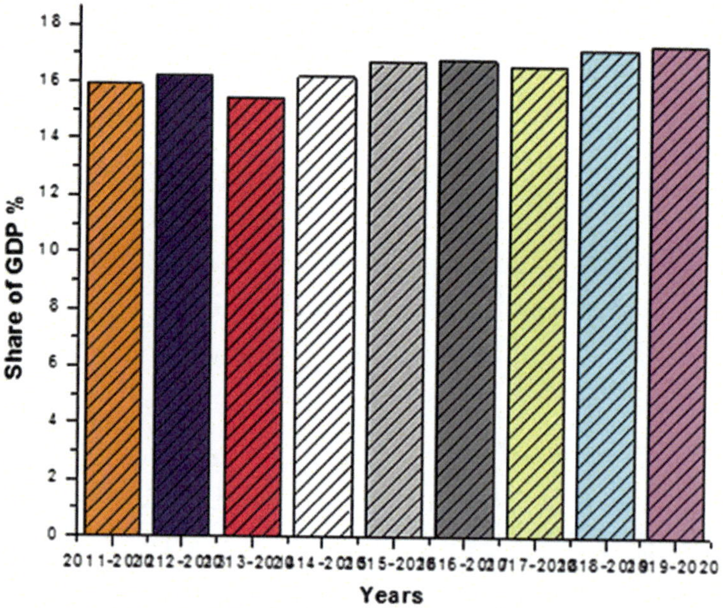

Fig. 1 Shows manufacturing in India as a percentage of GDP between 2011–12 and 2019–20

share. Packaging makers are the focus of an Asian corporation with a joint venture to develop a facility in Mumbai [4, 5].

The Make in India initiative is receiving increased attention from Indian Railways. It has a high-speed locomotive engine that it made on its own. The established Indian manufacturing sector is more linked with international trade, notably with overseas intermediate products for export. There are more jobs created in manufacturing in India. India has the chance to develop centers for multinational corporations. Numerous investments have been made in numerous economic areas throughout this period of time. With US $77.6 billion in mergers and acquisitions in 2017, this number grew 53.3% [6–8].

2 India's Automobile Market is Now in a Precarious State

Worldwide, private automobile travel is more common in this fast-paced environment. Global automakers are investing in technologies to minimize emissions and fuel consumption as part of a unified strategy with international associates to develop eco-friendly engines and diminish dangerous gases. Modern automobiles are more efficient and emit less pollution. Better possibilities for automobiles to progress in the coming era are electric and hybrid vehicles [9, 10].

Due to rising disposable income, Indian consumers are extremely aware of their purchasing habits in order to adapt to the new lifestyle. The cheapest automobile in the world that is ideal for the middle class is the Tata Nano type. Twelve percent of automobiles sold worldwide are exported from the Indian market.

The Indian vehicle sector is now ranked fourth in the world, but by 2030, it will have surpassed third place. 37 million people work in the automotive sector, which together with India's fourth-largest automaker accounts for 7.1% of the country's GDP in Fig. 2.

The price reduction to reduce output and the danger of job loss has left automakers in the current situation with financial difficulties. Automakers are being encouraged by the government to adopt policies that will help them thrive. A challenging period has been experienced by the global passenger automobile sector business. Due to the global economic downturn, rising fuel prices, and the corona virus (COVID19) impact, sales of cars are slowing down, particularly in India. India has 22 automobiles per 1000 people as of today, based on the country's 1.37 billion population.

An important part of the Indian economy is the automobile sector. The manufacture of automobile items is essential to this sector's prosperity. The development of an engine serves as the heart beat of a vehicle, generating the preponderance of the vehicle's power. It continues to play a large part in the Indian economy and is a major force behind the country's technical advancement [11].

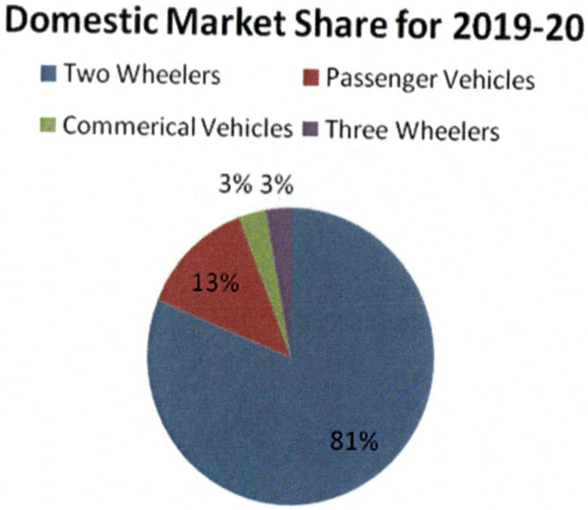

Fig. 2 Shows the domestic market share of various automobiles from 2019 to 2020

3 The Indian Mission Plan is a Developing Vehicle Market

According to the Automotive Mission Plan (AMP 10 years) 2016–2026 of the Indian government, it will increase from 7 to 12%. By 2021, India aspires to grow as the third-largest traveler car market in the world. In India, the automobile sector supports 19 million jobs. It contributes 40% to worldwide R&D and 4.30% of India's exports.

According to the industry, 25 million of the 3.5 million automobiles produced in 2018 were exported. India aimed to become the third-largest automobile market in the world in the future. It is currently ranked fourth in the entire globe. More than 6 million hybrid and electric vehicles are anticipated to be sold in India's automotive market in 2020 mainly due to the "Make in India" plan's goal of having the majority of its components made in India.

The strategy has been crucial in boosting the world economy as a whole. In order to meet future demand for a secure world, the Indian automobile industry is concentrating on producing more hybrid and electric vehicles. The automotive sector may be divided into passenger, marketable, three-wheeled, and two-wheeled vehicles. Due to strong demand from middle-caste families and other groups, 2-wheeled vehicles also hold a prominent position in the market.

The automotive sector has a straightforward business strategy: sell as many cars as possible in line with demand and offer after-sale services like repair and maintenance. Since automakers must compete fiercely to produce cutting-edge features to make money in the Indian market, the auto industry is growing dramatically every day. Government initiatives have an effect on the automobile sector. Newcomers have historically sought out markets with strong growth rates, liberal policies, and tax breaks. A greater standard of life is the second motive, which is to attract a new generation. Due to tax breaks and new government regulations, the Indian automobile market has grown during the past several years.

4 Development and Future Prospects of the Automotive Industry

Humans may continually see new advances and updated inventions in this industry, such as cutting-edge technologies and ecosystems in the automobile sector, which is a global, day-to-day presence in this global market. SUVs are quite popular in this market, and electric SUVs, in particular, are currently contributing to the global expansion of this market category.

Worldwide sales of passenger automobiles reached 45.4 million in 2005. From 2007 to 2009, it went from selling 50.8 million to 49.6 million units to 55.8 million to 70.6 million units each year from 2010 to 2017, with a little decline to 68.6 million units in 2018. With 64.3 million automobiles sold globally in 2019, there has been a decline in sales over the previous three years.

According to a German auto lobby group, sales this year were down by 4 million vehicles particularly in comparison to last year. The global automotive industry has more tendency to lose jobs in the next 12 months due to the global recession, so the worldwide auto market will not be in a prime place at the peak of the crisis [12].

5 Analysis of the Indian market's Attractiveness for the Automotive Industry

The automobile industry's profitability is shown here using the five forces analysis. In order to advance the automotive sector in the Indian market, according to this research, a low-moderately profitable automobile section of the business is needed.

5.1 Threat from Newcomers (Barriers to Entry): Low

In Fig. 3 this industry is not open to new entrants from any company. High levels of cash, expertise, technology, size, and market knowledge are necessary for new entrants. Due to the vast volume of international automakers and the high demand for novel product concepts, the global market has the greatest impact on the domestic market. Auto manufacturers have a reputation and brand. To invest in the new entrance in India, brand equity and money are needed [13].

5.2 A Fierce Rivalry in Competition

There are several rivals in the market, including Ford, Maruti Suzuki, Hyundai, Tata Motors, Toyota, Mahindra & Mahindra, GM, and Volkswagen, among others. As a

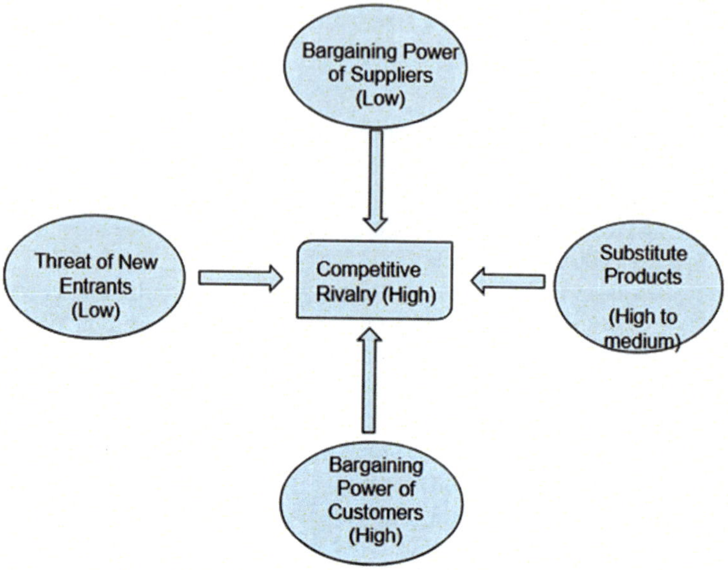

Fig. 3 Porter's study of the Indian automobile industry's five factors

result, a highly competitive market often has poor profits because of high costs and price-based rivalry in the car sectors. Because there are so many new designs on the market, it is problematic that the car's design is the most crucial aspect of this competition because any alterations would bring delays, risky proposals, and slower revenue development. Many automakers are lowering their prices while maintaining the same level of features due to competition, which is intensified by entrants from international automakers.

5.3 Replacement Threat: High to Medium

There are various alternatives in the auto business, including SUVs, micro cars, and utility vehicles, depending on the need. For electric hybrid, CNG, LPG, petrol, and diesel automobiles, there are several alternative vehicle possibilities. It is the furthermost prevalent two-wheeler petition based on segment availability and the vehicle's lower price. Buses, trains, and airlines are all excellent alternatives for long distance travel.

5.4 *Customer Bargaining Power: High*

Customers in India are particularly sensitive to pricing since there are so many options available on the market, and efficient automobiles provide value for money in accordance with customer demand due to the industry's high level of competitions [14].

5.5 *Low Supplier Negotiating Strength*

Due to India's large number of suppliers, both domestic and international suppliers of car components have little negotiating strength. The majority of vehicle manufacturers rely on big suppliers and a single or two clients for each specialist area.

6 Future Investment Trends in India's Automotive Industry

The advantages of this market sector in India should be supported for a few key reasons. First off, the country's robust middle-class family income growth and youthful population growth rate are the key causes. To expand exports for the next mission plan 2026, the Indian car industry's GDP is expected to grow by a factor of five.

Furthermore, it enjoys a significant competitive edge due to its lower manufacturing costs than those of Europe and the USA, as well as its strong access to resources and trained labor. Last but not least, the Indian government intended to expand the number of electric automobiles in order to minimize emissions of hazardous gases and to create a massive R & D industry in India.

The new GST policy, which supports government policy, would lower the tax from 12 to 5% for electric vehicles, providing greater incentives to buy them. Suppliers are developing battery technology to increase the capacity for long-distance trips in response to the current demand for electric automobiles. For instance, a number of automakers are working on electric cars, including Mahindra & Mahindra, Tata, MG Motors, Hyundai, Toyota, BMW, and Maruti Suzuki. The new first EV model will be tested in 2020. In 2020, Maruti Suzuki India Ltd. will start making electric vehicle batteries at its Hansalpur manufacturing facility. India has a competitive edge in that it ranks second globally in terms of steel production, giving it a cost advantage.

7 The Indian Automobile Industry's Main Competitive Advantage

A crucial success aspect mostly has to do with the customer-seller resolution process. The main success factors for the Indian automobile industry are labor productivity and flexibility, capital efficiency, qualified labor, the availability of raw materials, and infrastructural upgrades. The primary KSF to gain an inexpensive edge under government initiatives to benefit the Indian auto sector is technology and procedures. Other important success elements include the purchasing process, after-sales support, location, a technical edge, and low-cost ownership and upkeep. The cheapest little automobile on the domestic market is made by Maruti Suzuki.

8 SWOT Analysis is a Strategic Planning Technique Used in the Indian Automobile Sector

Strengths: The popular auto industry is changing as a result of higher export levels, more affordable labor, and low-cost automobile ownership. Both upper middle class and working class income groups make up the population. Higher market skill levels and high engineer availability in the compact vehicle category. With a better distribution network and more output, the automobile sector has a stronger brand and market position.

Weaknesses: Comparatively speaking to European nations with low labor productivity, inferior goods are readily available. In comparison to China, the interest rate is high and the cost of production is greater. Low levels of R&D spending and weak demand in the market for high-end vehicles are observed.

Opportunities: Infrastructure changes in tandem with population increase. The next generation's status is increasing as a sign of lifestyle, which is changing consumer behavior. Women of working age are becoming more prevalent in this auto market as innovation and technology are in more demand. Future electric/hybrid eco-friendly automobile demand presents a chance for both international and national businesses.

Threats: Other market sectors hold a larger stake, such as the two-wheeler segment, which has an 81% market share. Due to recent competition and natural disasters that have affected the performance of sales in the automotive industry from international competitors, the sector has a less trained labor force, which will result in higher tariffs and maintenance costs.

9 Government Policy in India is Changing in Anticipation of New Initiatives

The Indian government is adhering to this four-phase in the automobile and four-wheeler industries (vans, Cars and motor utility vehicles).

Phase 1

B7 grievance By 2020, passenger cars and utility vehicles with compression ignition engines will be manufactured and offered in India; throughout this period, the convenience of biodiesel would increase from 0.001 to 5%.

Phase 2

All new cars must be produced with the BS-VI system between 2020 and 2025, and the oil sector will boost B7 fuel's ability to supply 5% availability of biodiesel.

Phase 3

Between 2025 and 2030, B7 would adopt new diesel vehicles, and the market for hybrid and electric vehicles would become increasingly popular.

Phase 4

Electric vehicles will be widely deployed between the years of 2030 and 2047, and they will be expected to provide urban transportation mobility and in order to decrease the import of crude oil, the government might also boost the use of biodiesel fuel in the automotive industry.

10 The Eco-Business Revolution: Electric Vehicle Mobility

Due to Indian consumers' preference for zero-emission hybrid and electric vehicles, the market for electric cars is expected to experience consistent expansion in the coming years. The primary issue is the environment's impact and the depletion of crude oil reserves. Global battery manufacturing capacity is growing daily to take advantage of economies of scale in the electric sector, making batteries a key force behind the development of this industry in India.

The Society of Manufacturers of Electric Vehicles (SMEV) reports that sales of electric cars in India decreased by 6% in 2020 to 3400 units from 3600 units in 2019. According to SMEV, India now imports 87% of its fossil fuels and produces just 13% of its own; by 2020, this percentage will climb to 92 due to an increase in population and increased demand. Therefore, it is vital to improve the new, environmentally friendly section. When compared to cars with internal combustion engines, electric, hybrid, and electric vehicles should emit 35 to 45% fewer greenhouse gases (Fig. 4).

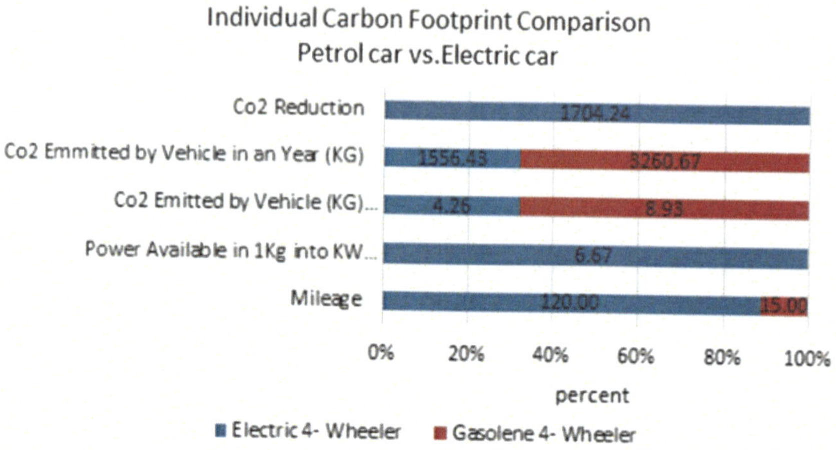

Fig. 4 Shows a personal carbon footprint comparison between gasoline and electric vehicles

11 Conclusion

In a summary, the study of the Indian auto industry presents fresh chances and market demand difficulties in India with plenty of resources. The Indian auto industry is helping the fourth-largest manufacturer's market share to increase in 2019 in this quickly evolving industry.

With a market share of 12.9% for passenger vehicles in 2020 and a CAGR of 14.2% from 2016 to 2020, small and midsized automobiles provide the majority of profit for the automotive industry.

In addition, Mahindra & Mahindra is the second-best corporation in India for overall performance based on high car sales and high revenues in order to maintain a dominant position on the current market. Maruti Suzuki is the greatest firm overall in India. It will lead to the electric car category and has a good liquidity position, but the business must improve its working capital management system.

Future transportation needs eco-friendly vehicles for a variety of reasons. With a 1% market share for electric vehicles, the Indian automotive industry is concentrating on lowering technological costs, battery costs for long distances, and the availability of charging stations for the country's electric-hybrid vehicle market. With CO_2 emissions in electric vehicles ranging from 8 to 24% lower than those with internal combustion engines, and with battery output predicted to rise by 2030, electrified automobiles will have a beneficial macroeconomic and environmental influence on the automotive industry in the future.

The automotive business model for vehicles in the past was a straightforward plan since it depended on selling as many cars as possible before relying on after-sales services like service, maintenance, and repair.

Due to the sophisticated technical value chain and ecosystem structure, the business model has undergone substantial modifications nowadays. The business model has three components, such as a value and product network, rules and laws, and so on. Environment, law, and government are rules and regulations' three primary pillars.

Funding The research and/or authoring of this paper were not supported financially by the author(s).

Declaration of Conflicting Interests In relation to the research, writing, and/or publishing of this paper, the author(s) reported no possible conflicts of interest.

References

1. Kale D (2011) Co-evolution of policies and firm level technological capabilities in the Indian automobile industry. In: Atlanta conference on science and innovation policy, Atlanta, GA, USA
2. Kale D (2017) Sources of innovation and technology capability development in the Indian automobile industry. Inst Econ 4:121–131
3. Okada A, Siddharthan NS (2007) Industrial clusters in India: evidence from automobile clusters in Chennai and the national capital region (Discussion Paper No. 103). Institute of Developing Economies, JETRO (2007)
4. Meena A, Dhir S, Sushil (2021) An analysis of growth-accelerating factors for the Indian automotive industry using modified TISM. Int J Product Perform Manag 70(6):1361–1392. https://doi.org/10.1108/IJPPM-01-2019-0047
5. Zhang X, Rao R, Xie J, Liang Y (2014) The current dilemma and future path of China's electric vehicles. Sustainability 6(3):1567–1593
6. Zheng X, Lin H, Liu Z, Li D, Llopis-Albert C, Zeng S (2018) Manufacturing decisions and government subsidies for electric vehicles in China: a maximal social welfare perspective. Sustainability 10(3):672
7. Smiti (2019) India preps its own EV charging station standards. Clean Techn. https://cleantechnica.com/2019/05/30/india-preps-its-own-ev-charging-stationstandards
8. Shalender K, Yadav RK (2018) Promoting e-mobility in India: challenges, framework, and future roadmap. Environ Dev Sustain 20(6):2587–2607
9. Leviäkangas P, Kinnunen T, Kess P (2014) The electric vehicles ecosystem model: construct, analysis and identification of key challenges. Manag Global Trans: Int Res J 12(3)
10. IBEF Indian Automobile Industry Analysis (2019). https://www.ibef.org/industry/automobiles-presentation
11. Hübner Y, Blythe P, Hill G, Neaimeh M, Austin J, Gray L, Wardle J (2013) 49,999 electric car journeys and counting. In: 2013 world electric vehicle symposium and exhibition (EVS27). IEEE, pp. 1–9
12. Adheesh SR, Vasisht SM, Ramasesha SK (2016) Air-pollution and economics: diesel bus versus electric bus. Curr Sci 110(5):858–862

13. Bergek A, Hekkert M, Jacobsson S, Markard J, Sandén B, Truffer B (2015) Technological innovation systems in contexts: Conceptualizing contextual structures and interaction dynamics. Environ Innov Soc Trans 16:51–64. 445
14. Sturgeon TJ, Memedovic O, Van Biesebroeck J, Gereffi G (2008) Globalisation of the automotive industry: main features and trends. Int J Technol Learn Innov Dev 2(1–2):7–24

Relevance of Industry 4.0 in Manufacturing and Financial Systems Engineering (MFSE)—Connected Factories and Networks

K. Ravi and P. Naveenchanran

Abstract Today's industrial organizations face issues such as resource utilization, a shorter time to market, the creation of personalized products, and the connectivity of infrastructural pieces. This article discusses Industry 4.0 and Internet-connected technology's impact on corporate and societal value creation. Industry 4.0 is a solution to these challenges; it is the evolution of industrial systems that enables them to remain competitive. Industry 4.0 includes cyber-physical systems, additive manufacturing, virtual reality, predictive maintenance, and other features. Industry 4.0 is discussed in scholarly articles, conference proceedings, and social media sites. The authors analyse data gleaned from social network posts about Industry 4.0 in order to providing novel insights into the implementation of Industry 4.0. The data was gathered over time from the Twitter social network and then assessed in terms of the visibility and existence of Industry 4.0 and its components. The suggested roadmap is important because it accelerates the implementation of Industry 4.0 technologies for smart manufacturing and bridges the gap between sophisticated technologies and their implementation in manufacturing, especially for MFSEs.

Keywords Industry 4.0 · Networks · Manufacturing · Financial systems

1 Introduction

Today's industrial systems face significant challenges in the areas of resource efficiency, accelerated time to market, customized product production, and the integration of system infrastructure components. The purpose of every industrial system and industrial revolution is to maximize productivity. Most industrial systems are still ill-equipped to manage big data due to a lack of intelligent analytic tools, and businesses are confronting obstacles in addressing big data issues related to swift

K. Ravi (✉) · P. Naveenchanran
Department of Automobile Engineering, Bharath Institute of Higher Education and Research, Chennai, Tamil Nadu, India
e-mail: kraviphd2020@gmail.com

H. Kumar et al. (eds.), *Recent Advances in Intelligent Manufacturing*, Lecture Notes in Mechanical Engineering,
https://doi.org/10.1007/978-981-99-1308-4_12

decision-making for improved efficiency. Along with these obstacles, trade value formation necessity is focused toward long-term sustainability, which is now influenced by the transition to the fourth stage of industrialization, known as Industry 4.0. [1].

In the case of KM 4.0, the trend is toward establishing a communication channel for the constant exchange of information regarding needs and individual situations in real spending time with online retailers, healthcare professionals, producers, housekeepers, coworkers, customers themselves, and energy providers, among others. Information is typically sent between devices in the vast majority of situations.

Full automation, the digitalization of processes, and the use of electronics and information technology (IT) in manufacturing and providing services in the private sector will all be characteristics of the fourth industrial revolution. SMBs will be significantly impacted by technological developments like 3D printing as well as the expansion of online sales services like auto services, medical exams from home, ordering real food from the store to the refrigerator, etc. Smart factories, smart products, and smart services incorporated into the internet of things are the cornerstones of Industry 4.0. Industry 4.0 includes elements like virtual reality, additive manufacturing, cyber-physical systems, and predictive maintenance. New business models are emerging to take these qualities into account. The earlier industrial rebellion had a major influence on the shop floor, while this one is focusing on increasing productivity through improving brainwork and decision-making processes. Given the infancy of Industry 4.0's development, it is crucial to disseminate information about its benefits and raise awareness about the relevance of incorporating its components. Social media platforms, on the other hand, are employed for social contact and can be an extremely effective and potent information distribution tool. Although information distributed at a rapid pace through social networks might be correct or wrong, helpful or worthless, there is evidence that social networks prefer true information over rumours. The purpose of this article is to investigate the use of Industry 4.0 in Manufacturing and Financial Systems Engineering (MFSE)—Connected factories and networks.

2 Background Information About Theory

2.1 Industry 4.0

The goal of this essay is to assess what has been learned so far about the significance of Industry 4.0 and its influence on organizational and human environment changes that result in value added. The German governments, as well as several of its firms, colleges, and research institutes, are working to construct completely automated, Internet-based "smart" factories. The German government is promoting a political economic endeavour to assist industrial production preserve its competitive edge in the face of developing-country labour-cost advantages and a recovery in U.S. manufacturing. The European Union has decided to support research in the field of smart technology as a result. A research and innovation programme called Horizon 2020 provides funding for initiatives like gathering data on smart cities and communities, examining the strategic roles that smart cities can play in addressing energy and transportation issues, determining the viability of implementing integrated smart city and community solutions on a large scale, and more (European Commission, 2015).

2.2 *Industry 4.0's Theoretical Foundation*

Although the Internet-driven digital revolution is still ongoing, big data, artificial intelligence, and networking all point to a new cycle of digital disruption. Sector 4.0 is approaching and will have a big impact on the overall transformation of the industry since it reflects progress in three areas.

1. Digitization of manufacturing—administration and production scheduling information systems;
2. Automated Data Interchange (A.D.I.) connects industrial locations in a full supply chain.
3. Automation—systems for gathering data from manufacturing lines and putting it to use with equipment.

Improved attractiveness through smart apparatus, the utilization of knowledge regarding high-wage locales, population shifts, resources, energetic efficiency, and urban manufacturing are all hallmarks of Industry 4.0.

In this extent, we anticipate automating operations that will necessitate some level of worker automation. People will still be required to use their intellect. New goods and solutions will provide added value—handling numbers is not a productive task (Table 1).

3 Customer Comportment 4.0 in the Case of Future Connected Factories and Networks

The automobile sector was one of the first to recognize the benefits of the Internet and device connectivity. The availability of Long-Term Evolution (5G) communication between the car and in-car electronics in mid-range automobiles makes it simpler to communicate with the Internet and the outside world while driving. Connecting the smartphone to the car's control system allows the driver to view data on a screen that they may adjust to their liking. When a smart key is in close proximity to a car, the car will start and unlock when the owner approaches. Smart keys can alter seat as well as steering wheel settings as starting the car depending on previously saved data. The relevance of the technical condition of the goods as well as the quality and

Table 1 Basic perception of Industry 4.0

S. No.	Expression	Explanation
1	In the creation of products and services, new systems are being used	Individualized product and service development will be implemented. Approaches like open innovation and product intelligence, as well as product memory, are critical in this setting
2	Smart factory, smart manufacturing, intelligent factory, and factory of the future are all terms that can be used to describe a smart factory	The intelligent, flexible, and dynamic smart factory has more intelligent, flexible, and energetic. Sensors, actors, and self-driving systems will be installed in manufacturing
3	Self-organization	Processes vary across the supply and production chains in manufacturing. Manufacturing processes will be intimately linked across organizational borders as a result of all of these developments
4	Digital sustainability	The design of smart cities and smart industries is increasingly focusing on sustainability and resource efficiency. When exploiting private information, it is vital to follow ethical guidelines. These are the foundational criteria for a successful product
5	New distribution and procurement mechanisms are being implemented	Individualized distribution and procurement will become more common

dependability of the information acquired and given will be more widely understood by customers. This will have an impact on the real-time collection and analysis of data and, as a result, on future value generation guidelines for consumers [2]. It raises the issue of how customers will adjust to new technology associated with items.

The adoption of Internet technologies into industry is the primary technological backdrop of Industry 4.0, this methodological foundation is frequently coupled with related future ideas [3]. The Internet has had a significant impact on the world we live in and our way of life in general. In today's wireless telecommunications world, a new paradigm known as the Internet of Things (IoT) is gaining ground swiftly. The concept is based on the commonplace existence of several different items or objects, such as Radio-Frequency Identification (RFID) tags, sensors, actuators, mobile phones, etc. that are able to communicate with one another and work together toward a common goal by virtue of their individual addressing systems. Research at the intersection of fields, including telecommunications, information systems, electronics, and the social sciences, is essential to developing practical applications for the Internet of Things. With the addition of two-way communication between smart objects, the concept of "anytime, anywhere, any medium, any object" communications can be realised. The creation of useful insights from large data sets is one potential outcome of the Internet of Things becoming ubiquitous in the near future. The Internet of Things (IoT) provides information from the environment, while social

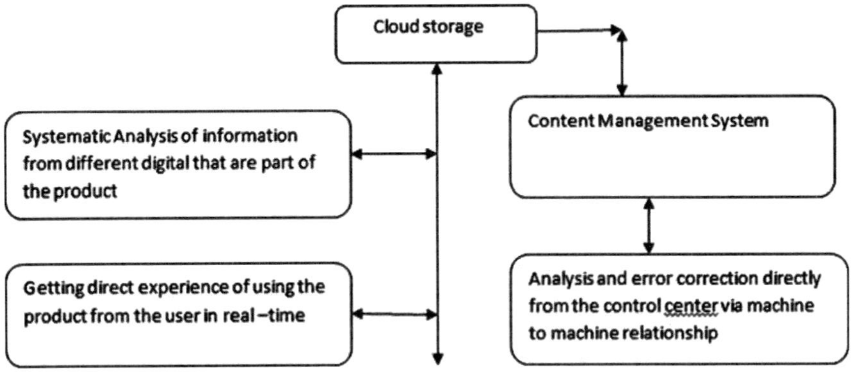

Fig. 1 Process of relevance of Industry 4.0

networks offer the glue that enables human-to-device interactions, forming a cluster that enables individuals to connect to the ubiquitous computing universe (Fig. 1).

4 Results Discussion

Public are discussing about Industry 4.0 on Twitter, according to a study done in three working days. Most people connect this problem with the phrases "job" and "hiring," which may be the most intriguing finding of the study. This is also the most talked-about topic in the United States (especially the East Coast) and Europe (especially Germany). That is most likely due to the fact that the majority of Twitter users are in the United States, while Industry 4.0 was originally introduced in Germany. The time of day when this issue is discussed the most is in the middle of the day. Another intriguing statistic is that most tweets about this issue have only one hashtag.

5 Conclusions

This paper set out to show how strong social media sites are for knowledge and information exchange, and how important Industry 4.0 is for monitoring industrial systems. This was just the beginning of our investigation into the presence of Industrial 4.0 on Twitter; it's most referenced and crucial components, and the regions most covered by the topic. The burgeoning social media mining community will benefit from the expansion of research and development, as well as the discovery of both online and offline human behaviour and interaction patterns, thanks to the preliminary success of social media mining research activities. Given that there are around 320 million monthly active Twitter users and 600 million tweets posted every day, Twitter may be a highly strong and valuable medium for promoting the Industry

4.0 idea. Future studies will concentrate on SNA, a sociological method for analysing patterns of relationships and interactions between social actors. Smart factories, smart products and smart logistics, and so on are only some of the other facets of Industry 4.0 that will be taken into account.

Funding The research and/or authoring of this paper were not supported financially by the author(s).

Declaration of Conflicting Interests The author(s) of this study disclosed no potential conflicts of interest with regard to the research, authorship, and/or publication of this work.

References

1. Bonzanini M (2016) Mastering social media mining with Python. Packt Publishing Ltd
2. Espejo R, Dominici G (2016) Cybernetics of value cocreation for product development. Syst Res Behav Sci. https://doi.org/10.1002/sres.2392
3. Lee I, Lee K (2015) The IoT (IoT): applications, investments, and challenges for enterprises. Bus Horiz 58:431–440. https://doi.org/10.1016/j.bushor.2015.03.008
4. Almada-Lobo F (2016) The Industry 4.0 revolution and the future of manufacturing execution systems (MES). J Innov Manage 3:16–21
5. Dominici G, Roblek V, Abbate T, Tani M (2016) "Click and drive": consumer attitude to product development. Towards future transformations of driving experience. Bus Process Manage J 22:420–434. https://doi.org/10.1108/BPMJ-05-2015-0076
6. Gamarra C, Guerrero JM, Montero E (2016) A knowledge discovery in databases approach for industrial microgrid planning. Renew Sustain Energy Rev 60:615–630. https://doi.org/10.1016/j.rser.2016.01.091
7. Nanry J, Narayanan S, Rassey L (2015) Digitizing the value chain: challenges remain for "Industry" 4.0, but the buzz is growing. McKinsey Quarterly. http://www.mck-insey.com/insights/manufacturing/digitizing_the_value_chain
8. Rocco RA, Bush AJ (2016) Exploring buyer-seller dyadic perceptions of technology and relationships: implications for sales 2.0. J Res Interact Market 10:17–32. https://doi.org/10.1108/JRIM-04-2015-0027
9. Scheer AW (2012) Industrierevolution 4.0 ist mit weitreichen- den organisatorischen Konsequenzen verbunden! (Industrial Revolution 4.0 is associated with far-reaching organizational con- sequences!). Inf Manage Consult 3:10–11
10. Sommer L (2015) Industrial revolution—industry 4.0: are German manufacturing SMEs the first victims of this revolution? J Ind Eng Manage 8:1512–1532. https://doi.org/10.3926/jiem.1470
11. Weber E (2015) Industrie 4.0—Wirkungen auf Wirtschaft und Arbeitsmarkt (Industry 4.0 Impact on the economy and labor market). Wirtschaftsdienst 95:722–723
12. Stock T, Seliger G (2016) Opportunities of sustainable manufacturing in Industry 4.0. In: Procedia CIRP 40 on 13th global conference on sustainable manufacturing—decoupling growth from resource use, pp 536–541
13. KTN. KTP helps embed Industry 4.0 techniques to transform business visibility and efficiency; case study: KTN; June 2020. https://www.ktp-uk.org/case-study/ktp-helps-embed-industry-4-0-techniques-to-transform-business-visibility-and-efficiency-for-beverston-engineering/. Accessed 1 June 2020
14. Gurela T (2019) Industry 4.0 and the factory network. I4.0 Today (1) February 2018. http://i40today.com/magazines/

15. PlatformIndustrie 4.0. Digitization of Industrie Platform Industrie 4.0; Federal Ministry for Economic Affairs and Energy (BMWi), Berlin, Germany (2016)
16. Rüßmann M, Lorenz M, Gerbert P, Waldner M, Justus J, Engel P, Harnisch M (2015) Industry 4.0: the future of productivity and growth in manufacturing industries. The Boston Consulting Group (BCG), Boston, MA, USA

Investigations on Performance of Crater Film Cooling with Different Film Hole Configurations at Compound Angles

V. G. Krishna Anand, Atul Babbar, and M. Mohammed Mohaideen

Abstract In crater assisted cooling the holes are implanted within the crater-shaped structure developed on the gas turbine surface that need to be cooled. This study aimed to analyze the performance spherical shaped crater implanted with different film hole configurations (circular, rectangular, square, semicircular 1, and semicircular 2) oriented at 45° and 180°. Numerical investigations were carried out using the finite volume-based Reynolds averaged Navier strokes approach. Experiments were performed on a heated suction-type film cooling facility to validate the numerical data. The study show that the semicircular 2 model film hole configuration at 180° orientation angle has produced higher cooling performance for the cratered cooling surface. It has also delivered an impressive lateral spreading with higher local and spanside averaged cooling effectiveness on the cratered test surface model.

Keywords Film cooling · Crater · Compound angle · Cooling effectiveness · Blowing ratio

1 Introduction

The gas turbines have wide applications in the areas of electricity production and vehicle propulsion. Advanced cooling methods are essential to further enhance the performance of gas turbines. The crater assisted film cooling (FC) is a type of FC where the FC holes are implanted within the crater depression developed on the FC surface. Bunker [1] reported that the crater assisted FC has delivered a 50% improved FC effectiveness on the test model for an unity blowing ratio (M). Also, the performance of crater FC improved with the increase in M. Lu and Ekkad [2] performed a computational study to examine the adiabatic FC effectiveness for the

V. G. Krishna Anand (✉) · M. Mohammed Mohaideen
Department of Aeronautical Engineering, MRCET Campus, Hyderabad, Telangana, India
e-mail: kakrishnaanand@gmail.com

A. Babbar
Department of Mechanical Engineering, SGT University, Gurugram, Haryana, India

H. Kumar et al. (eds.), *Recent Advances in Intelligent Manufacturing*,
Lecture Notes in Mechanical Engineering,
https://doi.org/10.1007/978-981-99-1308-4_13

crater embedded geometry. The studies were carried out for three different crater geometry with a fixed crater depth. The film holes were inclined at 30° to the crater wall surface with an L/D ratio of 4 and a hole diameter of 12.7 mm. The results of the numerical investigation show that FC effectiveness delivered with crater assisted geometry were effective than the non-crated film hole model. The cratered coolant injection leads to the development of weaker opposite rotating vortex structure pairs at the lowerside location of the film coolant injection region. Kim and Kim [3] optimized the various design parameters of the crater geometry and analyzed its influence on area-averaged FC effectiveness. The studies were performed on a model with FC hole inclined at 30° and having a diameter of 12.7 mm. The unstructured mesh system was employed for the computational domain and SST turbulence model was used for numerical modeling. The mainstream velocity and temperature were maintained at 13.8 m/s and 321 K respectively. The analysis were carried out for an M of 0.5. The numerical analysis concludes that crater geometry has delivered improved area-averaged FC effectiveness. Tran et al. [4] numerically studied the effects of rectangular crater geometry, design parameters, and its influence on test surface FC effectiveness. The investigations were aimed to identify the major crater parameters that influence the spatially averaged FC performance. The output of the study showed that the crater geometry has produced better film cooling effectiveness than a non-cratered cylindrical film hole configuration. Kalghatgi and Acharya [5] carried out computational studies to analyze the behavior of contoured crater geometry on the FC effectiveness of test surface. The variables analyzed are viz., depth of the crater, the effects of increasing injection ratios, and the impact of contouring crater wall on FC effectiveness. The crater depth investigated was in the range of 0.2 D–0.75 D and the tested blowing ratios were in the range of 1.0–2.0. The cooling holes have a diameter of 12.7 mm with an inclination angle of 35° to test the surface and the L/D ratio is maintained at 1.75. The study reported that the contoured crater configuration has produced an effective FC performance on the test model in comparison with the non-contoured crater geometry. All tested crater depth configurations showed an effective enhancement in FC performance at all tested M. The increase in crater depth showed an enhanced spanwise averaged FC performance on the test surface. The crater depth of 0.75 D has delivered a better FC performance compared to other crater depth at higher M. Vortical structures were also observed in the intermediate crater depths. A higher level of span side averaged FC performance is observed close to the FC hole.

Lu et al. [6] performed experimental studies on different crater configurations with IR measurement technique. They reported that in comparison with the conventional FC model, the crater configuration have delivered a 50% enhanced FC performance on the test surface. Rhee et al. [7] conducted experimental studies on different film hole configurations for flat surface film cooling. All tested film hole configurations were maintained at the same hydraulic diameter for comparison. They concluded that the rectangular film hole model wase effective than circular hole configuration. Gartshore et al. [8] reported that the exit geometry of the FC hole has a considerable influence on the FC performance. They also showed that the square-shaped FC hole has produced slightly better FC effectiveness in comparison with the circular film hole model. Takahashi et al [9] performed experimental studies on different FC hole configurations embedded on a conventional flat surface FC and showed that the rectangular FC hole configuration has produced higher FC performance in comparison with circular hole configuration. Aga and Abhari [10] showed that the higher oriented FC hole (90°) has produced better film coolant distribution along the spanside direction of the test model. Goldstein and Jin [11] concluded that the oriented angle cooling hole has produced better FC performance than the simple angle FC hole.

Most of the studies reported on crater film cooling were performed with circular film hole configurations. However, the studies performed on non-crater surfaces with different film hole configurations showed higher FC performance than the circular hole model. Also, it is outlined in the literature that the oriented hole angle can deliver higher performance than the simple angle film cooling hole. Hence this study aimed to investigate the crater FC performance of different FC hole models (circular, rectangular, square, semicircular 1, and semicircular 2) oriented at 45° and 180°. The study can deliver the flow and FC insights of cratered FC hole and can help to understand the FC behavior of cratered test surface with different tested FC holes at two different orientation (45° and 180°).

2 Computational Methodology

The computational model for the cratered film cooling utilized for this investigation is presented in Fig. 1. The repeating pattern of film cooling holes in computational studies is simulated with the application of periodic boundary condition (BC) in the lateral vertical regions of the computational boundaries. The origin of the computational domain is positioned at the trailing edge of the crater. The computational domain is based on the dimensions of the FC experimental facility developed in the laboratory.

The longitudinal extension of the computational region between the and outlet is 87D. Also D is the FC hole diameter. The X-axis forms the free stream distance along the wall surface, Y-axis forms the column distance concerning the test surface and Z-axis represents the width of the test wall surface in the spanwise axis. The injection angle of the FC hole (γ) is 30°. The L/D ratio and P/D ratio of the FC hole are maintained at 6 and 3 respectively. The hydraulic diameter of all the studied FC holes is maintained at 4 mm for purpose of comparison. The computational domain was sized based on the dimensions of the experimental test section module. The pitot-tube module was utilized for the measurement of free stream flow velocity at a distance of 25 D ahead of slot leading position of test wall surface and the numerical domain was sized and modeled correspondingly. The free stream inlet location for the computational study is modeled at a position of 25 D ahead of the slot leading edge. Further, the complete experimental section is uniform concerning the principal axis of the slot and FC holes of the test wall model and it results in the utilization of one FC hole integrated with the slot for the numerical domain. The

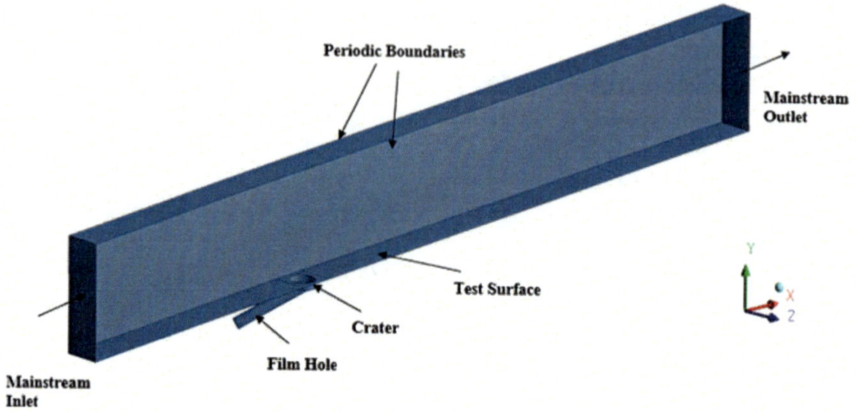

Fig. 1 Three dimensional views of the domain

entire FC holes of the manufactured test model have an identical cross-sectional area at entry and exit. Designing the complete experimental module was infeasible with our current computational facilities. The investigations are extended to investigate the influence of different film hole shapes (circular, rectangular, square, semicircular 1 and semicircular 2) embedded in the spherical crater model for FC effectiveness. The schematic details of the different hole shapes embedded within the spherical crater are shown in Fig. 2. In the present investigation, a structured mesh system with prism layers on the wall surface is used for the mainstream domain and an unstructured mesh system is adopted for the film hole. The boundary layer mesh system ensures Y+ value close to unity on the wall surfaces. The mesh independence study is performed with three different meshes viz., 0.9, 2.1 and 3.9 million cells. The final mesh system chosen after the mesh independence study is 2.1 million cells.

The results of the FC effectiveness predicted with the k-ε realizable model shows a close accordance with the experimental results. The other turbulence models (RNG k-ε and SST k-ω) investigated have overestimated the results of FCE. Therefore for the present investigation, the realizable k-ε model with EWT and activated thermal effects were employed for all cases of the simulation. The inclusion of proper boundary conditions in the computational domain is an important part of the computational investigation. The boundary condition defined during the modeling phase of the domain includes adiabatic walls, velocity inlet, pressure outlet, periodic lateral boundaries, and symmetric boundaries. Free stream flow inlet is defined as inlet BC with a constant velocity of 13 m/s and the thermal condition of the free stream hot air was retained at 343 K. All the spanwise column boundaries were defined with repeating BC (periodic). The faces of the computational regions were defined with the adiabatic and no-slip wall condition. The level of intensity of turbulence for free stream inlet and FC hole inlet was 5 and 2% respectively. The hydraulic diameter computed based on free stream inlet and FC hole inlet were defined at their relevant boundaries. The SIMPLE scheme is employed for the coupling of pressure

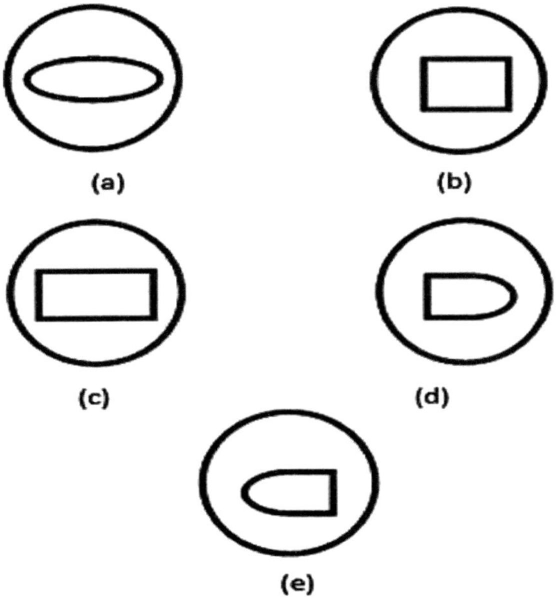

Fig. 2 Schematic of the spherical crater with **a** circular film hole **b** rectangular film hole **c** square film hole **d** semicircular 1 film hole **e** semicircular 2 film hole

velocity terms. The 2nd order upwind scheme is adapted for spatial discretization. The convergence criteria are set to fall of residuals below 10^{-5} for continuity, momentum, turbulence equations and 10^{-8} for the energy equation. The temperature at the location of X/D 10 along the spanside axis of the test wall surface was monitored and it was ensured that after reaching the set residual limits there is no change in solution even after the iterations are continued.

3 Results and Discussion

The simulation results need to be compared with the experimental data for their acceptance. The experimental study for crater FC is performed on a FC facility developed in the fluid dynamics laboratory. The test facility have different sections viz., diffuser, heater module, settling chamber, converging section, test section, constant section, divergent section, and blower assembly. The experimental facility developed is a heated wind tunnel film cooling facility. The developed facility has provisions to measure the freestream flow velocity. The flat plate test models can be inserted on to the test section of the test facility. The static pressure and thermal measurements were carried out on the inserted test plates. The velocity of the heated airflow can be measured with the pitot-static probe. The static pressure measurements can be carried out with a scani-valve device. The scani-valve used in the experimentation is DSA 3218-PTP pressure scanner. It is a 16 channel pressure scanner with an accuracy of ±0.05%. The temperature measurement at the surface of the test models we carried

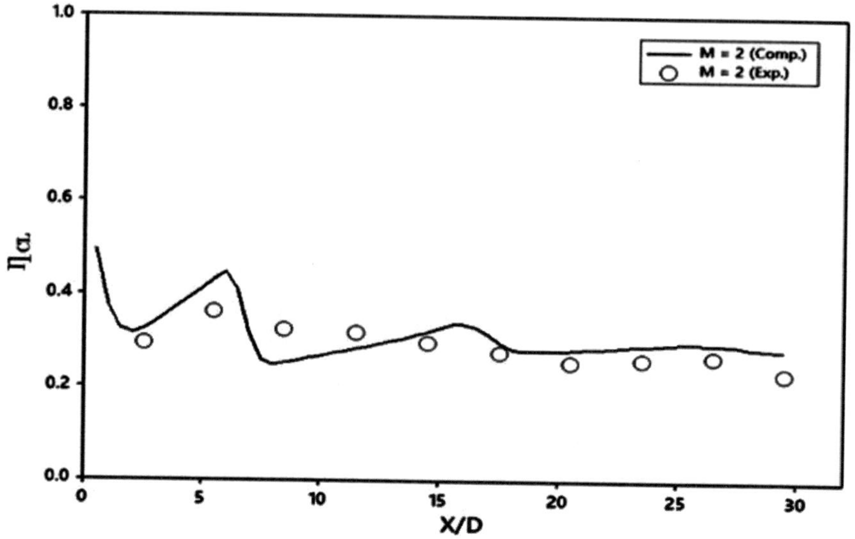

Fig. 3 Simulation and experimental centreline FCE for crater FC at M of 2.0

out with the thermocouples is embedded on the test plate. The k-type thermocouples were used in this study for the temperature measurement. Figure 3 shows the comparison plot for centreline effectiveness along the central axis of the wall surface for CFC at M of 2. Here the computational results are mostly predicted higher in comparison with experimental values. However, there is a slight under-prediction observed between X/D of 8–12 along the central axis of the test wall surface.

The influence of different film hole shapes at 45° compound angles (circular, square, rectangle, semicircle 1, semicircle 2) is compared for FC effectiveness. The distribution of FC performance on test surface and on the region around the crater model for different film hole shapes is shown in Fig. 4. The FC effectiveness distribution on the test surface were in line with the orientation angle of the FC hole i.e. 45° and irrespective of the FC hole shape. However, the semicircular 2 film hole model has produced an higher film coolant distribution on the test surface. The rectangular shape FC hole at a compound angle of 45° has produced an ineffective film coolant distribution on the test wall surface in comparison with other investigated compound angle FC hole.

The local effectiveness distribution on the test surface for different film hole shapes (circular, square, rectangle, semicircle 1, semicircle 2) with a backward oriented (180°) film hole embedded in the spherical crater model is shown in Fig. 5. In comparison with other FC hole shapes, the semicircular 2 shaped film hole at 180° compound angle has produced higher coolant distribution on the test surface model. This effect is due to the lower jet lift-off and better spanwise distribution of the ejected coolant jet on the test wall surface.

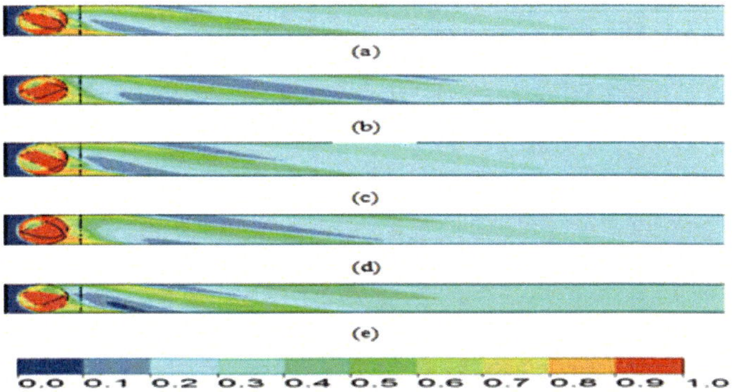

Fig. 4 Distribution of η for 45° oriented **a** circular film hole **b** rectangular film hole **c** square film hole **d** semicircular-1 film hole **e** semi-circular-2 at M of 1.5

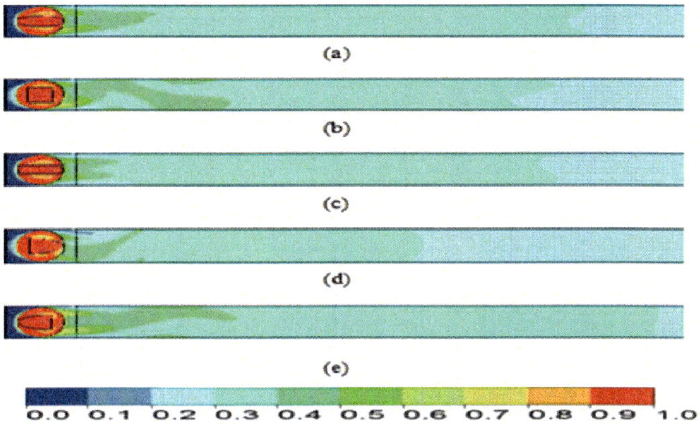

Fig. 5 Distribution of η for 180° oriented **a** circular film hole **b** rectangular film hole **c** square film hole **d** semicircular-1 film hole **e** semi-circular-2 at M of 1.5

The spanside averaged FC effectiveness along stream-wise direction for different film hole shapes (CIR—circular film hole, REC—rectangular film hole, SQR—square film hole, SC1—semi-circular model 1, SC2—semicircular model 2) embedded in spherical crater model is presented in Fig. 6.

The SC2 model film cooling hole has delivered the highest FC effectiveness at M of 1.5 for a spherical crater model implanted with 45° compound angle FC hole. The highest FCE of the SC2 film hole model was due to the higher lateral side distribution of the coolant film on the wall surface

The lateral averaged FC effectiveness for different film hole shapes embedded in the spherical crater with backward oriented film hole is shown in Fig. 7. The rectangular and semi-circular 2 film hole models have delivered higher and almost similar FCE. However, after X/D > 10, the SC2 film hole model with backward

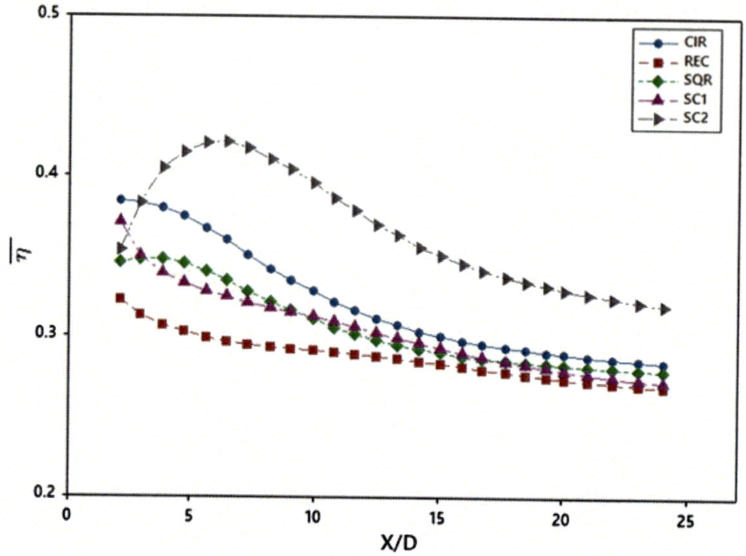

Fig. 6 Variation of $\bar{\eta}$ for different film shapes with 45° orientation film hole embedded in the spherical crater at M of 1.5

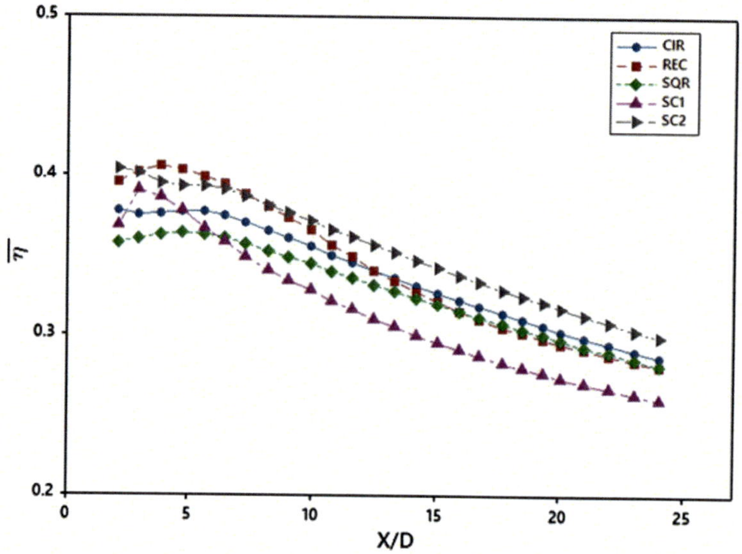

Fig. 7 Variation of $\bar{\eta}$ for different film shapes with 180° orientation film hole embedded in the spherical crater at M of 1.5

injection has delivered higher cooling effectiveness distribution on the test surface in comparison with the other film hole shape investigated.

4 Conclusion

Numerical investigations were carried out to study the cooling performance of a spherical crater configuration implanted with different film hole configurations (circular, rectangular, square, semicircular 1, and semicircular 2) oriented at two different orientation angles (45° and 180°). Experimental studies for film cooling measurements were carried out on a film cooling facility to validate the numerical data. RANS-based simulations were performed with the selected Realizable k-ε turbulence model. Results of the investigation showed that the semicircular 2 model film hole has produced an effective performance with 180° oriented FC hole. Also, the semicircular 2 model FC hole has delivered a better spanwise spreading of the film coolant with higher local and spanside averaged FCE on the cratered test surface.

Acknowledgements The first author thankfully acknowledges DST, Government of India, for financial support through DST INSPIRE fellowship.

References

1. Bunker RS (2005) A review of shaped hole turbine film hole technology. ASME J Heat Transf 127:441–453
2. Lu Y, Ekkad SV (2006) Film cooling for cratered cylindrical inclined holes. In: Proceedings of international mechanical engineering congress and exposition, Illinois, USA, IMECE2006–14406
3. Kim J, Kim K (2016) Surrogate based optimization of a cratered cylindrical hole to enhance film cooling effectiveness. J Therm Sci Technol 11(2):1–14
4. Tran NT, Nguyen CQ, Ho SH, Kapat JS (2010) Prediction of adiabatic effectiveness of various cratered film hole configurations: sensitivity analysis for rectangular shaped mask. In: AIAA aerospace science meeting, Florida, AIAA 2010-404
5. Kalghatgi P, Acharya S (2015) Improved film cooling effectiveness with a round film cooling hole embedded in a contoured crater. ASME J Turbomach 137:101006-1-10
6. Lu Y, Dhungel A, Ekkad SV, Bunker RS (2009) Film cooling measurements for cratered cylindrical inclined holes. ASME J Turbomach 131:011005-1-12
7. Rhee DH, Lee YS, Cho HH (2002) Film cooling effectiveness and heat transfer of rectangular-shaped film cooling holes. In: Proceedings of ASME turbo expo: power for land, sea, and air. ASME Paper No. GT2002-30168
8. Gartshore M, Salcudean I, Hassan I (2001) Film cooling injection hole geometry: hole shape comparison for compound cooling orientation. AIAA J 39:1493–1499
9. Takahashi H, Nuntadusit C, Kimoto H, Ishida H, Ukai T, Takeishi K (2001) Characteristics of various film cooling jets injected in a conduit. Ann N Y Acad Sci 934:345–352

10. Aga V, Abhari RS (2011) Influence of flow structure on compound angled film cooling effectiveness and heat transfer. ASME J Turbomach 133:031029-1-12
11. Goldstein RJ, Jin P (2001) Film cooling downstream of a row discrete holes with compound angle. ASME J Turbomach 123(2):222–230

Machine Configuration Based on Machine Reliability and Production Rate Criteria Through Line Balancing Algorithm in Reconfigurable Manufacturing System (RMS)

Ashutosh Singh, Mohammad Asjad, Yash Vardhan Singh, and Shahnawaz Alam

Abstract A reconfigurable manufacturing system (RMS) is specially designed around a part family of manufacturing products because of its core characteristic such as Diagnose-ability, Convertibility, Scalability, and Machine adaptability, Customization, Modularity, Mobility, Integrability, Sustainability, and Flexibility etc. In RMS, the selection of machine equipment, machine arrangement, plant layout arrangement, material handling arrangement and assessment of operations etc. has a significant impact on their performance. In this paper, various machine configurations of RMS are investigated in terms of different machine reliability, production rate and cycle time. In the first phase, line balancing technique is deployed to find out optimal machine configuration and the second phase includes different values of machine reliability (like 0.85 and 0.75) for calculating the minimum number of machines of different machine configurations in RMS. The developed approach is demonstrated and validated using a case study. This proposed approach can provide decision support for management decision makers for selecting optimal machine configuration at the design stage in RMS.

Keywords Cycle time · Dedicated manufacturing system (DMS) · Line balancing technique · Machine reliability · Production rate · Reconfigurable manufacturing system (RMS)

A. Singh (✉) · Y. V. Singh
Department of Mechanical Engineering, JIMS Engineering Management Technical Campus, Greater Noida 201308, India
e-mail: ashurajput86@gmail.com

M. Asjad · S. Alam
Department of Mechanical Engineering, Faculty of Engineering and Technology, Jamia Millia Islamia, New Delhi 110025, India

H. Kumar et al. (eds.), *Recent Advances in Intelligent Manufacturing*, Lecture Notes in Mechanical Engineering,
https://doi.org/10.1007/978-981-99-1308-4_14

1 Introduction

Generally, the manufacturing system can be categorised into three types as Dedicated manufacturing system (DMS), Flexible manufacturing system (FMS) and Reconfigurable manufacturing system (RMS) [1–8]. In DMS, all CNC machines are arranged in a series manner with fixed automation (for both software and hardware modules) and fixed machine configuration [9]. They are designed for mass production of specific types of product at a minimum cost and a very high production rate [3]. It is worthwhile to note that DMS cannot face large fluctuation in market demand [10]. By contrast, FMS consists of NC and CNC machine and programmable machine [11]. The investment cost of FMS is very high and has a relatively low throughput as compared to DMS. FMS is used in the Ford Windsor Engine Plant containing 120 CNC machines [12]. In comparison of these two manufacturing systems, RMS is ready to modify its scalability (production volume) and functionality by adding or removing machine hardware and software module in the machine tool or by changing its machine configuration. The world-first patent on RMS was filled in 1998 by Koren and Ulsoy [25]. In RMS, there are many fields of reconfiguration at different levels such as changing machine controller, changing factory communication software, changing the material handling, changing configuration of machine tool, changing machine configuration, buildings blocks to face market globalization [1, 13, 14]. Production rate, reconfigurability, responsiveness, investment cost of machine and tool, capacity scalability, throughput, no of product type etc. that are highly influenced by machine configuration [15, 16] in which different machines (like lathe, milling, shaper, drilling, slotter etc.,) are arranged in series, parallel or mixed configurations [2]. Zhong et al. [17] studied the different machine configuration lines (series and pure parallel). Figure 1 helps to illustrate the four stage machine configuration. It is trivial to realize that designing of the machine line is not a simple task because it contains a very large number of issues such as line balancing, production rate, reliability of machines etc. [14]. Keeping all these factors in mind, Lu et al. [18] designed the RMS on the basis of system reliability parameter. Further, Goyal et al. [16] discussed a NAGA-2 and TOPSIS algorithm for finding optimal machine configuration in RMS. He also discussed machine capability and reconfigurabilty. Therefore, machine reliability is an important performance parameter for selecting the optimal machine configuration. According to Ebeling [19] reliability is defined as the probability that a component or system will perform a required function for a given period of time when used under stated operating condition. Since reliability is a probability, system reliability may be determined from the component reliability. Reliability is not just a direct reflection of the quality indicators, but also on the success or failure of the entire system operation [18]. System reliability is very important parameter for any manufacturing system for calculating cost benefits analysis, life cycle costing, optimal machine configuration and replacement decision of machine, machine parts etc. [20]. Therefore, selecting machine, machine configuration, layout configuration and proper sequencing of manufacturing operation are the crucial performance parameters for the reconfigurable manufacturing system.

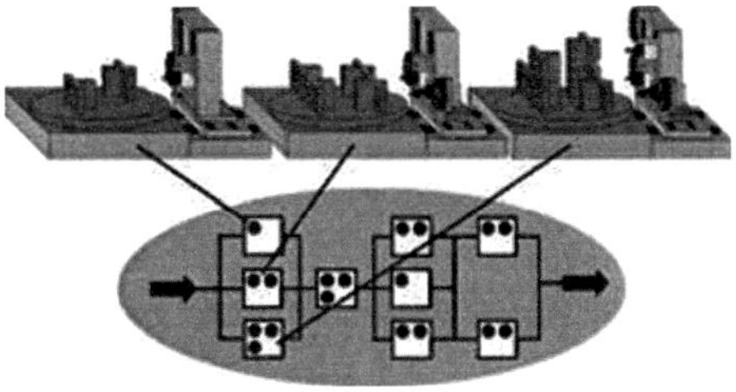

Fig. 1 Four stage machine configuration in RMS [20]

Cost, reliability, operational capability and production time are four performance parameters used for the selection of machine alternatives across all stages for the product flow line [21]. But, how to measure the performance of machine configuration and to develop the optimal machine configuration of RMS is still an issue till now. Despite numerous advantages of RMS over DMS and FMS, less research work has been carried out in the area of selection of optimal machine configuration and line balancing. Being motivated from these research gaps, this research was systematically planned and conducted to achieve the following objectives: (i) to explore line balancing algorithm and production rate criteria in machine configuration in RMS (ii) to analyse the effect of machine reliability and cycle time concepts for calculating the minimum number of machines in the manufacturing plant. Further, this research work elaborates here different types of machine configurations on the basis of varying machine reliability (like 0.85 and 0.75) and different cycle times (like 2.0 min/ stage and 2.5 min/stage) for a particular type of product. In RMS, multiple parallel identical machines are arranged in several stages (like two stages, three stages, four stages, five stages etc.). The present paper aims at reviewing the current practices, approaches and suggests the possible future directions based for future research. Further, a case study is presented in which we have assumed different machine reliabilities, cycle time and production rate to determine optimal machine configuration of two, three, four and five stages. Finally in the conclusion the overall research and future scope of the overall research is given.

2 Methodology: Case Study—I

In order to demonstrate the application of the line balancing, reliability to find out RMS Configuration, a case study is adapted from Koren and Shpitalni [22] based on an example part. The developed machine configuration approach is applied to the test part; assume an example: Design a machining system to machine a part that requires t = 12.2 min of machining time in five machining tasks. The execution times for the five tasks are given in Fig. 2 where the required daily volume is Q = 500 parts per day. The total working time per day is 1000 min. Machine reliability is assumed to be 0.85 and 0.75 in to two different cases respectively. This is the engine part after

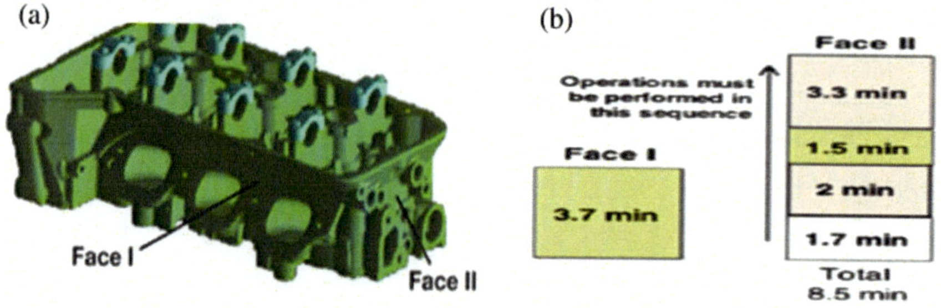

Fig. 2 **a** An engine part after machining. **b** Machining times [22]

casting operation. In this problem, we assume only two faces as shown in the figure with their machining time.

Step No. 1: According to Reginato et al. [23]. The cycle time (T_c) is the time when a product is released from station to station, defined by the following Equation,

$$T_c = \text{available time in period/demand in period(no of components)} \quad (1)$$

Step No. 2 Find out the minimum number of machines for performing this manufacturing operation:

Let N be the number of machines required to meet the demand requirement. The minimum value of N can be calculated. According to [22]

$$N_{min} = Q * Tt/\text{min time per day} * RM \quad (2)$$

Q = Daily volumes in parts/day.

Tt = Total machining time for manufacturing one product.

R_M = Machine reliability.

Step No. 3: Find out the minimum no of machine for both manufacturing face I and face II.

Step No. 4: Developed different optimal machine configuration based on machine reliability and cycle time.

Step No. 5: Identify those machine configurations that are satisfying cycle time constraints and have the highest production rate only.

Different machine configurations for two, three, four and five stages (consider machine reliability is 0.85 and cycle time $\leq$ 2.0 min):

By using equation no 2, we can calculate the minimum number of machines for any manufacturing condition.

Minimum no. of machine (for both face I & II) N = 500*12.2/1000* ≈ 8 machine.

For machining face I, Minimum no of machine N1 = 500*3.7/1000*0.85 = 2.17 ≈ 3 machine.

For machining face 2, Minimum no of machine N2 = 500*8.5/1000*0.85 = 5.0 ≈ 5 machine.

Therefore, total no of machine N = N1 + N2 = 3 + 5 = 8 machines, further we can arrange these machine in series, parallel or mixed configuration by using line balancing technique. Line balancing has been an important technique for manufacturing system design, because a completely balanced system can provide maximum resource utilization at the designed capacity [24]. By contrast, to achieve a balanced RMS configuration only the following line balancing relationship needs to be satisfied [22].

$$Ts1/Ns1 = Ts2/Ns2 = Tsi/Nsi \tag{3}$$

where Nsi = total number of machines in stage i, Tsi = total processing time per machine in stage i.

It should be noted that the sequence of the machining of different operations is fixed throughout the calculation. For two stages there is only one possible machine configuration shown in Fig. 3. The machining time for each stage (two, three, four and five) is discussed in Table 1. In three, four and five stages, only four machine configurations are possible, as shown in Figs. 3, 4, 5 and 6. On the basis of this result, we can find out easily the optimal machine configuration for maximum throughput or production rate. We can only consider that machine configuration where the cycle time is less than or equal to two minutes only. Machine configurations of Figs. 4d, 5d and 6b–d are not accepted because all of them don't fullfill the cycle time. Further, Table 2 shows an overview summary of machining time of the proposed machine configuration for two, three, four and five stages. It is cleare from Table 2 the machine configurations that are best among all. The machine configuration which gives higher production per day (i.e. 500 parts per day) is assumed to be of best configuration. For example in three stage, machine configuration Fig. 4c is the best configuration because the production rate per day is very high (i.e. 588 parts per day) and machine configuration Fig. 4d has a very low production rate per day (i.e. 300 parts per day). As similarly, n four stages, the production rate per day for Fig. 5b machine configuration is very high (i.e. 571 parts per day) and in case of five stages, only Fig. 6a is the best machine configuration. After calculating the machining time and production rate per day, we are plotting the graph in between different machine configurations in horizontal axis (for two stage, three stage, four stage and five stage) and production rate of plant per day in Y-axis. We will get the following graph (Fig. 7) for machine reliability 0.85. Finally, the number of configuration per stage in this case is reduced for the entire possible machine configuration.

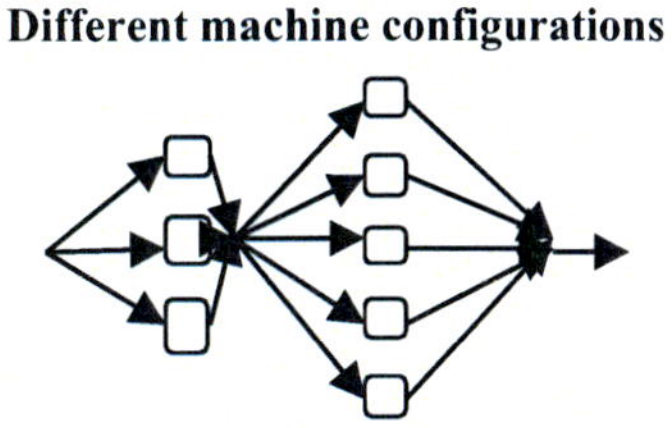

Fig. 3 Two stage machine configuration

Table 1 Machine stage cycle time for each stages

Machine stage cycle time for two stage configuration					
Machine configuration	First stage	Second stage			
Fig. 3	3.7/3 = 1.23 min	8.5/5 = 1.70 min			
Machine stage cycle time for three stage configuration					
Machine configuration	First stage	Second stage	Third stage		
Fig. 4a	3.7/3 = 1.23 min	3.7/2 = 1.85 min	4.8/3 = 1.60 min		
Fig. 4b	3.7/3 = 1.23 min	5.2/3 = 1.73 min	3.3/2 = 1.65 min		
Fig. 4c	3.7/3 = 1.23 min	1.7 min	6.8/4 = 1.70 min		
Fig. 4d	3.7/3 = 1.23 min	5.2/4 = 1.3 min	3.3 min		
Machine stage cycle time for four stage configuration					
Machine configuration	First stage	Second stage	Third stage	Fourth stage	
Fig. 5a	3.7/3 = 1.23 min	3.7/2 = 1.85 min	1.5 min	3.3/2 = 1.65 min	
Fig. 5b	3.7/3 = 1.23 min	1.7 min	3.5/2 = 1.75 min	3.3/2 = 1.645 min	
Fig. 5c	3.7/3 = 1.23 min	1.7 min	2.0 min	4.8/3 = 1.60 min	
Fig. 5d	3.7/3 = 1.23 min	3.7/3 = 1.23 min	1.5 min	3.3 min, Bottle neck	
Machine stage cycle time for five stage configuration					
Machine configuration	First stage	Second stage	Third stage	Fourth stage	Fifth stage
Fig. 6a	3.7/3 = 1.23 min	1.7 min	2.0 min	1.5 min	3.3/2 = 1.65 min
Fig. 2b, (Not accepted configuration)	3.7/3 = 1.23 min	1.7/2 = 0.85 min	2.0 min	1.5 min	3.3 min, Bottle neck
Fig. 6c,(Not accepted configuration)	3.7/3 = 1.23 min	1.7 min	2.0/2 = 1.0 min	1.5 min	3.3 min, Bottle neck
Fig. 6d, (Not accepted configuration)	3.7/3 = 1.23 min	1.7 min	2 min	1.5/2 = 0.75 min	3.3 min, Bottle neck

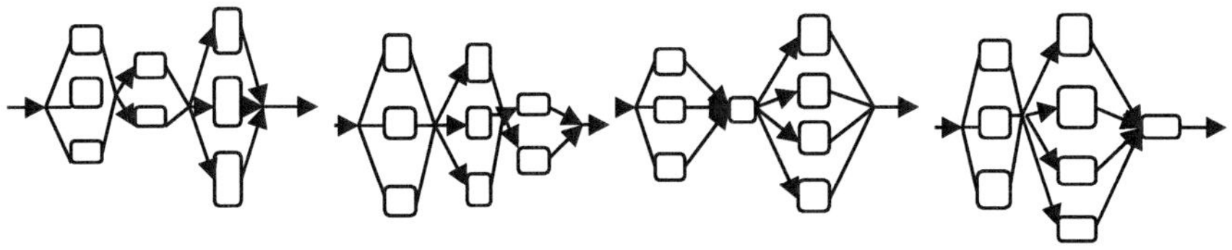

(a) (b) (c) (d)

Fig. 4 Illustration of different three stage machine configurations

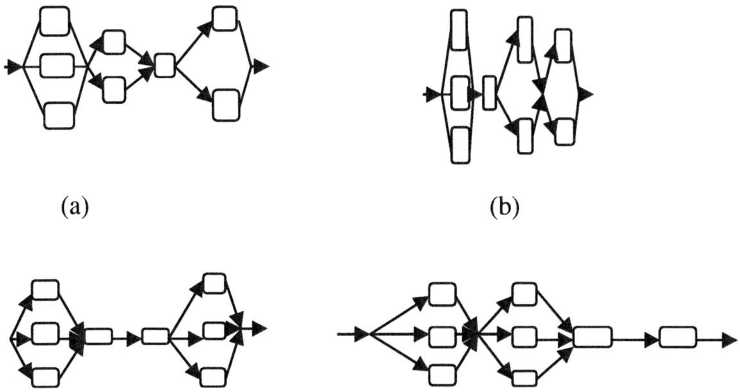

(a) (b)

(c) (d)

Fig. 5 Illustration of different four stage machine configurations

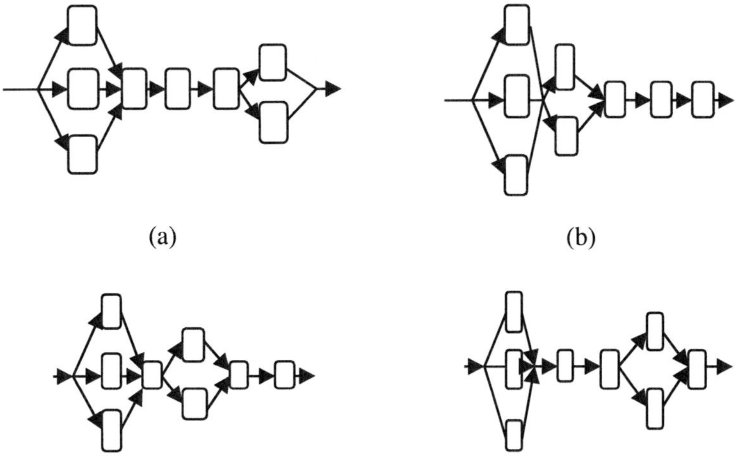

(a) (b)

(c) (d)

Fig. 6 Illustration of different five stage machine configurations

Table 2 Throughput per day of different machine configurations (for two stage, three stage, four stage, and five stage when machine reliability is 0.85 and cycle time is 2.0 min)

	Two Stage	Three Stage				Four stage				Five Stage			
	3a	4.1a	4.2b	4.3c	4.4d	5.1a	5.2b	5.3c	5.4d	6.1a	6.2b	6.3c	6.4d
Production per day when machine reliability = .85 Q=1000/t_{max}	588	540	578	588	303	540	571	500	303	500	303	303	303

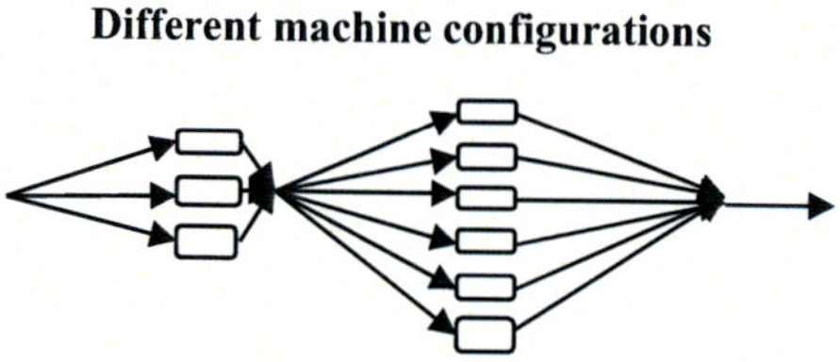

Fig. 7 Machine configuration for 2 stages

Different machine configurations for two, three, four and five stages (consider machine reliability is 0.75 and cycle time Tc ≤ 2.0 min):

Minimum no. of machine N = 500*12.2/1000*0.75 ≈ **9** machine.

For machining face I, Minimum no of machine N_1 = 500*3.7/1000*0.75 = 2.4666 ≈ **3** machine.

For machining face 2, Minimum no of machine N_2 = 500*8.5/1000*0.75 = 5.667 ≈ **6** machine.

Therefore, total no of machines N = $N_1 + N_2 = 3 + 6 =$ **9** machine.

By using equation no. 3 for calculating the minimum number of machines to complete the task, will be 9 machines (3 machines for face 1 and 6 machine for face 2). By following the same steps, just change the machine reliability 0.85 to 0.75. We will find out that the different machine configurations for two, three, four and five stages as shown in below. In three, four and five stages, only four machine configurations are possible, as shown in Figs. 7, 8, 9 and 10 and manufacturing cycle time outlined in Table 3. On the basis of this result, we can find out easily the optimal machine configuration for maximum throughput or production rate. We can only considered the machine configuration whose cycle time is less than or equal to two minutes only. It is clear from the below table, that machine configurations of Figs. 8d, 9d and 10d are not accepted because one of them does not fulfil the cycle time. Table 4 shows an overview summary of machining time of the proposed machine configuration for two, three, four and five stages. It is clear from the below table which machine configurations are best. For three stage, machine configuration Fig. 8c is the best configuration because the production rate per day is very high and machine configuration Fig. 9d has a very low production rate per day. In four stages, production rate per day for Fig. 9c machine configuration is very high and

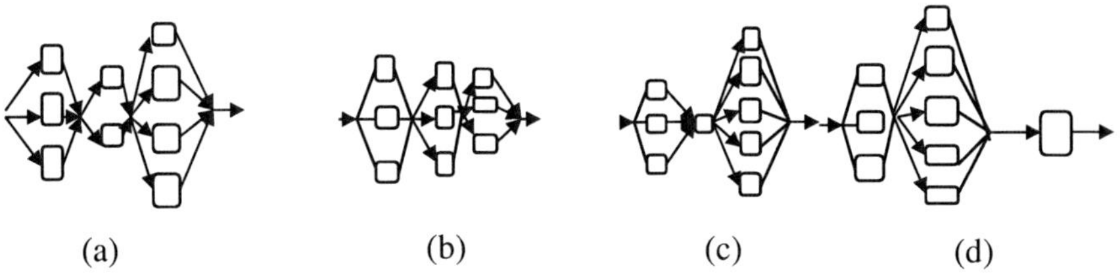

Fig. 8 Illustration of different five stage machine configurations (when machine reliability is 0.75)

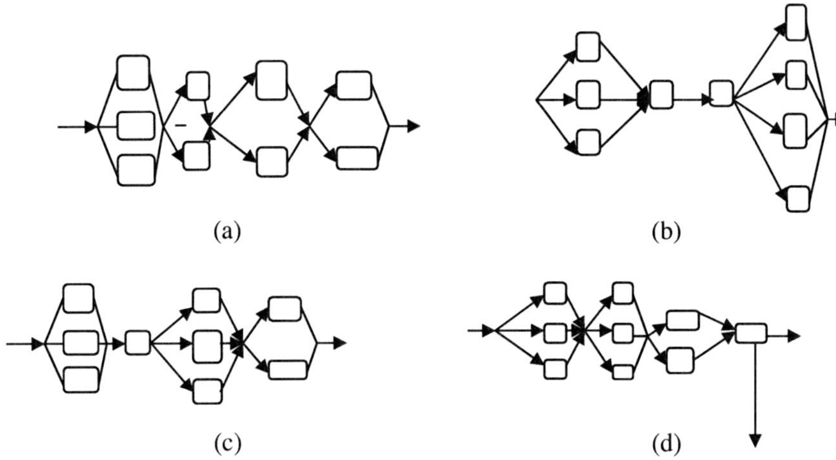

Fig. 9 Illustration of different four stage machine configurations (when machine reliability i.e. 0.75)

in case of five stages. Table 4 shows an overview summary of production per day of the proposed machine configuration for two, three, four and five stages (Table 5).

2.1 Case Study—II

Different machine configurations for two, three, four and five stages (when machine reliability is 0.85 and cycle time ≤ 2.5 min per stage): In this problem, the total machining time (1000 min) and sequence of manufacturing task for face 1 and face 2) are same. But, we are just changing the production per day i.e. 400 products per day in place of 500 parts per day. In the previous case study, it was 500 parts per day. By using equation no 1, the cycle time for this problem is 1000/400 = 2.5 min per cycle. Therefore, T_c **≤ 2.5** min for each stage of manufacturing operation. After calculating the cycle time, we will find out the different machine configurations for different stages (i.e. Two, three, four and five stage) by using the

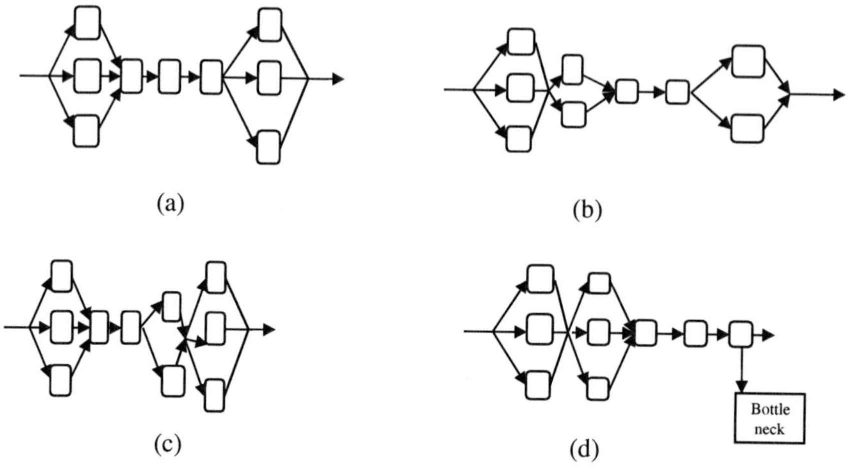

Fig. 10 Illustration of different five stage machine configurations (when machine reliability i.e. 0.75)

equation no 3. Minimum no. of machine N = 400*12.2/1000*0.85 therefore N = 5.741≈ **6** machine.

For machining face I, Minimum no of machine N_1 = 400*3.7/1000*0.85 = 1.741 ≈ **2** machine.

For machining face 2, Minimum no of machine N_2 = 400*8.5/1000*0.85 = 4.000 ≈ **4** machine.

Therefore, total no of machines $N = N_1 + N_2 = 2 + 4 =$ **6** machine.

In three stages, there are only two possible machine configurations as shown in Fig. 12. We can only considered that machine configuration where the cycle time is less than or equal to 2.5 min. It is clear from the Table 6, that machine configuration Fig. 12c is not accepted because one of them does not fulfil the cycle time. Which one becomes a bottleneck with a cycle time of 3.3 min in third stage and cannot meet the required demand (400 parts per day)? The machine configurations shown in Fig. 12a–c help to illustrate the three stage configuration. In four stage, there is only one possible accepted machine configuration as shown Fig. 13b. Table 7 represents the production per day for each machine configuration (two, three and four stages) and Fig. 16 shows the graph between productions per day versus no. of machine configuration for different stages.

Two stage machine configuration (consider machine reliability 0.85) (Figs. 11, 12 and 13).

Different machine configurations for two, three, four and five stages (when machine reliability is 0.75 and cycle time ≤ 2.5 min per stage):

Minimum no. of machine N = 400*12.2/1000*0.75 ≈ **7** machine.

For machining face I, Minimum no of machine N_1 = 400*3.7/1000*0.75 = 1.9733 ≈ **2** machine.

Table 3 Machine stage cycle time for each stages

Machine stage cycle time for two stage configuration					
Machine configuration	First Stage	Second Stage			
Fig 8	3.7/3= 1.23 minute	8.5/6=1.46666 minute			
Machine stage cycle time for three stage configuration					
Machine configuration	First Stage	Second Stage	Third Stage		
Fig 9.1a	3.7/3= 1.23 minute	3.7/2= 1.85 minute	4.8/3=1.60 minute		
Fig 9.2b	3.7/3= 1.23 minute	5.2/3= 1.73 minute	3.3/2=1.65 minute		
Fig 9.3c	3.7/3= 1.23 minute	1.7 minute	6.8/4=1.70 minute		
Fig 9.4d	3.7/3= 1.23 minute	5.2/5= 1.04 minute	3.3 minute bottle neck		
Machine stage cycle time for four stage configuration					
Machine configuration	First Stage	Second Stage	Third Stage	Fourth Stage	
Fig 10.1a	3.7/3= 1.23 minute	3.7/2= 1.85 minute	1.5/2= 0.75 minute	3.3/2= 1.65 minute	
Fig 10.2b	3.7/3= 1.23 minute	1.7 minute	2 minute	4.8/4=1.2 minute	
Fig 10.3c	3.7/3= 1.23 minute	1.7 minute	3.5/3=1.16 minute	3.3/2= 1.65minute	
Fig 10.4d	3.7/3= 1.23 minute	1.7/3=0.56 minute	3.5/2=1.75 minute	3.3bottle neck	
Machine stage cycle time for five stage configuration					
Machine configuration	First Stage	Second Stage	Third Stage	Fourth Stage	Fifth Stage
Fig 11.1a	3.7/3= 1.23 minute	1.7 minute	2.0 minute	1.5 minute	3.3/3 = 1.1 minute
Fig 11.2b	3.7/3= 1.23 minute	1.7/2= 0.85 minute	2.0 minute	1.5 minute	3.3 /2= 1.65minute
Fig 11.3c	3.7/3= 1.23 minute	1.7 minute	2.0 minute	1.5/2=0.75 minute	3.3/3= 1.1 minute)
Fig 11.4d, Not accepted configuration	3.7/3= 1.23 minute	1.7/3=0.56 minute	2 minute	1.5 minute	3.3 minute, Bottleneck

Table 4 Throughput per day of different machine configurations (for two stages, three stages, four stages and five stages when machine reliability is 0.75 and cycle time is 2.0 min)

	Two Stage	Three Stage				Four stage				Five Stage			
	8	9.1 a	9.2 b	9.3 c	9.4d	10.1 a	10.2 b	10.3 c	10.4d	11.1 a	11.2 b	11.3 c	11.4 d
Production per day when machine reliability = .75 Q=1000/t_{max}	709	540	578	588	303	540	500	588	303	500	500	500	303

Table 5 Throughput per day of different machine configurations (for two stage, three stage, four stage, and five stage when machine reliability is 0.85 and 0.75 but cycle time is 2.0 min only

Machine configuration	Two Stage	Three Stage				Four stage				Five Stage			
	3a	4.1 a	4.2 b	4.3 c	4.4 d	5.1 a	5.2 b	5.3 c	5.4 D	6.1 a	6.2 b	6.3 c	6.4 d
Production per day when machine reliability = .85 Q=1000/t_{max}	588	540	578	588	303	540	571	500	303	500	303	303	303
Machine configuration	Two Stage	Three Stage				Four stage				Five Stage			
	8a	9.1 a	9.2 b	9.3 c	9.4 d	10.1 a	10.2 b	10.3 c	10.4 D	11.1 a	11.2 b	11.3 c	11.4 d
Production per day when machine reliability = .75 Q=1000/t_{max}	709	540	578	588	303	540	500	588	303	500	500	500	303

*** ** Yellow box stands for high production rate per day and red box stands for not accepted machine configuration

For machining face 2, Minimum no of machine $N_2 = 400*8.5/1000*0.75 = 4.533 \approx$ **5** machine.

Therefore, total no of machines $N = N_1 + N_2 = 2 + 5 =$ **7** machine.

In three stages, there are only three possible machine configurations as shown in Fig. 15. We can only considered that machine configuration where the cycle time is less than or equal to 2.5 min. It is clear from the below table that machine configurations of Figs. 15d, 16d and 17b–d are not accepted because one of them does not fulfil the cycle time. Which one becomes a bottleneck with a cycle time of 3.3 min and cannot meet the required demand (400 parts per day)? In four stages, there are only two possible configurations which have highest production rate. All production rate values are mapped in Fig. 21. Table 9 represents the production per day for each machine configuration (two, three, four and five stages) when machine reliability are 0.85 and 0.75 (Table 8).

Two stage machine configuration (consider machine reliability as 0.75) (Figs. 14, 15, 16, 17 and Table 10).

Table 6 Machine stage cycle time for each stage

Machine stage cycle time for each two stages				
Machine configuration	**First stage cycle time(t≤ 2.50 minute**	**Second stage cycle time(t≤ 2.50 minute**		
Fig 13	3.7/2= 1.85 minute	8.5/4 = 2.125 minute		
Machine stage cycle time for each three stages				
Machine configuration	**First stage cycle time(t≤ 2.50 minute**	**Second stage cycle time(t≤ 2.50 minute**	**Third stage cycle time(t≤ 2.50 minute**	
Fig 14.1a	3.7/2 = 1.85minute	1.7 minute	6.8/3= 2.266 minute	
Fig 14.2b	3.7/2 = 1.85minute	3.7/2= 1.85 minute	4.8/2= 2.4 minute	
Fig 14.3c(not accepted)	3.7/2 = 1.85minute	5.2/3= 1.73 minute	3.3minute	
Machine stage cycle time for each four stages				
Machine configuration	***First stage cycle time (t ≤ 2.50 minute)***	***Second stage cycle time (t ≤ 2.50 minute)***	***Third stage cycle time (t ≤ 2.50 minute)***	***Fourth stage cycle time (t ≤ 2.50 minute)***
15.1a (not accepted)	3.7/2 = 1.85minute	3.7/2= 1.85 minute	1.5 minute	3.3 minute
15.2b	3.7/2 = 1.85minute	1.7 minute	2 minute	4.8/2=2.4 minute
15.3c (not accepted)	3.7/2 = 1.85minute	1.7 minute	3.5/2=1.85 minute	3.3minute

Table 7 Throughput per day of different machine configuration (for two stages, three stages, and four stages, when machine reliability is 0.85 and cycle time is 2.50 min)

	Two Stage configuration	**Three Stage configuration**			**Four stage configuration**		
	13	14.1 a	14.2 b	14.3 c	15.1 a	15.2 b	15.3 c
Production per day when machine reliability = .85 **Q=1000/tmax**	470	442	416	303	303	416	303

Fig. 11 Machine configuration for 2 stages

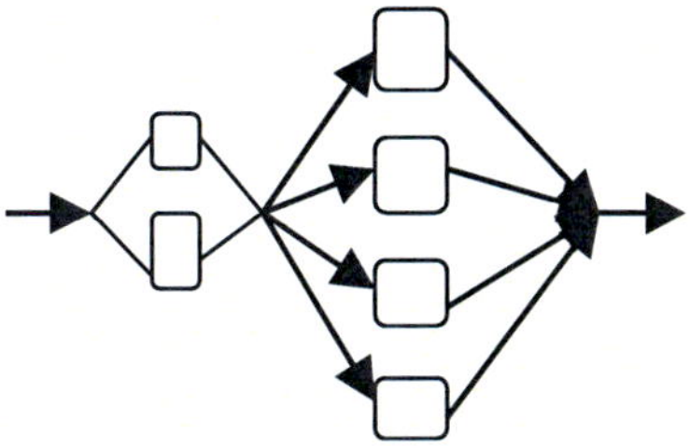

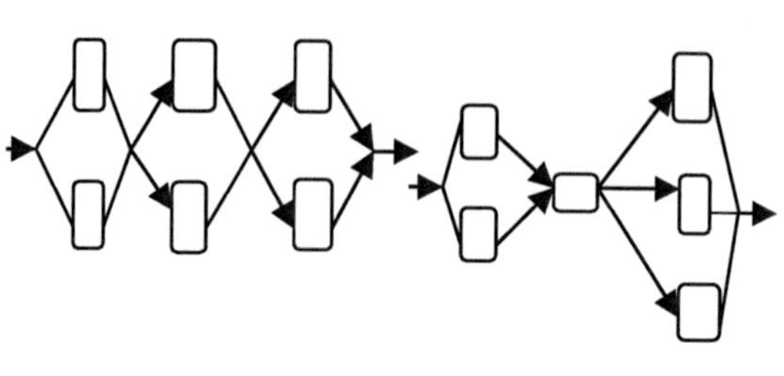

(a) (b) (c)

Fig. 12 Illustration of Different three stage machine configurations (when machine reliability i.e. 0.85)

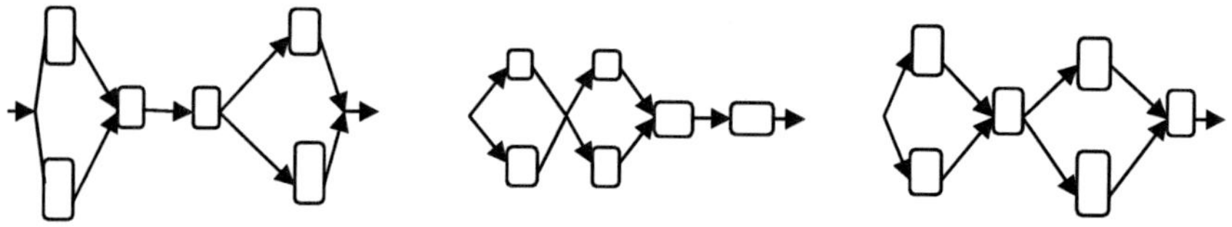

(a) (b) (c)

Fig. 13 Illustration of Different four stage machine configurations (when machine reliability i.e. 0.85)

3 Conclusion and Future Scope

In the present study, a thorough literature review of the current state of research regarding RMS machine configuration selection has been provided. However, there are many challenges in this RMS domain such as assignments of operation, layout configuration, material handling selection etc., A new line balancing algorithms and machine reliability approach was introduced for the selection of optimal machine configuration (for two, three, four and five stages) having maximum production rate per day (i.e. 500 parts per day). A case study was presented in this paper, which validates the proposed approach. Because of cycle time limitation, the number of machine configuration is reduced. The conclusions that might be drawn from this illustration are:

- We can easily calculate the minimum no. of machines in the manufacturing system based on the total manufacturing time and daily requirement of products.
- We can easily reduce the different machine configurations from 46 to 20 (As shown in Fig. 18) when line balancing algorithms and cycle time concepts are taken into consideration.
- This methodology provides a precursor for future research on quantifying system reliability, availability, scalability, minimum number of machines, and maximum possible configuration of the reconfigurable manufacturing system. There are

Table 8 Machine stage cycle time for each two stage

Machine stage cycle time for two stage configuration					
Machine configuration	First Stage	Second Stage			
Fig 17	3.7/2= 1.85 minute	8.5/5=1.7 minute			
Machine stage cycle time for three stage configuration					
Machine configuration	First Stage	Second Stage	Third Stage		
Fig 18.1a	3.7/2= 1.85 minute	3.7/2= 1.85 minute	4.8/3=1.60 minute		
Fig 18.2b	3.7/2= 1.85 minute	5.2/3= 1.73 minute	3.3/2=1.65 minute		
Fig 18.3c	3.7/2= 1.85 minute	1.7 minute	6.8/4=1.70 minute		
Fig.18.4d not accepted	3.7/2= 1.85 minute	5.2/4 = 1.30 minute	3.3 minute		
Machine stage cycle time for four stage configuration					
Machine configuration	First Stage	Second Stage	Third Stage	Fourth Stage	
Fig 19.1a	3.7/2= 1.85 minute	3.7/2= 1.85 minute	1.5 minute	3.3/2= 1.65 minute	
Fig 19.2b	3.7/2= 1.85 minute	1.7 minute	3.5/2=1.75 minute	3.3/2= 1.65 minute	
Fig 19.3c	3.7/2= 1.85 minute	1.7 minute	2.0 minute	4.8/3= 1.6 minute	
Fig 19.4d not accepted	3.7/2= 1.85 minute	3.7/2=1.233 minute	1.5 minute	3.3 minute	
Machine stage cycle time for five stage configuration					
Machine configuration	First Stage	Second Stage	Third Stage	Fourth Stage	Fifth Stage
Fig.20.1a	3.7/2= 1.85 minute	1.7 minute	2.0 minute	1.5 minute	3.3/2 = 1.65 minute
Fig. 20.2b	3.7/2= 1.85 minute	1.7/2= 0.85 minute	2.0 minute	1.5 minute	3.3minute
Fig. 20.3c	3.7/2= 1.85 minute	1.7 minute	2.0/2=1 minute	1.5 minute	3.3minute
Fig. 20.4d	3.7/2= 1.85 minute	1.7 minute	2.0/2=1 minute	1.5/2=.75 minute	3.3minute

Table 9 Throughput per day of different machine configuration (for two stages, three stages, and four stages, when machine reliability is 0.75 and cycle time is 2.50 min)

	Two Stage	Three Stage				Four stage				Five Stage			
	17a	18.1 a	18.2b	18.3 c	18.4d	19.1a	19.2b	19.3c	19.4d	20.1a	20.2b	20.3c	20.4d
Production per day when machine reliability = .75	540	540	540	540	303	540	540	500	303	500	303	303	303

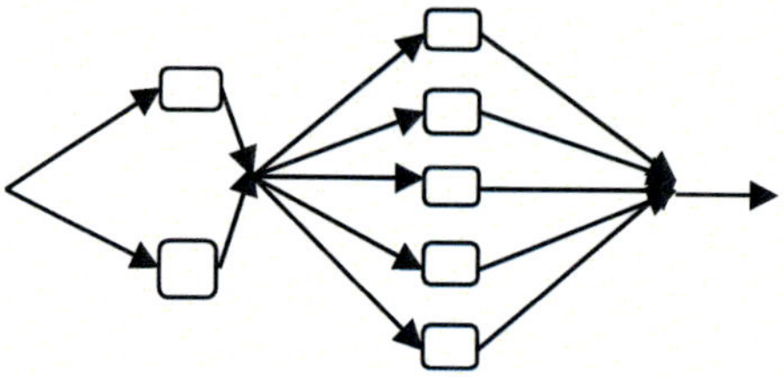

Fig. 14 Two stage machine configuration

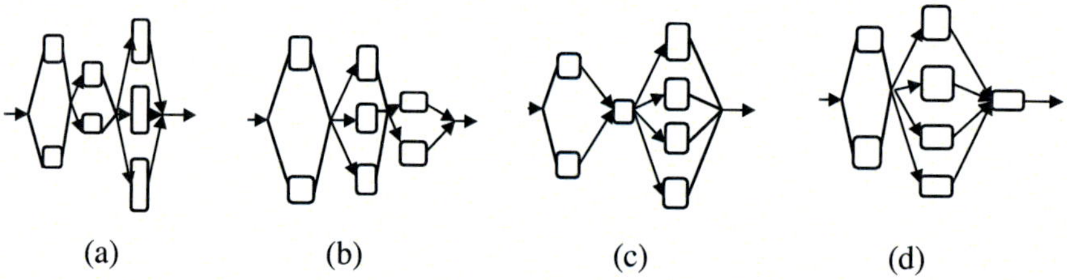

Fig. 15 Illustration of different three stage machine configurations (when machine reliability 0.75)

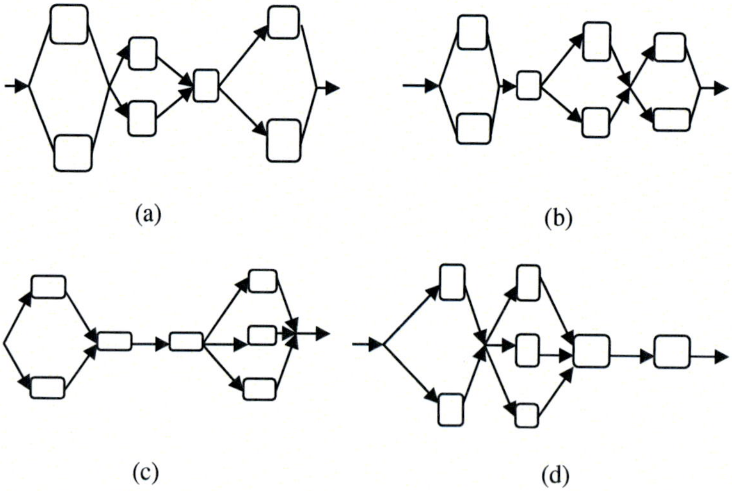

Fig. 16 Illustration of different four stage machine configurations (when machine reliability i.e. 0.75)

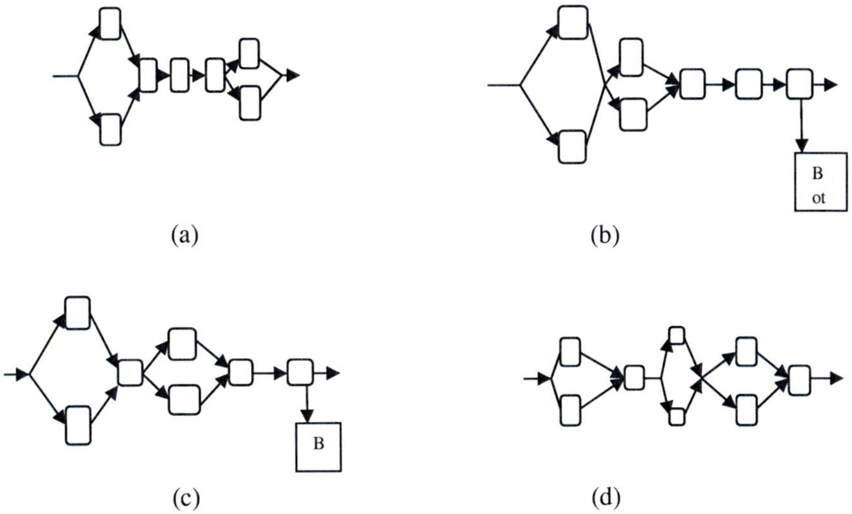

Fig. 17 Illustration of different five stage machine configurations (when machine reliability i.e. 0.75)

Table 10 Throughput per day of different machine configurations (for two stage, three stage, four stage, and five stage when machine reliability is 0.85 and 0.75 and cycle time is 2.0 min and 2.5 min)

Machine configuration	Two Stage	Three Stage				Four stage				Five Stage			
	13	14.1 a	14.2 b	14.3 c		15.1 a	15.2 b	15.3 c					
Production per day when machine reliability = .85 $Q=1000/t_{max}$	470	442	416	303		303	416	303					
Machine configuration	Two Stage	Three Stage				Four stage				Five Stage			
	17a	18.a	18.2b	18.3c	18.4d	19.1 a	19.2 b	19.3c	19.4d	20.1a	20.2b	20.3 c	20.4d
Production per day when machine reliability = .75 $Q=1000/t_{max}$	540	540	540	540	303	540	540	500	303	500	303	303	303

Fig. 18 Graph between total possible machine configurations (46) versus optimal machine configurations (20)

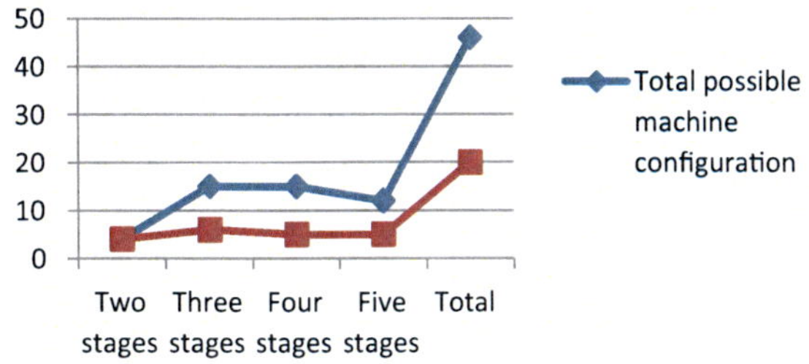

several ways to explore this work in the future. Effect of modularity, flexibility, maintainability, scalability etc. can also be integrated in the process of judging the production rate, no of manufacturing stages of the system etc., The problem can be extended to account for more complex architectures of reliability function. This helps support the management decisions regarding system configurations for a reconfigurable manufacturing system by providing the best alternatives at the beginning of each configuration period.

References

1. Mehrabi MG, Ulsoy AG, Koren Y (2000) Reconfigurable manufacturing systems: key to future manufacturing. J Intell Manuf 11(4):403–419
2. Singh A, Gupta S, Asjad M, Gupta P (2017) Reconfigurable manufacturing systems: journey and the road ahead. Int J Syst Assur Eng Manag 8(Suppl. 2):1849–1857
3. Koren Y, Gu X, Guo W (2017) Reconfigurable manufacturing systems: principles, design, and future trends. Front Mech Eng. https://doi.org/10.1007/s11465-018-0483-0
4. Singh A, Asjad M, Gupta P, Quamar J (2019) An approach to develop shaper cum slotter mechanism: a reconfigurable machine tool. South Asian J Bus Manage Case 8(2):195–206
5. Singh A, Gupta P, Asjad M (2019) Reconfigurable manufacturing system (Rms): accelerate towards industries 4.0 (March 18, 2019). In: Proceedings of international conference on sustainable computing in science, technology and management (SUSCOM-2019), Amity University Rajasthan, Jaipur, India. SSRN. https://ssrn.com/abstract=3354485
6. Singh A, Asjad M, Gupta P (2019) A conceptual development of reconfigurable drill machine tool (RDMT). SN Appl Sci 1(10):1141
7. Singh A, Asjad M, Gupta P (2019) Reconfigurable machine tools: a perspective. Life Cycle Reliab Safety Eng 8(4):365–376
8. Singh A, Asjad M, Gupta P (2019) Enhancing the reconfigurability issues of machine tool for reconfigurable manufacturing system (RMS). J Ind Integr Manage. https://doi.org/10.1142/S2424862220500037
9. Mehrabi MG, Ulsoy AG, Koren Y, Heytler P (2002) Trends and perspectives in flexible and reconfigurable manufacturing systems. J Intell Manuf 13(2):135–146
10. Wang W, Koren Y (2012) Scalability planning for reconfigurable manufacturing systems. J Manuf Syst 31(2):83–91
11. Gupta A, Jain PK, Kumar D (2015) Configuration selection of reconfigurable manufacturing system based on performance. Int J Ind Syst Eng 20(2):209–230
12. Krygier R (2005) The Integration of flexible, reconfigurable manufacturing with quality. In: Proceedings of CIRP 3rd conference on reconfigurable manufacturing. Ann Arbor
13. Youssef AM, ElMaraghy HA (2007) Optimal configuration selection for reconfigurable manufacturing systems. Int J Flex Manuf Syst 19(2):67–106
14. Tang L, Yip-Hoi DM, Wang W, Koren Y (2004) Concurrent line-balancing, equipment selection and throughput analysis for multi-part optimal line design. J Manuf Sci Prod 6(1–2):71–82
15. Spicer P, Koren Y, Shpitalni M, Yip-Hoi D (2002) Design principles for machining system configurations. Ann CIRP 51(1):275–280
16. Goyal KK, Jain PK, Jain M (2012) Optimal configuration selection for reconfigurable manufacturing system using NSGA II and TOPSIS. Int J Prod Res 50(15):4175–4191
17. Zhong W, Maier-Speredelozzi V, Bratzel A, Yang S, Chick SE, Hu SJ (2000) Performance analysis of machining systems with different configurations. In: Proceedings of the 2000 Japan–USA flexible automation conference, Michigan, July, pp 783–790

18. Lu C, Li A, Xu L (2010) Research and optimization of reconfigurable manufacturing system configuration based on system reliability. Kybernetes 39(6):1058–1065
19. Ebeling CE (2000) An introduction to reliability and maintainability engineering. McGraw-Hill. ISBN-13: 978-0070188525. New York
20. Spicer P, Carlo HJ (2006) Integrating reconfiguration cost into the design of multi-period scalable reconfigurable manufacturing systems. J Manuf Sci Eng 129(1):202–210
21. Hasan F, Jain PK (2015) Genetic modelling for selecting optimal machine configurations in reconfigurable manufacturing system. Appl Mech Mater 789–790:1229–1239
22. Koren Y, Shpitalni M (2001) Design of reconfigurable manufacturing systems. J Manuf Syst 29(4):130–141
23. Reginato G, Anzanello MJ, Kahmann A, Schmidt L (2016) Mixed assembly line balancing method in scenarios with different mix of products. Gestão & Produção 23(2):294–307
24. Son SY, Lennon OT, Yip-Hoi D (2001) An approach to scalability and line balancing for reconfigurable manufacturing systems. Integr Manuf Syst 12(7):500–511
25. Koren Y, Ulsoy G, US Patent 6349237, 2002-02-9

Effect of Oil Temperature, Hydro-Motor Speed, Hydro-Motor Size and the Load Torque on the Loss Coefficients of the Bent Axis Hydro-Motor

Alok Vardhan and Ajay Pratap Singh

Abstract In this paper, the effect of oil temperature, hydro-motor speed, hydro-motor size, and load torque are investigated in relation to the loss coefficients (i.e. external leakage, internal leakage, and inlet valve-port coefficients) of the bent axis hydro-motor. In this regard, a governing system has been made, and the bond graph simulation method has been used to represent it. For a broad range of speed and load-torque at different oil temperatures and motor size, the loss coefficients of the hydro-motors are investigated. The results demonstrate that loss coefficients significantly affect the performance of the hydro-motors.

Keywords Bent axis hydro-motor · Loss coefficients · External leakage coefficient · Internal leakage coefficient · Inlet valve-port coefficient

1 Introduction

The hydro-motors' performances are significantly impacted by the loss coefficients. Several loss modelling techniques for hydrostatic pumps and hydro-motors have been investigated by eminent experts. In light of this, Watton [1] created a closed-loop design technique that makes predictions based on test data about the steady-state speed characteristics of an axial piston hydro-motor. Knezevic et al. [2] have researched the pressure dependency of oil viscosity on hydraulic system analysis. To simulate a hydrostatic transmission system, Kohmascher et al. [3] have developed a novel modelling technique. In-depth losses, their components, and the tribological behaviour of the contact surfaces have all been examined by Murrenhoff et al. [4]. He drew the conclusion that the primary source of loss in hydrostatic units is due to tribological interactions from there. Rahmfeld et al. [5] have given investigation of "efficiency measurement and loss modeling for hydrostatic systems" that has resulted in a valuable database for HST system design by demonstrating that accurate

A. Vardhan (✉) · A. P. Singh
Department of Mechanical Engineering, Ajay Kumar Garg Engineering College, Ghaziabad, India
e-mail: alok.or.monu@gmail.com

H. Kumar et al. (eds.), *Recent Advances in Intelligent Manufacturing*,
Lecture Notes in Mechanical Engineering,
https://doi.org/10.1007/978-981-99-1308-4_15

measurements when combined with trustworthy modelling are ineffective. Bergada et al. [6] have given a collection of flow loss performance models. The "slipper swash plate pair or the valve plate cylinder pair" are the primary causes of the leakage in the pump and motor, accounting for around 94% of the overall leakage, according to the models. Kumar et al. [7] have investigated how operating pressure affects the leakage flow properties of different-sized piston type bent axis hydro-motors. Inaguma [8] has examined how temperature and pressure affect the leakage flow characteristic of positive displacement pumps that are not piston-type. A useful method for forecasting the impact of "operating pressure and temperature on leakage losses in various kinds of hydrostatic pumps" was demonstrated in the study. The analyses' findings are only applicable to positive displacement pumps that are not piston-type pumps. The hydrostatic system is made up of pumps and motors. The effectiveness of the hydrostatic system depends significantly on the motors. Loss coefficients, however, affect motor performance. Studying how different operational factors affect loss coefficients is crucial. The loss coefficients (also known as external leakage, internal leakage, and the inlet valve-port coefficients) of the bent axis hydro-motor are examined in this article in relation to oil temperature, hydro-motor speed, hydro-motor size, and load torque. Using the bond graph simulation approach, a governing system has been chosen for this situation. For a wide range of speed and load torque at various oil temperatures, the hydro-motors' loss coefficients are examined.

2 The Governing System

The governing system shown in Fig. 1 has been used to examine the relationship between different loss coefficients and oil temperature, load torque, speed, and motor size. The paper Vardhan et al. [9] discusses a detailed explanation of the aforementioned system.

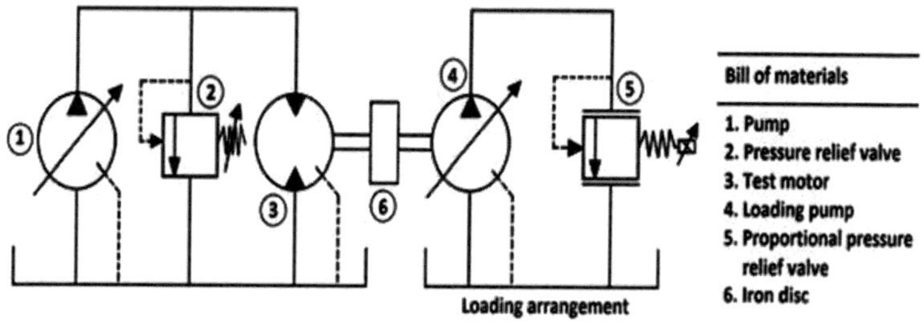

Fig. 1 The governing system

3 Modelling of the Governing System

The steady-state model of the bent axis hydro-motor is created using the bond graph simulation approach and is illustrated in Fig. 2 to analyse the impact of oil temperature, hydro-motor speed, hydro-motor size, and the load torque on the loss coefficients. The external leakage coefficient, inlet valve-port coefficient, and internal leakage coefficient are represented in the model by R_{el}, R_1, and R_2, respectively. R_{el} and R_2 coefficients contribute to the motor's flow loss, whereas R_1 accounts for the motor's torque loss.

The governing equations obtained from the model are:

The flow loss due to fluid compressibility:

$$Q_{cp} = Q_s - \frac{P_p}{R_{el}} - \frac{P_p - (T_{ld}/D_m)}{R_1} \tag{1}$$

Ignoring the flow loss due to fluid compressibility ($Q_{cp} = 0$), the inlet valve-port coefficient is:

$$R_1 = \frac{P_p - (T_{ld}/D_m)}{Q_s - (P_p/R_{el})} \tag{2}$$

The speed of the hydro-motor is:

$$\omega_{ma} = \frac{\left(\frac{P_p - (T_{ld}/D_m)}{R_1} - \frac{(T_{ld}/D_m)}{R_2}\right)}{D_m} \tag{3}$$

From Eq. (3), the internal leakage coefficient of the hydro-motor is:

$$R_2 = \frac{(T_{ld}/D_m)}{-\omega_{ma} D_m + \left(\frac{P_p - (T_{ld}/D_m)}{R_1}\right)} \tag{4}$$

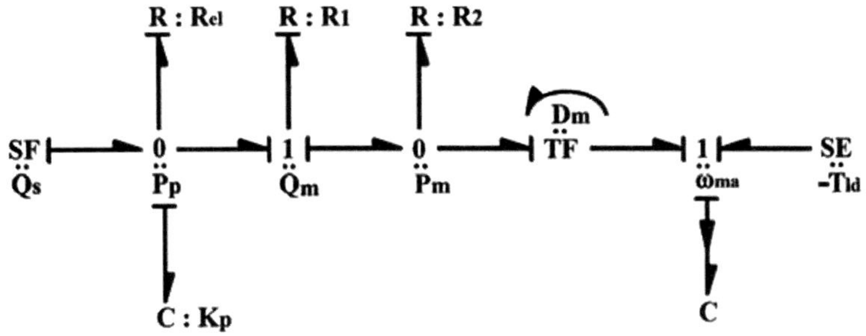

Fig. 2 The governing system's steady-state bond graph model

The external leakage coefficient is:

$$R_{el} = \frac{P_p}{Q_{el}} \tag{5}$$

where,

$$P_p = K_p \int Q_{cp} dt \tag{6}$$

4 Results and Discussions

The values of R_1, R_2, and R_{el} are determined at three different oil temperatures using the test data of ω_{ma}, P_p, Q_s, Q_{el}, and T_{ld} in Eqs. (2), (4), and (5) for five different sizes of the hydro-motors. These loss-coefficients' features are derived using the best-fit curves to data points at various T_{ld}. Figures 3, 4, 5 show changes in such loss coefficients as a function of hydro-motor speed (ω_{ma}).

4.1 External Leakage Coefficient (R_{el}) of the Hydro-Motors

The features of R_{el} of five different hydro-motor capacities at three different temperatures are shown in Fig. 3. These are displayed at different load torques and hydro-motor speeds.

We may draw the following conclusions from Fig. 3a through 3c:

- For a certain hydro-motor speed (ω_{ma}), increasing the load torque (T_{ld}) results in a reduction in the external leakage coefficient (R_{el}). The pressure across the motor rises as a result of the higher load torque, which increases the pressure-induced leakage flow through the clearance of the motor. As a result, the leakage flow increases and the external leakage coefficient drops.
- An increase in hydro-motor speed (ω_{ma}) tends to increase the external leakage coefficient (R_{el}) at a given load torque (T_{ld}). This is owing to the fact that flow losses decrease as speed increases.
- The external leakage coefficient (R_{el}) increases as the hydro-capacity motor's (D_m) increase with a constant oil temperature. This is due to the reduced leakage flow clearance, which causes the piston's stroke length to lengthen.
- For a fixed hydro-motor size (D_m), the external leakage coefficient (R_{el}) decreases as the oil temperature rises. This happens mostly as a result of the oil viscosity decreasing with rising oil temperature.

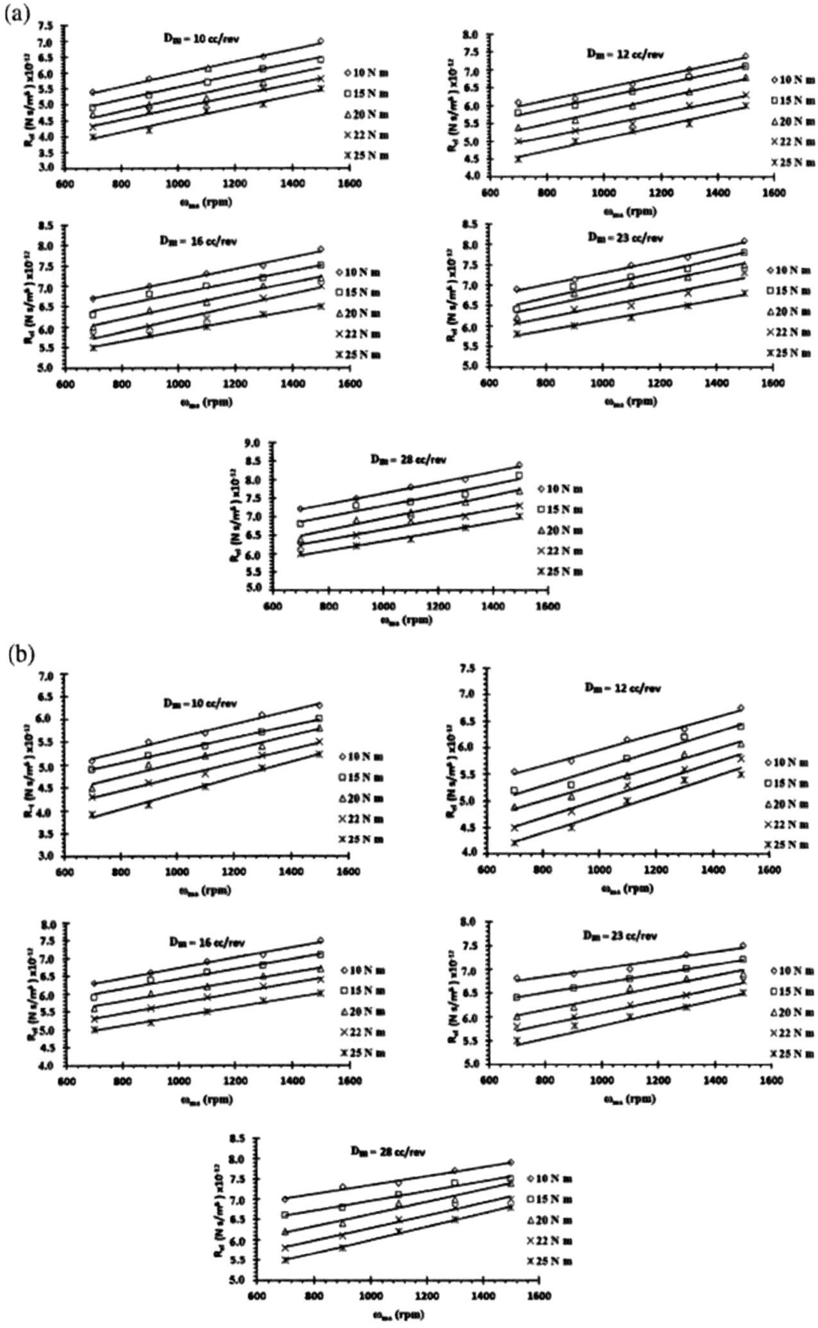

Fig. 3 **a** External leakage coefficient characteristics of the bent axis hydro-motor at 50 °C. **b** External leakage coefficient characteristics of the bent axis hydro-motor at 60 °C. **c** External leakage coefficient characteristics of the bent axis hydro-motor at 70 °C

(c)

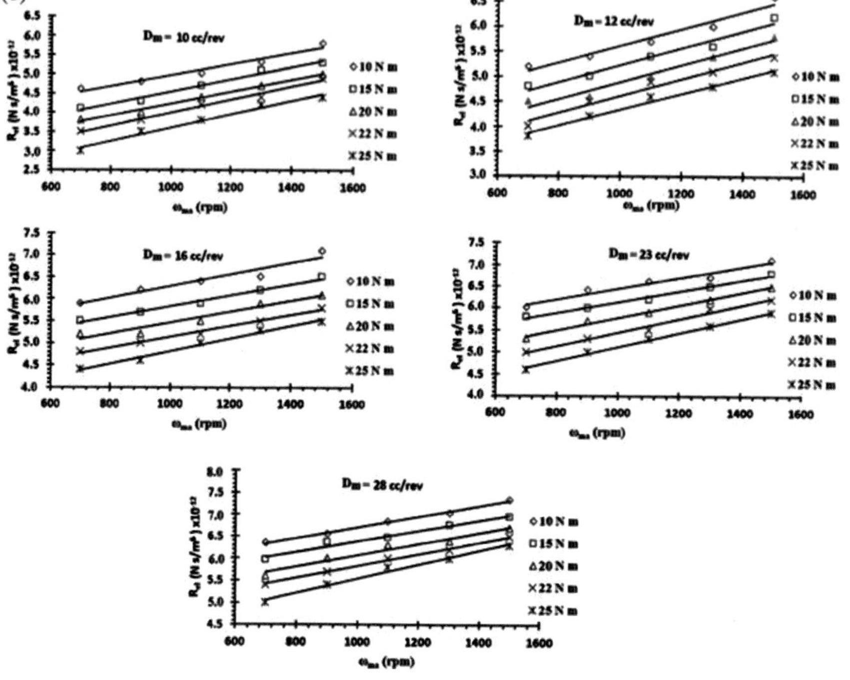

Fig. 3 (continued)

4.2 Inlet Valve-Port Coefficient (R_l) of the Hydro-Motors

Figure 4 shows the characteristics of R_l for a hydro-motor with five different capacities and three different oil temperatures. These are displayed at different load torques and hydro-motor speeds.

From Fig. 4a through 4c, the following inferences are possible:

- With a fixed T_{ld}, the R_l falls as the ω_{ma} increases. This is due to Eq. (2) demonstrating that increasing speed causes an increase in the flow through the motor's inlet valve port. R_l decreases as a consequence.
- For a certain ω_{ma}, the increasing load torque causes the inlet valve-port coefficient (R_l) to increase (T_{ld}). This is so because according to Eq. (2), R_l increases as a result of an increase in leakage flow brought on by an increase in load torque (T_{ld}).
- The hydro-motor inlet valve-port coefficient (R_l) increases at a constant oil temperature (T).
- As the oil temperature (T) increases, the inlet valve-port coefficient (R_l) decreases for a certain hydro-motor size (D_m). As temperature rises, oil viscosity falls, causing a corresponding drop in viscous resistance that reduces the coefficient of the valve port.

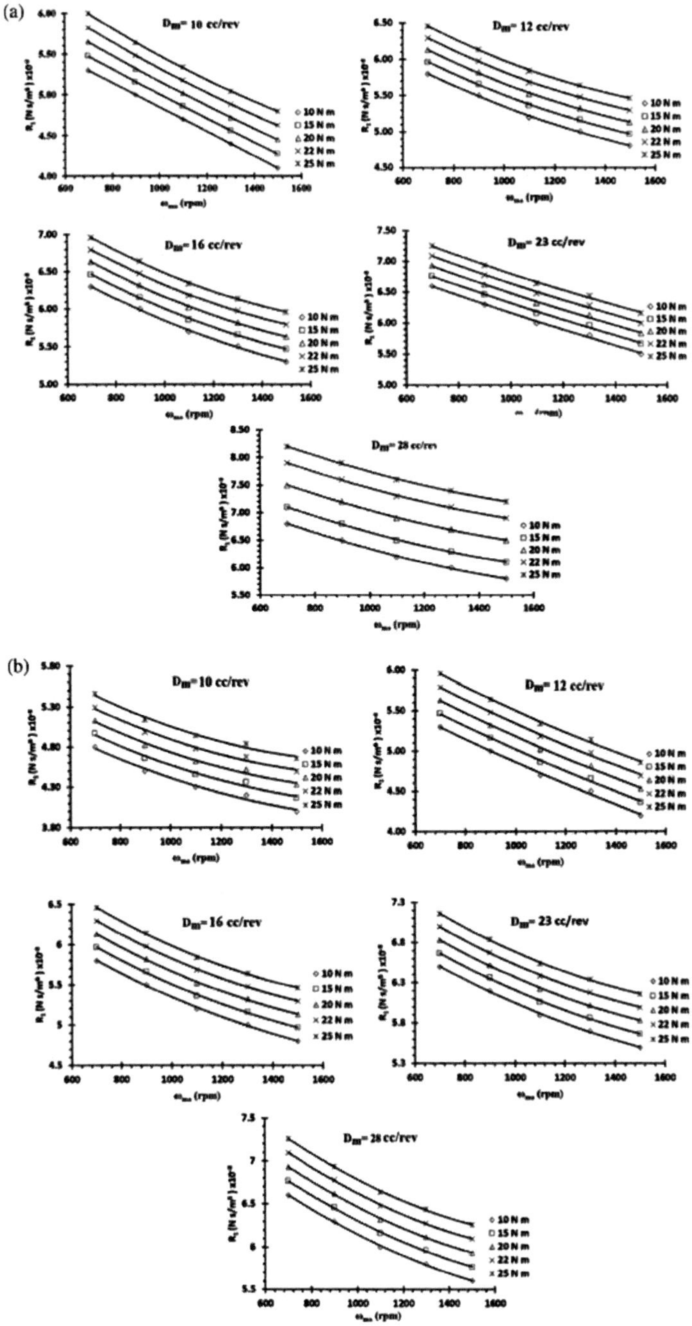

Fig. 4 **a** Inlet valve-port coefficient characteristics of the bent axis hydro-motor at 50 °C. **b** Inlet valve-port coefficient characteristics of the bent axis hydro-motor at 60 °C. **c** Inlet valve-port coefficient characteristics of the bent axis hydro-motor at 70 °C

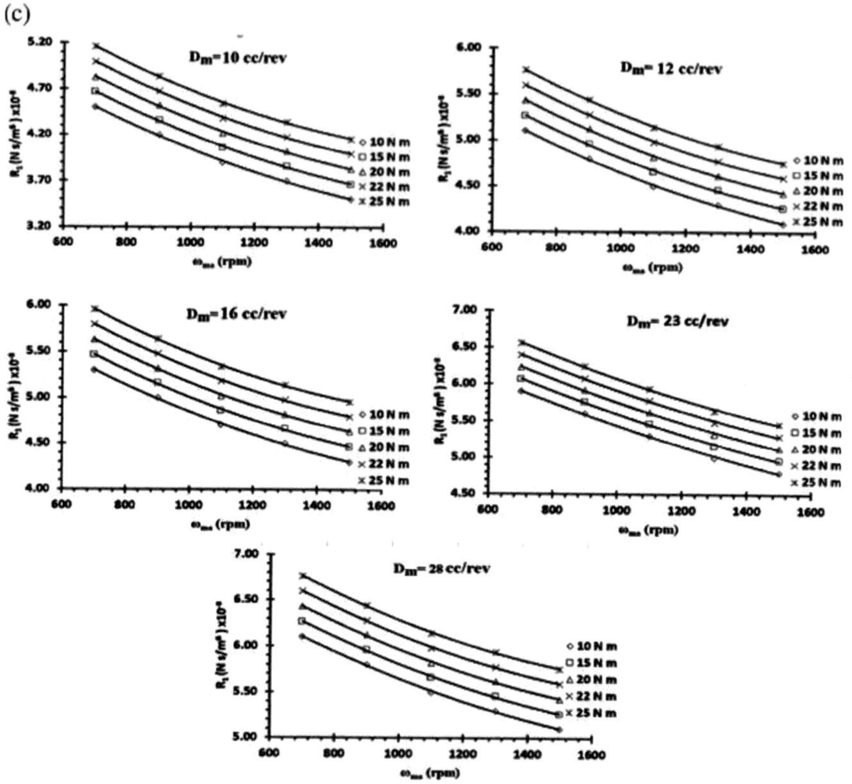

Fig. 4 (continued)

4.3 Internal Leakage Coefficient (R_2) of the Hydro-Motors

Figure 5 depicts the R_2 characteristics of the hydro-motors with five different capacities and three different oil temperatures. Plots of them at various load torques are made in relation to hydro-motor speed.

We may draw the following conclusions from Figs. 5a through 5c:

- Increasing the hydro-motor speed (ω_{ma}) at a certain T_{ld} improved the internal leakage coefficient (R_2). This is a result of the fact that leakage flow reduces as speed increases.
- For a certain hydro-motor speed (ω_{ma}), the internal leakage coefficient (R_2) decreases with increasing load torque (T_{ld}). This is caused by an increase in pressure-induced leakage flow through the motor's clearance with increasing load torque, leading to an increase in leakage flow and a drop in leakage coefficient (R_2).

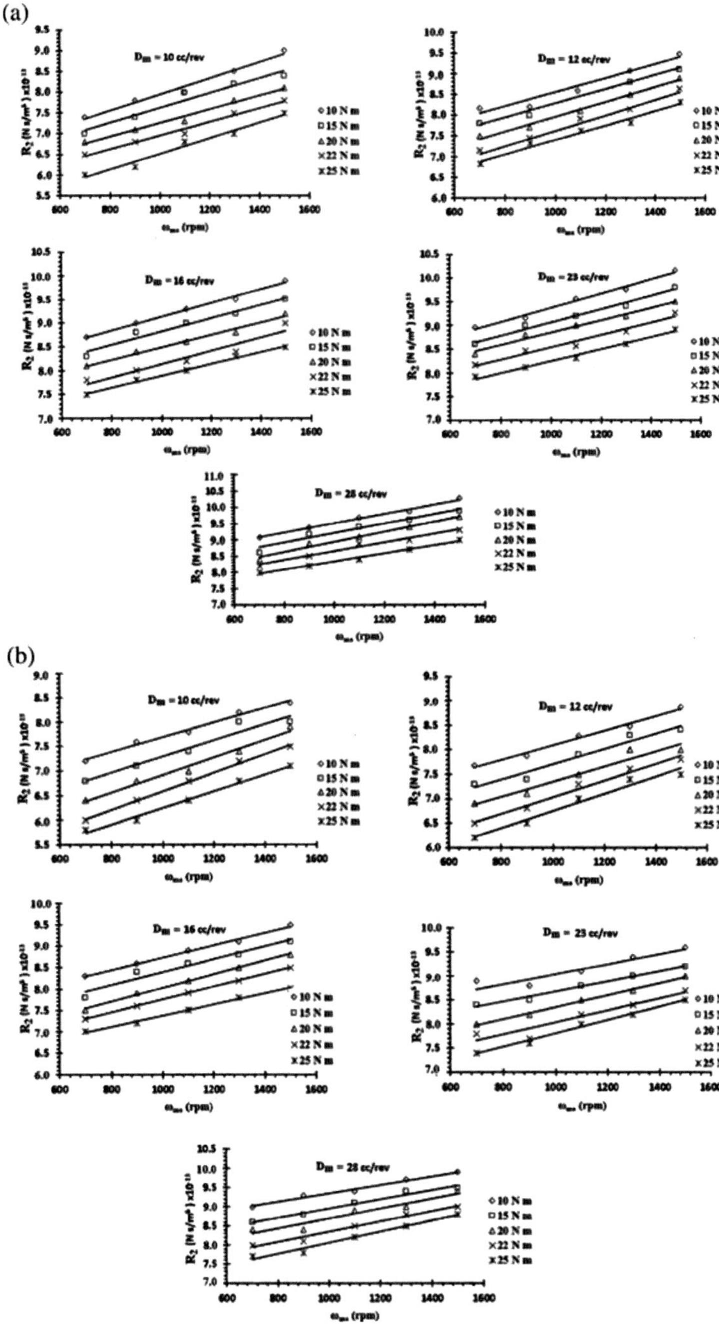

Fig. 5 **a** Internal Leakage coefficient characteristics of the bent axis hydro-motor at 50 °C. **b** Internal Leakage coefficient characteristics of the bent axis hydro-motor at 60 °C. **c** Internal Leakage coefficient characteristics of the bent axis hydro-motor at 70 °C

(c)

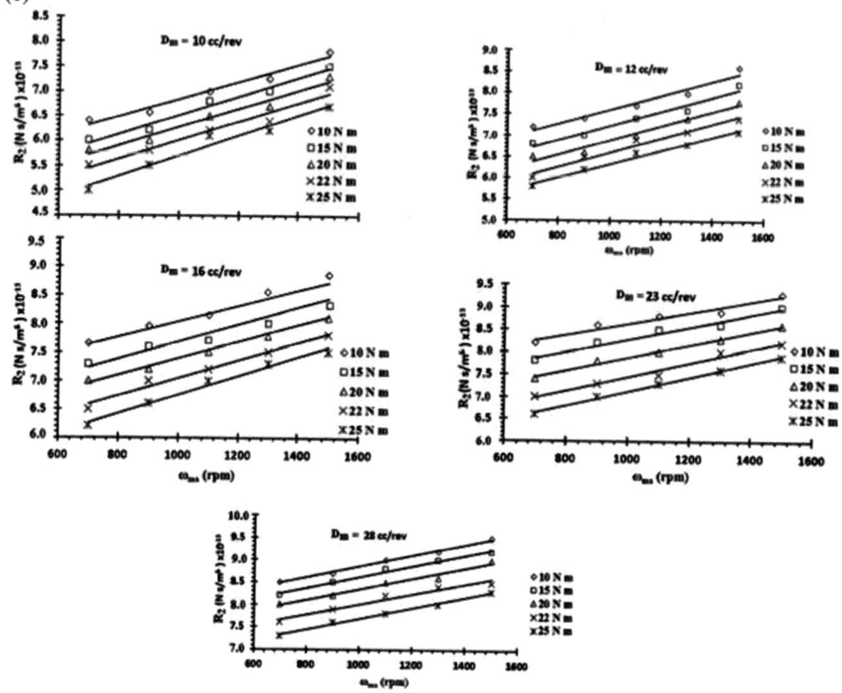

Fig. 5 (continued)

- For a fixed oil temperature (T), increasing the hydro-capacity of the motor causes the internal leakage coefficient (R_2) to increase (D_m). This occurs because as the piston's stroke length increases, the leakage flow through clearances decreases.
- For a given hydro-motor size (D_m), the R_2 falls as the oil temperature (T) rises. This happens mostly as a result of the oil viscosity decreasing with rising oil temperature.

5 Conclusions

The loss coefficients of the bent axis hydro-motor are explored in this study along with the effects of oil temperature, hydro-motor speed, hydro-motor size, and load torque. In this regard, a governing system has been made, and its modelling is done through the bond graph simulation technique. The loss coefficients of the hydro-motors are investigated throughout a wide-range of speed and load-torque at varied oil temperatures and motor size. The results demonstrate that the performance of the hydro-motors is significantly influenced by the loss coefficients. The decrease in the R_{el} and R_2 values causes flow loss resulting in the decrease in the volumetric efficiency of the hydro-motor, whereas the increase in the value of R_l translates higher torque loss resulting in the decrease in the mechanical efficiency of the hydro-motor.

References

1. Watton J (2006) An explicit design approach to determine the optimum steady state performance of axial piston motor drives. Proc Inst Mech Eng, Part I: J Syst Control Eng 220(2):131–143
2. Knežević D, Savić V (2006) Mathematical modeling of changing of dynamic viscosity, as a function of temperature and pressure, of mineral oils for hydraulic systems. Facta Univ-Ser: Mech Eng 4(1):27–34
3. Kohmascher T, Rahmfeld R, Murrenhoff H, Skirde E (2007) Improved loss modeling of hydrostatic units: requirement for precise simulation of mobile working machine drivelines. In: ASME international mechanical engineering congress and exposition, vol 42983. pp 195–206
4. Murrenhoff H, Piepenstock U, Kohmäscher T (2008) Analysing losses in hydrostatic drives. In: Proceedings of the JFPS international symposium on fluid power, vol 2008, no 7–1. The Japan Fluid Power System Society, pp 103–108
5. Rahmfeld R, Skirde E (2010) Efficiency measurement and modeling-essential for optimizing hydrostatic systems. In: Proceedings of the 7th international fluid power conference. Aachen, Germany, pp 22–24
6. Bergada JM, Davies DL, Kumar S, Watton J (2012) The effect of oil pressure and temperature on barrel film thickness and barrel dynamics of an axial piston pump. Meccanica 47(3):639–654
7. Kumar N, Dasgupta K, Ahmad F (2013) Analysis of leakage flow characteristics in bent axis motors. In: Proceedings of the 1st international and 16th national conference on machines and mechanisms (iNaCoMM2013). pp 318–323
8. Inaguma Y (2013) A practical approach for analysis of leakage flow characteristics in hydraulic pumps. Proc Inst Mech Eng C J Mech Eng Sci 227(5):980–991
9. Vardhan A, Pandey AK, Singh AP (2022) Effect of various operating parameters on the steady-state performance of the bent axis motor. Proc Inst Mech Eng, Part E: J Process Mech Eng 236(3):1034–1046.

Design an Electric Vehicle with Advanced Power Transmission

Nitin Kumar Waghmare, Akhil Raj, Anubhav Batra, and Harsh Gupta

Abstract The field of wireless power transfer has expanded significantly with recent improvements in power electronics, Speed and capacity simulations of magnetic fields, among other fields. The future of the transportation sector, electric automobiles are being looked into as a potential replacement for internal combustion engine-powered vehicles, especially for CO_2 reduction and alternative energy uses. However, a number of serious issues need to be resolved due to the vehicle's heaviness, big capability, and short distance to travel. The innovative on-road dynamic wireless charging technology for electric automobiles is described in the article. Electric car charging methods can be classified as fixed or dynamic, sluggish or rapid, conductive or wireless. In this essay, an idea for charging an electric car while it is moving—without having to stop—is presented.

Keywords Power transmission · Arduino · Charging · Accident · Monitoring

1 Introduction

The act of transmitting substantial amounts of energy between the transmitter and the receiver without moving or utilizing mass is known as wireless power transmission. Power is transmitted throughout space via electromagnetic fields. It can eliminate the need for wires and batteries, making energy delivery more convenient, transportable, and secure. Increasingly used and well-liked, this technology has a wide range of uses.

Power is transferred from a source to an electrical load without the usage of connectors. When physical wiring is inconvenient, WPT can be used to power electrical devices. The mutual inductance theory serves as the technique's foundation.

N. K. Waghmare (✉) · A. Raj · A. Batra · H. Gupta
Department of Automobile Engineering, Manav Rachna International Institute of Research and Studies, Faridabad, Haryana, India
e-mail: nitinwaghmare123@gmail.com

H. Kumar et al. (eds.), *Recent Advances in Intelligent Manufacturing*, Lecture Notes in Mechanical Engineering,
https://doi.org/10.1007/978-981-99-1308-4_16

One of the potential applications is in the automotive sector, specifically with regard to electric vehicles.

This article's goal is to cover the investigation and creation of without wire transmission-operated mechanism for electric vehicles. Building charging systems and transferring energy via resonance coupling are the main goals. The systems consist of a battery as the electric load and an alternating current source, as well as transmission and receiving coils, a converter, and other components.

More than 2,36,802 electric automobiles were sold in the year 2020–2021 related to 27,11,457 IC engine vehicles. A normal combustion vehicle costs about 7–10 lacs, however the regular price of an electric automobile is over 22 lacs. Electric automobiles take 98 percent longer to fully charge than their gasoline-powered counterparts. The future of transportation is electric cars (EVs). Instead of its standard induction vehicle substitute that is both better for the environment and more long-term. The market for electric vehicles has grown significantly over time for a number of obvious reasons, including [1–4] government assistance and lesser emissions of greenhouse gas, with one of the most significant being decreased emissions of CO_2 and CO_2 [5–7]. Numerous countries in Europe and the US carried out extensive research and observed that the automobile industry contributed more than 25% of emissions. EV would not be sufficient to give a hassle-free drive on the highway; automobile owners continue to purchase traditional induction vehicles [8–10]. The charger has to charge the vehicle with the use of pads which are placed on the car as well as charging station, and can be used to charge electric automobiles instead of utilizing a cable. Later, current is generated using the core scientific concept of EMI in order to charge the vehicle's battery. The block diagram for this project's concept is shown in Fig. 1.

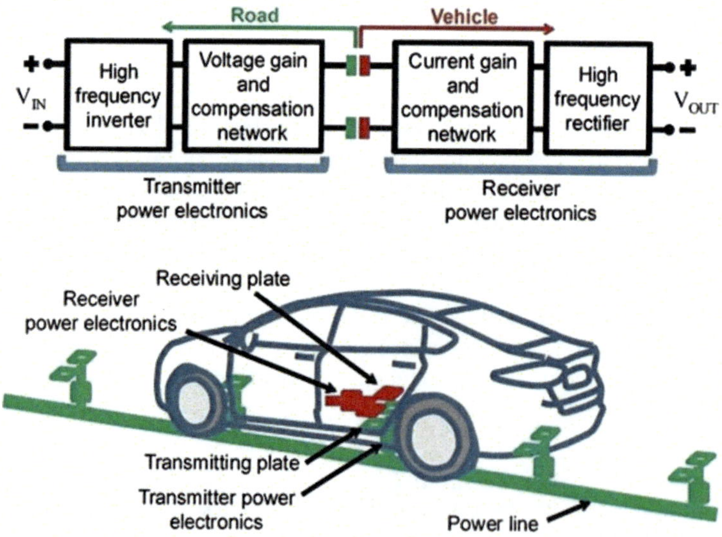

Fig. 1 Block diagram and demonstration

1.1 Present Scenario of Electric Vehicles

The world is heading toward electric mobility at the moment to reduce the amount of pollution produced by nonrenewable fossil fuel-powered vehicles and to provide a cost-effective substitute for expensive petroleum. The driving range and charging process for electric automobiles, on the other hand, are the two main obstacles restricting their acceptability in comparison to conventional vehicles. In recent years, performance and range of electric cars (EVs) have improved significantly. Several commercially available models are presently available and more electric vehicles on the roads. Most of EVs on the road today are charged by electric lines, but companies like Tesla, BMW, and Nissan have started working on wirelessly charging electric cars that do not require complicated connections [11, 12]. The wireless (inductive) link effectively reduces sparking brought on by plugging and disconnecting rather than depending on a physical cable connection. Charging, during running a vehicle, is another benefit of without wire charging. When this is accomplished, the demand for battery storage will be much reduced because EVs won't be restricted by their limited electric driving range.

With the development of wire charging technology, it is easy to charge at the time of driving as well as in parking area. Previously, you had to wait for hours at charging stations. Why can't we transmit electricity over the air when we are well familiar with wireless data, audio, and video transmission?

2 Fundamentals of Power Transmission in Electric Vehicle

An induction motor with two air-core coils is the latest technology for WPT in EVs. Ampère's Law states that when a time-varying current is used to energize the transmitter coil, also known as the main coil, it creates a dynamic magnetic field around the structure [13, 14]. When a secondary coil or receiver is able to concatenate some of the flux, induced voltage is observed. A thorough enlightenment of Quasi Dynamic, Static and Dynamic Wireless Charging is given [15, 16] after conductive and wireless charging are contrasted. The article explores potential solutions for problems like misalignment, frequency, power level constraints, and coil design for power pads. The standards are then added together to create a consistent representation of the current (Fig. 2).

Again nano-manufacturing has been categorized into three different generations with their nature of complexity that comes during the time of production and metrology. The first generation behaves as an economic stimulator that consists of an integrated form of nanostructures into the micro and macroscopic systems [17, 18]. The second generations consist of nano, submicron, quantum dot, lasers, and LEDs, and the last generation consists of the system which uses the nanoscale (less than 100 nm) in size e.g. sensors and device systems.

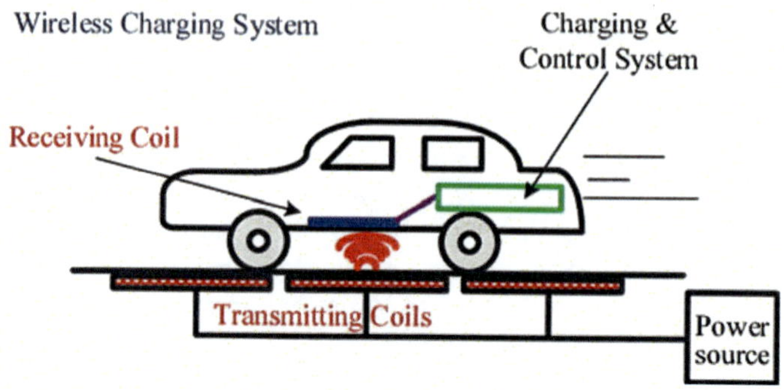

Fig. 2 Basic of transmission

2.1 Electric Vehicle Charging Requirement

Consumers like electrical vehicles because they are easy to use. The need for convenient parking and charging space is one essential requirement. A suggested approach aims to offer an effective solution by merging these two systems to the issue. This article's aim is to provide a description of the layout of a system that can control pricing plans and the availability of parking spaces. All vehicle types cannot be accommodated under the current parking arrangement. For electric cars, parking spaces and charging facilities are necessary. Using a smartphone to make reservations for space is possible with the provided method. The system's three main components are the lot, map and vehicle manager, and it then manages all related processes based on details like the battery level and time of the vehicle's arrival [19–21].

3 Working of Circuit Diagram

Electric car charges continuously while moving through this technology. Moreover, it presents an innovative and sustainable way to provide energy for these autos. The principal components are IR sensors and coils; as a car approaches, the IR sensor triggers, activating the coil below the automobile through a relay, which then begins the vehicle's charging (Fig. 3).

3.1 Electromagnetic Induction

Electric vehicle recharging is handled via dynamic WPTS. The technique induces current through two copper coils using electromagnetic induction, a cornerstone of magnetism. One is known as the main coil and is situated on the bottom chassis of

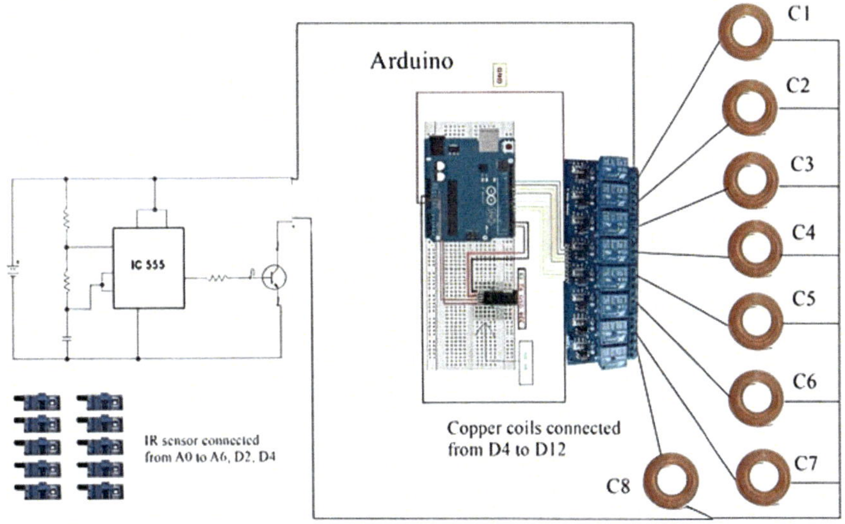

Fig. 3 Circuit diagram of safety system

the vehicle, while the other, known as the secondary coil, is situated on the road just below its surface. Conductive wires connect the main coil to the power source. When electricity passes through the primary coil, a magnetic field is created all around it. The secondary coil generates a magnetic field due to the two coils' nearness to one another. The electricity produced by the secondary coil at any one time is converted from a magnetic field created by the primary coil [22]. The charging procedure is then completed without the need for the car to stop or slow down at any time by transferring this electrical energy to the battery of the car via conductive connections. The system must be ensured not to use any energy while there are no vehicles on the street. In order to address this, infrared sensors have been deployed on both sides of the street [23]. Any form of motion can be detected using these sensors, which is their main job. Either the sensor senses movement of the car, the system's circuit closes and supplies electricity to the car; or no movement is detected, the circuit is left open and no energy consumption. A reliable and consistent wireless power transfer requires inductors with high quality factors.

3.2 Resonant Frequency and Energy

The resonant frequency must be found before the quality factor can be computed, as indicated in the equation below

Resonant frequency formula

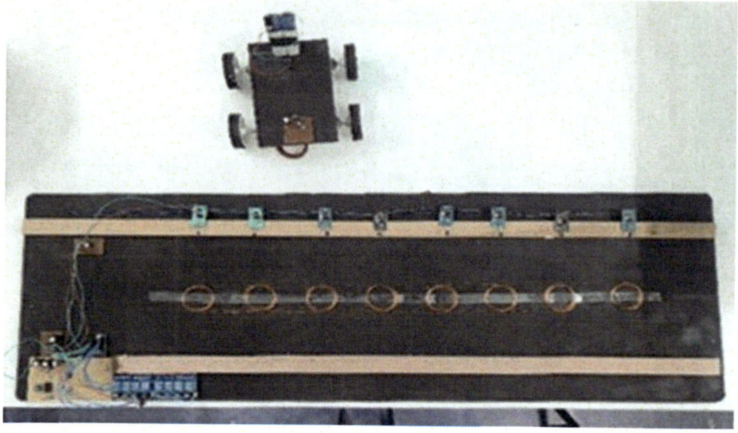

Fig. 4 Show the wireless charging model

$$f = \frac{1}{2 \cdot \pi \cdot \sqrt{L \cdot C}}$$

After calculating the resonance frequency, evaluate the inductor coils level of quality to understand if the specified standards of inductance, capacitance, and resonant frequency are viable for the circuit.

A typical WPT system requires a quality factor of at least 100. Direct energy from the car itself might be used to power the charging process. The use of a flywheel, ratchet, axle, and dynamo can all produce energy. Only around 2/5 of the flywheel is apparent from outer of the road or speed braker since the remaining three-fifths are buried beneath the surface. The ratchet marks straight contact with the flywheel and when a vehicle passes over the flywheel, it causes movement. Ratchet and dynamo are joined by an axle [24]. The dynamo uses the rotation of the axle caused by the ratchet to transform energy from mechanical into electrical energy using the electromagnetism induction process.

Through the use of conductive wires, the produced energy is put in storage in a battery. Hydropower facilities and solar panel installation are additional ways to produce energy. The microcontroller used is an Arduino Uno. Relays are attached to it, and coding is used to run the relay that corresponds to the IR sensor that is activated by a moving vehicle (Figs. 4 and 5).

4 Result Output

The dimension of the RX coil limits the efficiency of WPT system (WSE, Following regulation, what is the net power increase from the transmitted power at the TX coil to the RX load?) and how much power is supplied to the load (Table 1 and Figs. 6, 7).

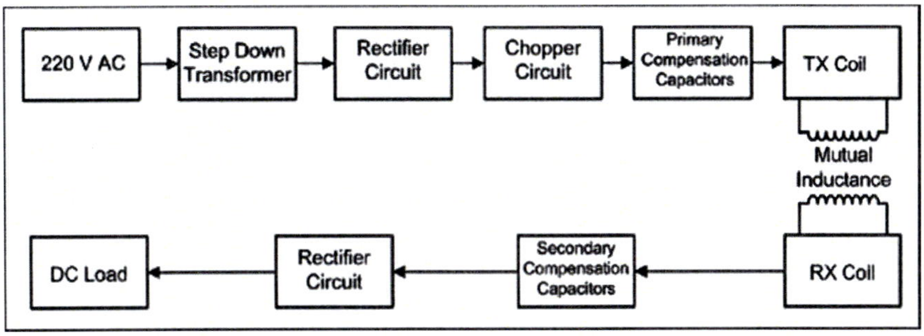

Fig. 5 Complete flow chart working of the model

Table 1 Readings

Parameters	Inductive coupling
Human safety	Depends on energy transferred
Efficiency	73%/1 cm
Max voltage RX	Up to 20v
Frequencies	50 kHz
Max voltage TX	Upto 5v

Fig. 6 Output in receiver coil RX

5 Outcomes and Key Findings

Arduino, an IR sensor, and relays are used to illustrate a wireless charging system from the road to the vehicle. The expanding use of EVs calls for the development of novel technologies that make charging simpler through increasing its autonomy and reducing the need for human interaction. This article discusses the technology involved in wirelessly charging electric cars.

Fig. 7 Output in transmitter coil TX

Key findings are system is easy to use and gives ease of mind, smaller batteries will be used as compared to traditional EVs, Extended life of the vehicle, minimal emissions and no maintenance required as such.

6 Future Perspective and Conclusion

Magnetic induction charging mats were employed in the early development of wireless transmission technology, which needed touch and precise alignment of the device, such as a phone. As a result, technology was created that allows smartphones and other gadgets to be charged from meters away using a powerful wireless system.

Up to a distance of 5 cm from the project, WPT for short range or close field occurred, after which the power transmitted began to diminish substantially. In this experiment, we were able to charge a 9 V battery using electricity that was supplied wirelessly. Research on a better, more efficient circuit employing phased locked loops (PLL) and power amplifiers is needed to obtain greater frequency stability, more power, and efficiency.

References

1. Machura P, Li Q (2019) A critical review on wireless charging for electric vehicles. Renew Sustain Energy Rev 104:209–234. https://doi.org/10.1016/j.rser.2019.01.027
2. Khan I, Sharma H, Kumari A (2020) Multi-output fast wireless charging for electric vehicle electric vehicle charging. Int J Eng Tech Res 9
3. Jang YJ (2018) Survey of the operation and system study on wireless charging electric vehicle systems. Transp Res Part C Emerg Technol 95. https://doi.org/10.1016/j.trc.2018.04.006

4. Mohiuddin K (2020) Wireless power transfer system for charging of electric vehicles. https://doi.org/10.13140/RG.2.2.16333.46569
5. Jin Y, Xu J, Wu S, Xu L, Yang D (2021) Enabling the wireless charging via bus network: route scheduling for electric vehicles. IEEE Trans Intell Transp Syst 22:1827–1839. https://doi.org/10.1109/TITS.2020.3023695
6. Liang X, Chowdhury M (2018) Emerging wireless charging systems for electric vehicles—achieving high power transfer efficiency: a review 1–14. https://doi.org/10.1109/IAS.2018.8544484
7. Yakala R, Pramanick S, Nayak D (2021) Optimization of circular coil design for wireless power transfer system in electric vehicle battery charging applications. Trans Indian Nat Acad Eng 6.https://doi.org/10.1007/s41403-021-00224-z
8. Qiu C, Chau KT, Ching T, Chunhua L (2014) Overview of wireless charging technologies for electric vehicles. J Asian Electric Veh 12:1679–1685. https://doi.org/10.4130/jaev.12.1679
9. Tavakoli R, Pantic Z (2017) Analysis, design and demonstration of a 25-kW dynamic wireless charging system for roadway electric vehicles. IEEE J Emerg Select Topics Power Electron 1. https://doi.org/10.1109/JESTPE.2017.2761763
10. Ushijima-Mwesigwa H, Khan Z, Chowdhury M, Safro I (2017) Optimal placement of wireless charging lanes in road networks. J Ind Manage Optimizat 13.https://doi.org/10.3934/jimo.2020023
11. Panchal C, Stegen S, Lu J (2018) Review of static and dynamic wireless electric vehicle charging system. Eng Sci Technol Int J 21. https://doi.org/10.1016/j.jestch.2018.06.015
12. Stamati T-E, Bauer P (2013) On-road charging of electric vehicles 1–8. https://doi.org/10.1109/ITEC.2013.6573511
13. Balde B, Sardar A (2019) Electric road system with dynamic wireless charging of electric buses 1–4. https://doi.org/10.1109/ITEC-India484572019.ITECINDIA2019-251
14. Bi Z, Kan T, Mi C, Zhang Y, Zhao Z, Keoleian G (2016) A review of wireless power transfer for electric vehicles: prospects to enhance sustainable mobility. Appl Energy 179:413–425. https://doi.org/10.1016/j.apenergy.2016.07.003
15. Yan L, Shen H (2021) Utilizing game theory to optimize in-motion wireless charging service efficiency for electric vehicles. ACM Trans Cyber-Phys Syst 5:1–26. https://doi.org/10.1145/3430194
16. Sun B, Tsang D (2017) Performance analysis of dynamic wireless charging system for electric vehicles: a queueing approach 168–178. https://doi.org/10.1145/3077839.3077849
17. Manshadi S, Khodayar M, Abdelghany K, Uster H (2017) Wireless charging of electric vehicles in electricity and transportation networks. IEEE Trans Smart Grid 1–1. https://doi.org/10.1109/TSG.2017.2661826
18. Mohammed Q, Jung J-W (2021) A comprehensive state-of-the-art review of wired/wireless charging technologies for battery electric vehicles: classification/common topologies/future research issues. IEEE Access 1. https://doi.org/10.1109/ACCESS.2021.3055027
19. Bagchi AC, Kamineni A, Zane R, Carlson R (2021) Review and comparative analysis of topologies and control methods in dynamic wireless charging of electric vehicles. IEEE J Emerg Select Top Power Electron 1–1. https://doi.org/10.1109/JESTPE.2021.3058968
20. Danping Z, Juan L, Yuchun C, Yuhang L, Zhongjian C (2019) Research on electric energy metering and charging system for dynamic wireless charging of electric vehicle 252–255. https://doi.org/10.1109/ICITE.2019.8880214
21. Zahiri V, Thrimawithana D, Covic G (2021) An inductive coupler array for in-motion wireless charging of electric vehicles. IEEE Trans Power Electron 1. https://doi.org/10.1109/TPEL.2021.3058666
22. Muharam A, Suziana A, Hattori R, Hapid A (2020) 13.56 MHz scalable shielded-capacitive power transfer for electric vehicle wireless charging 298–303. https://doi.org/10.1109/WoW47795.2020.9291299
23. Sun L, Ma D, Tang H (2018) A review of recent trends in wireless power transfer technology and its applications in electric vehicle wireless charging. Renew Sustain Energy Rev 91:490–503. https://doi.org/10.1016/j.rser.2018.04.016

24. Mohamed N, Flah A, Mouna B (2019) Wireless charging system for a mobile hybrid electric vehicle. https://doi.org/10.1109/ISAECT.2018.8618829
25. Triviño A, Gonzalez-Gonzalez J, Aguado J (2021) Wireless power transfer technologies applied to electric vehicles: a review. Energies 14:1547. https://doi.org/10.3390/en14061547

Experimental Study and Recent Scope on Drilling Electrochemical Discharge Machining Process: A Review

Gaurav Yadav, Riya Choudhary, Himanshu Bhardwaj, and B. K. Bhuyan

Abstract Among all growing machining, Electrochemical Discharge Machining (ECDM) is a new challenging hybrid machining process in which the machine is electrically non-conductive fragile or has rigid substances with precise dimensional accuracy which has also zero stress of residual. The mechanism of ECDM process is based on electro discharge principle and combines features of electric discharge machining (EDM) and electro chemical machining (ECM) process. Drilling-ECDM is the one of the configuration of the ECDM process which is apposited to make accurate holes on the difficult-to-machine materials compared to the traditional drilling process. This article is used to focus the study on the fundamental physical principle of the drilling-ECDM process and scrutinize the experimental study of past researchers in the tabular form. Moreover the process variables and responses of the drilling-ECDM process are symbolized through the Fish bone diagram and the influences of input parameters on machining characteristics are also reported in the present work. From the critical study of literature, the scopes of the materials, year wise publications and potential of research have been illustrated in the analytical manner through a pie-chart as well as line diagram.

Keywords ECDM · EDM · ECM · MRR · Overcut

1 Introduction

The Electrochemical Discharge Machining (ECDM) is a more desirable process which is used to make drill cavity on assorted materials regardless of their cracked nature. The ECDM process is complicated which is under exploration. There are so many experiments which are presided over numerous actions on output responses. This is also a transpire fusion process under the individualistic process. In ECDM, the tool wears initiate in collation for a technique of EDM (electro-discharge machining)

G. Yadav · R. Choudhary · H. Bhardwaj · B. K. Bhuyan (✉)
Department of Mechanical Engineering, FET, MRIIRS, Faridabad, Haryana, India
e-mail: bkbhuyan.fet@mriu.edu.in

H. Kumar et al. (eds.), *Recent Advances in Intelligent Manufacturing*,
Lecture Notes in Mechanical Engineering,
https://doi.org/10.1007/978-981-99-1308-4_17

and ECM (Electrochemical Machining). This task is very much important and due to it can do various machining materials, composites, glass, alumina, etc. There are so many processes in which drilling-ECDM is one of the processes which have the potential to machining effectively insulating fragile materials. The high electrical discharge involves a melting process above the tip of the electrode during the electrolysis which enables the machine to have difficulties with resistance materials [1]. It is a specific technology that is used to process the high productivity, good elasticity, and cut-price. This can help to hold the dielectric materials such as hard and brittle elements. Over a time the growth of EMS (electro mechanical system) fluidics automation, hi-tech requirements become higher which rose for fabrication of insulated materials. There are so many studies that are conducted for improving the machining accuracy, the motion control approach for changing the control on the motion approach, and optimise the process parameters using the response surface method for changing the motion control [2, 3].

There are so many scientific restrictions which can get the better of the use of some non-metallic substances. Thus the different types of applications in modern industries using Drilling-ECDM are playing a key role. The performance limit of the metal has reached due to engineering, the advantages of the use of good creep resistance which is not an alternative but to seek increased corrosion resistance, increased thermal shock resistance, increased temperature resistance, and increased hardness characteristics; the principle of ECDM is for the eroding effect of controlled ECDM on the electrodes [4]. ECDM process is also known as ECSM (electrochemical spark machining) which is a hybrid machining process and combines the features of the ECM (electrochemical machining) and EDM processes. Both ECM and EDM processes have the potential to machine electrically conductive composite materials. However, ECDM is more suitable to machine insulating hard materials [5–7]. Huang et al. proposed a new form of technology to enhance the machining efficiency of drilling ECDM of stainless steel using pure water as an electrolyte and drilled successfully on it [8]. The different electrolytes are initiated by various researchers during ECDM of several glass polymer based composite materials. The coalescence of hydrogen gas bubbles develops at the cathode and oxygen gas bubbles are formed at the anode respectively [9, 10]. Antil et al. [11] conducted the experiments during drilling-ECDM on SiC based polymer composite materials and found better machining results while optimizing the process variables using the Taguchi based grey relational analysis approach. Chak and Rao [12] confirmed the outstanding potential of the drilling ECDM process during machining of ceramic workpiece using abrasive based rotating tool compared with a hollow static tool electrode and identified the improvement of the slicing ability while considering the abrasive tool electrode. Laio et al. had adopted new technique for improving the machining quality by adding Sodium Dodecyl Sulfate (SDS) into the electrolyte and found more bubble generation while adding SDS as compared to without SDS addition into the electrolyte [13]. Moreover, the Aluminium based Silicon Carbide composite material was introduced by Singh and co-researchers during the rotating electrode (RE) based ECDM process and reported the remarkable results obtained from RE-ECDM compared to sinking ECDM [14]. In addition to this the borosilicate glass was introduced during the drilling-ECDM process for improving the machining efficiency with economical cost of operations and the results of MRR was improved while increasing the voltage and electrolyte concentration [15]. Therefore the significance of glass based workpiece is a worthwhile in manufacturing industries and its application is widely employed in various MEMS sectors [16, 17]. In 1968, Kurafuji and Suda developed a new hybrid machining process which was appropriate to drill on a glass workpiece during the ECDM process with less time compared to the conventional drilling process [18]. Since there are many research studies existing already, an attention to do the experimental work due to which drilling-ECDM is discussed so proficiently.

Razfar et al. had stated the experimental study on the glass workpiece using the drilling-ECDM process and found the better efficiency of MRR using the vibrational tool compared to the static tool [19]. Arab and Dixit had proposed the feed rate analysis of the rotating tool during drilling ECDM on glass workpiece and represented the accurate feed rate of the tool for experimental work [20]. Kang and Tang carried out the experiments with coating as well as supper alloy materials and confirmed that electric discharge is playing a challenging role for removing the substrate of the supper alloy as compared to coating during the drilling-ECDM process [21]. Kumar et al. conducted the experimental work on glass materials for analysing the significant process variable on machining responses and the identified pulse duration is impacting more effect followed by other process parameters of the drilling ECDM. The higher machining efficiency was obtained while using electrolyte of NaOH as compared to NaCl [22]. Further salt solution with low conductivities was used for improving the machining performances as well as the surface of machining during tube electrode based drilling-ECDM with a high speed process and compared its various mechanisms of material removal with the drilling-EDM [23]. Geng et al. had initiated a new arrangement of backflush for the drilling-ECDM process on Nickel materials and reported some positive advantages like increase the drilling depth, reducing the short circuit and minimizing the angle of taper [24]. Zhang et al. presented a new concept of recast layer free zone using various fluids of the ECDM process during drilling on Nickel based supperalloy materials and confirmed the information through experimental as well as theoretical analysis. The radius of stray corrosion was found to drop with plummeting of flushing externally [25]. Arab et al. stated the influence of process parameters on responses during the drilling ECDM process and deliberated different values of inter electrode gap (IEG) for NaOH and KOH electrolyte experimentally which was also validated theoretically. The machining depth and tool wear were recommended to be increased and decreased due to the variation of IEG [26]. A low cost cutting amount method was spread for the cavity in the drilling of size minimized to greater than 1 mm diameter of steel using the drilling-ECDM process. The spring arrangement mechanism is employed in this process to drill the tinny materials in a smooth manner and helps to remove the debris particles from the machining surface during the operation [27]. Pawar et al. [28] had conducted the experiments on sodalime glass during the drilling-ECDM process using Taguchi based L_{27} approach and demonstrated the applied voltage and rotational speed that were more significant parameters for perfect machining depth as well as hole diameter followed by others process parameters. Further the abrasive particles in powder form was also mixed in the electrolyte for enhancing the machining potential during drilling of a blind-hole on sodalime glass in ECDM process and the outstanding machining depth was reported while mixing of ceramics based abrasive particles in the electrolyte compared to the normal electrolyte [29]. The newly designed drilling-ECDM setup was developed with a better feeding mechanism for making of a deeper hole on the glass workpiece and the contact force was employed so effectively to increase the large depth of machining [30]. In the drilling-ECDM process, heat affected zone (HAZ) is creating the vital issue inside the hole surface for reducing the machining potential. To cope up such type of problem, pulse voltage is used to reduce the thermal damage and enhance machining efficiency [31]. In addition to the tool erosion is also creating a big challenge for tumbling the dimensional accuracy of quality of the hole during drilling-ECDM of Nickel alloy. The two electrolyte solutions (Deionised water and $NaNO_3$) are tested for the experiments to improve the machining performances and $NaNO_3$ is performed to provide a better hole quality with less tool erode compared to deionised water [32]. The motionless electrolyte (sodium silicate) is also worthwhile to make a small hole diameter using the cylindrical based tool electrode of size (0.5–1.0) mm along with the increasing the distance between the tool and workpiece frontal surface [33].

Saeed and Amir had expressed two significant observations for axial wear and hole length based on voltage, liquid density, capacitance and tool diameter which were evaluated by analysis of variance and main effect plots [34]. Chak and Rao compared the hole diameters using copper tool and abrasive based tool electrode during ECDM of Al_2O_3 workpiece and also identified the pulsed DC power supply that plays a key role compared to DC supply as a result avoiding the cracks from the machining surface during the trepanning process [35]. Moreover the rotational copper tool electrode has a potential to improve the MRR, machined depth and hole diameter by setting the input parameters of the voltage, electrolyte concentration and tool speed during the ECDD of sodalime glass workpiece [36]. Dong et al. proposed the two experimental observations during high speed ECDD and EDM process on beryllium copper alloy workpiece using deionised water as an electrolyte and proved the better machining potential with lesser tool wear using high speed ECDD compared to EDM [37]. The pure water was also used as an electrolyte for conducting the experiments on stainless workpiece during the ECDM process using a high speed tool electrode and improved the quality of the hole by reducing the tool wear [38]. To resolve the issues of taper on Pyrex glass during the ECDM process, an innovative shape of the rotating tool electrode has been designed for enhancing the machining quality using pulse voltage and found no thermal cracks during proper setting of process parameters [39]. Singh et al. carried out the experiments by orthogonal array of L_{27} on silicon wafers during the ECDM process using rotating based tool and improved the outstanding results of overcut (1.40%) as well as circularity of holes (2.44%) setting the optimum parameters [40]. However increasing and decreasing THE hole taper was deciding a key role with rise in voltage as well as feed rate of THE tool and it was under control at an optimum setting of process parameters and also the significant role of applied voltage was identified for enhancing the characteristics of surface quality during the rotating tool based ECDM process [41]. Xu et al. elaborated a new method for passing the electrolyte through the brass tube tool electrode used as cathode and superalloys used as anode during ECDD process. They produced the deep hole of diameter 0.5 mm on the workpiece with lesser tool wear as well as better surface quality compared to the EDM process [42].

In the current study focuses the experimental investigations on various electrically non-conductive workpiece materials as well as different electrolytes used in the drilling-ECDM process along with major findings and effect of process variables on different responses represented as the fish bone diagram. Moreover, research potential and scope of the future work have also been given the priority in the following sections.

2 Working Principles

In-house ECDM is a combined feature of the EDM and ECM processes which is suitable to machine electrically non-conductive fragile and rigid workpieces. In the ECDM process, there are two different sizes of electrodes dipped into the electrolyte and distance between the two electrodes is (30–55) mm. The ratio of the surface area of the smaller electrode (called cathode) to the larger electrode (called anode) surface area is 1:100. The cathode tool is submerged (2–3) mm into the electrolyte which accumulated the hydrogen bubbles on the surface and the larger electrode also accumulated the oxygen gas due to the electrolysis process [43–48]. There are two types of power supply (DC or pulsating DC power) that can be used to complete the circuit in the process. The negative and positive ends of the power supply are connected with the cathode and anode (also called auxiliary electrode) electrode with stated specification of high voltage (80–150) V and low current of upto 15A [49]. The schematic diagram of the ECDM process is represented in Fig. 1.

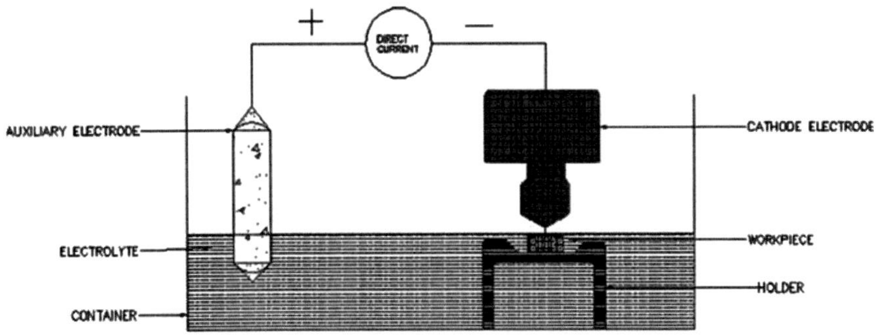

Fig. 1 Line diagram of ECDM process

Moreover, the current intensity increases at the smaller electrode due to lesser surface area compared to auxiliary electrode and the heating temperature of electrolyte is upto (40–60) °C owing to Ohm's law of principle. As breakdown voltage is below the applied voltage then the hydrogen gas bubbles generated at the surface of the cathode which creates a non-conductive bubble bridge as a result produces a high current resistance layer. Hence, the spark is initiated while the breakdown voltage is lesser than the applied voltage. The workpiece is placed at the vicinity of the spark zone between the tool and the electrolyte and the material is removed from the workpiece by melting and vaporization. Furthermore, the proper arrangement of the flushing method is needed to remove the by-products from the electrolyte and a suitable drainage system is required to exit the electrolyte to the reservoir. The reuse of the electrolyte is also possible to need an appropriate plan of filtration as well as pumping arrangement. The different types of electrolyte (like NaCl, NaOH, KOH, $NaNO_3$ and H_2SO_4) can be used during the experiments on various electrically non-conductive workpieces (like glass, ceramic, quartz and composites) [50–53]. The various names of the ECDM process (such as ECSM, ECAM and SACE etc.) create the process complexity but the material removal depends on the discharge phenomenon of the process. Hence, ECDM is a complex machining process where several disciplines of science and engineering are involved in its theory. It is not possible to present a simple and comprehensive theory, which explains the nature of the process with every detail.

The different configurations of the ECDM process is Sinking-ECDM, Drilling-ECDM, Wire-ECDM and Milling-ECDM [54–60]. For sinking-ECDM, a stationary tool is used, high-speed tool electrode is employed for drilling-ECDM, wire tool electrode is used for wire-ECDM and milling tool is employed for the milling-ECDM process to make several holes, cutting and slots etc. Various researchers have represented their work using sinking, wire and milling-ECDM but drilling-ECDM is a progressive and inspiring novel path for the twenty-second century. Thus, the present study of drilling-ECDM is an initiation in the direction of the research and its applications in various modern industries like aerospace, automobile, medical and space industries etc. The integration of the drilling-EDM and ECM process is known as the drilling-ECDM process under the hybrid machining process and the principle as well as mechanism of the drilling-ECDM process is similar to the ECDM process. The drilling-ECDM process has a strong machining potential to make a hole on the electrically non-conducting workpieces.

The common process variables namely voltage, current, pulse on-time, pulse off-time, electrolyte concentration and workpiece thickness and output characteristics like MRR (Material Removal Rate), TWR (Tool Wear Rate), Overcut and R_a (surface roughness) are playing a significant rule during the drilling-ECDM process for electrically non-conductive workpieces. The various possible process variables and output characteristics are also depicted through fish bone diagram which is represented in Fig. 2.

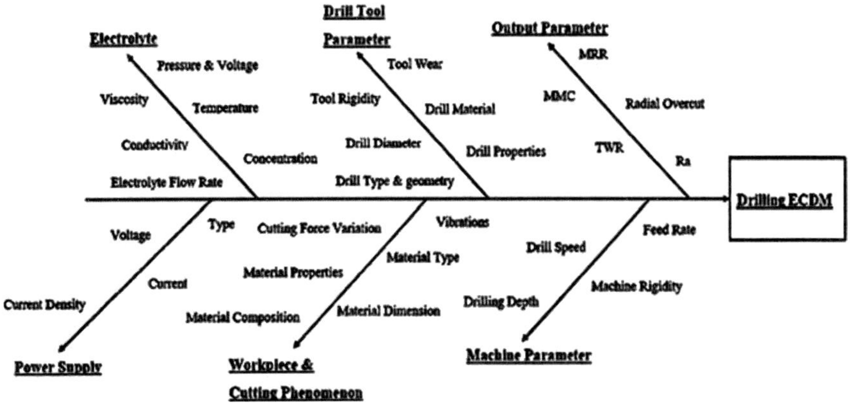

Fig. 2 Fish-bone diagram of drilling-ECDM process

3 Experimental Procedures

Several researchers have carried out the experimental and theoretical effort on electrically non-conductive materials during the ECDM process and investigated the possible applications in MEMS (Micro-electromechanical systems) as well as advanced automated industries. Many experimental works have been introduced by various researchers as compared to the theoretical study owing to the complexity of the ECDM process mechanism for the material removal [61–64]. In this article, a high speed rotational tool electrode based ECDM process (called drilling-ECDM) is studied to make a hole on non-conducting materials during the experiments at the setting of different input parameters. The drilling-ECDM process is based on the ECD principle and discharge takes place between the tool and electrolyte. The material is removed while the workpiece is placed at the vicinity of the spark region. The heat energy of the spark is responsible for damaging the tool and workpiece and higher MRR compared to ECG is observed due to potential difference between the electrodes [65]. Afterwards researchers have made productive use of ECD phenomenon during their research. Basic principle of the ECD is still under a laboratory study stage. Various researchers have put forth explanations of the ECD phenomenon based on their experimental studies. Some researchers have pointed out the mechanism of material removal on their investigations [30–35].

The ECDM process uses electrochemical discharge which generates heat for the purpose of removing of the work material [66]. If any electrically non-conducting or conducting material is placed in the closed vicinity of electrical discharge, material removal takes place. During the experimental work, different electrolytes (namely NaCl, NaOH, KOH, $NaNO_3$ and H_2SO_4) are used to enhance the machining capability and performance of the drilling-ECDM process and some attempts are also the initiation in the direction of the hybrid electrolyte which is mixing of more than two electrolytes [47]. The line diagram of the drilling-ECDM is illustrated in Fig. 3. The

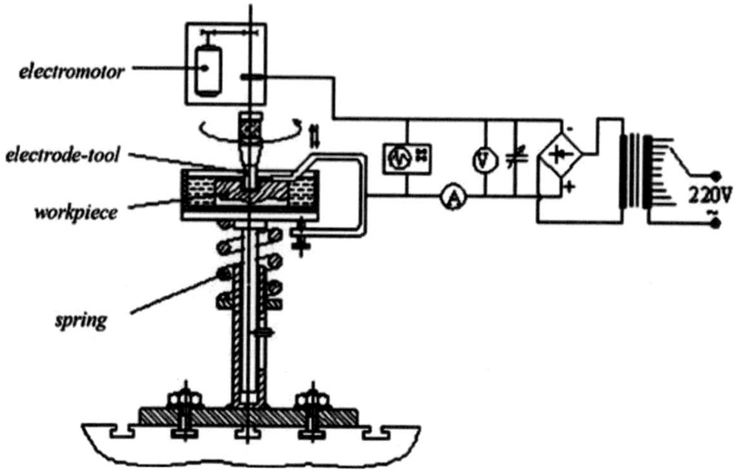

Fig. 3 Line diagram of drilling-ECDM process [20]

variation of electrolyte concentration on output responses is investigated thoroughly in different research works and studied the trend of influence on process responses through the graphs as increase in the electrolyte concentration enhances the MRR and R_a at certain points beyond that it is decreasing due to lack of spark intensity. During the drilling operation, the applied DC voltage comes from a programmable pulsating power supply with specification of high voltage and low current [23]. To shorten the stroke length of the work when the electrode tool is in the top position, then in order to avoid welding between the tool and the workpiece electrode during discharge, a screw pin was attached to the sleeve under the workpiece holder. The workpiece was then drilled due to the discharge energy and automatic feed arrangement was given to workpiece as spark generation is possible only while the tool electrode would be immersed (2–3) mm into the electrolyte. Moreover, the tool was also separated when a large amount of hydrogen gas was generated on the surface of the tool [32]. The process was analyzed by observing the process phenomena and determining voltage and current waveforms by using the sophisticated instruments. A suitable arrangement was set to control high speed rotating tool electrode in the machining chamber and the workpiece moved smoothly during the experiments until it touched the tool and then fixed properly [31].

4 Results and Discussion

From the critical study of literature, it has been seen that, output responses like MRR, Overcut, TWR and R_a during the drilling-ECDM process of electrically non-conducting workpieces is influenced by different key input process parameters namely voltage, current, pulse on-time, pulse off-time, electrolyte concentration and workpiece thickness. It has also been observed that complex, non-linear relationships exist between these process parameters with output responses. Moreover, the output characteristics are dominated by the variation of applied voltage, current and pulse on-time whereas these process parameters increase with enhance output responses and decline the input parameters with decrease in the responses [47, 52]. Pulse off-time and workpiece thickness are also playing a significant role during experiments as it is observed that increase and decrease of these process variables are inversely proportional with the abovesaid responses. From the experiments, it is seen that the electrolyte concentration increases with increase of output responses at a certain value but beyond that value it declines owing to the conductivity of the electrolyte that decreases with higher concentration of electrolytes during the drilling-ECDM process [67]. The experimental observations are summarized from various literatures which are tabulated in Table 1. This table includes various workpiece materials, electrolytes, process parameters with ranges, output parameters and major findings from the exhaustive study of literature review.

5 Research Potential and Scope for Future Work

Drilling-ECDM process is more suitable to make a symmetrical hole on electrically non-conductive materials with better dimensional accuracy. This process is also worthwhile to make micro-holes on rigid and hard materials which are not feasible on any conventional processes. The thickness of the processing material for drilling was chosen to be between 450 and 1350 μm [52]. As response parameters, permeability denotes the potential pathway by which energy might permeate a substance. The machining accuracy is characterized by overcut. It is preferable to use minimum values for proper editing. Surface damage, depth of cut, and aspect ratio as a result of drilling-ECDM were measured. The percentage of different output responses used in drilling-ECDM are 10% of MRR, 40% of R_a, 18% TWR, and 32% of Radial Overcut which is represented in Fig. 4.

The less experimental work is carried out on MRR and TWR and similar work followed by overcut and R_a. Thus, there is more opportunity available to work on MRR and TWR considered as output characteristics of the drilling-ECDM process for machining of several advanced materials. The number of publications in year wise from 2015 to 2021 is depicted in Fig. 5. In 2015, the maximum number of papers was published i.e. 25 papers were published and in 2021, 21 papers were published. In 2020, very less numbers of paper were published in the field of the drilling-ECDM process. The trend of the graph is represented to focus on publication rate in year wise. Therefore, more scope is open for the initiation of the drilling-ECDM process for making of hole on electrically non-conductive materials. The publication trend on composite materials is represented in Fig. 6. The 30% of carbon fibre reinforced polymer (CFRP), 25% of glass fibre reinforced polymer (GFRP), 20% of metal matrix and 25% of ceramic matrix have only been used in the past literature. Hence the drilling-ECDM process is less suitable for the metal matrix but it is more

Table 1 Summary of the Drilling-ECDM Process

Workpieces	Electrolytes	Input Parameters with range	Output Parameters	Major Findings	Ref.
Steel	Sodium Silicate	Voltage (35 V–45 V) Electrode diameter (0.5 mm-0.9 mm) Capacitance of discharge circuit (33 μF) Work liquid density (1.05) g/cm^3 Work piece Thickness (1.4 mm)	Electrode-tool axial wear	Analysis of combined machining methods and experimental studies resulted in an experimental relationship which highlights the effect of some of the analysed operating conditions on the axial wear of electrode tools	27
Single crystal nickel	$NaNO_3$	Voltage (80 V) Thickness (2 mm) Outside diameter (300 μm) Internal diameter (150 μm)	Radial overcut	Stray current in the non-machined area also decreases. Especially, when the conductivity of the external increases. However, for the average hole diameter, the nominal target hole is achieved by increasing the tool speed	25

(continued)

Table 1 (continued)

Workpieces	Electrolytes	Input Parameters with range	Output Parameters	Major Findings	Ref.
Soda lime glass	KOH	Voltage (55 V-75 V) Electrolyte Concentration (10%-20%) Tool Rotation (0 r/min-50 r/min)	Material removal rate (MRR)	The fabricated electrochemical discharge machine was utilized for the drilling of Soda-lime glass material through Copper tool material. From the present investigation, it is found that the average material removal rate increases with an increase in electrolyte concentration mostly	36
Pyrex glass	KOH	Voltage (40 V) Electrode Rotational Rate (500 rpm) Diameter (200 mm) Machine depth (450 μm) Pulse on Time (2 ms) Pulse off Time (0.1 ms- 2 ms)	Electrode-tool axial wear	The machining accuracy and time can be further enhanced through the rectangular pulse voltage and flat sidewall–flat front tool. From the research findings, it can conclude that the optimal parametric combinations for machining time and accuracy are sidewall–flat front tool and rectangular pulse voltage	39

(continued)

Table 1 (continued)

Workpieces	Electrolytes	Input Parameters with range	Output Parameters	Major Findings	Ref.
Silicon wafer	NaOH	Voltage (40–50 V) tool feed rate (100–250 μm/min) Inter-Electrode Gap (20 mm) Tool Electrode (ϕ 300 μm)	Overcut, taper and surface quality	The tool used for machining of silicon wafer exhibits axial, lateral and flank erosion. However, the axial erosion is higher as compared to lateral and flank erosion owing to high spark intensity at tool electrode tip	41
C17200 beryllium copper alloy	Deionized water	Peak Current (0.24A-0.97A) Pulse width (5 μs-30 μs) Gap voltage (250 V) Electrode (Tungsten)	MRR	Comprehensive analysis of the effects of peak current and pulse width on the machine time, electrode wear, dimensional accuracy, and the taper angle, a value of 0.34A peak current, and 10 μs pulse width can be considered	37
SiC reinforced polymer matrix composite	NaOH + HCL	Voltage (45–75 V) Electrolyte concentration (90–110 g/l) Inner-electrode gap (100–140 mm) Duty factor (0.5–0.75)	MRR	Among the treatment parameters, electrode spacing and voltage were found to be the most important parameters affecting MRR, overcut and taper during ECDM	11

(continued)

Table 1 (continued)

Workpieces	Electrolytes	Input Parameters with range	Output Parameters	Major Findings	Ref.
Al-6063 SiC_p	NaCl	Applied Voltage (62–70) Pulse on time (2.5–3.5) Tool rotation rate (700–1100) Electrolyte concentration (14–22) Current (3–5)	Overcut and penetration depth	Enabling tool rotation in ECDD successfully removes sludge/ debris from the treatment area and replenishes the electrolyte. This study of micro-hole machining in MMC using the RM-ECDD opens up new possibilities for the manufacturing industry	14
Stainless steel	Pure Water	Tool speed (24,000-42000 rpm), Tool diameter (0.25–0.40 mm) and Voltage (8–14 V)	Axial wear	An empirical mathematical model was created based on the experimental results, and the machining voltage has a large influence on the wear of the tool electrode, followed by the rotation speed of the electrode and the diameter of the tool electrode	38
Nickel-based superalloys	Deionized water and $NaNO_3$	Working Fluid Pressure (4 MPa) Electrode rotation (100 r/min) Pulse duration (12 μs) Pulse Interval (36 μs) Pulse voltage (80 V) Peak current (12A)	MRR and Tool Electrode Wear Rate (EWR)	The taper angle and average hole diameter increase with the increasing conductivity of the solution during the ECDD process	32

(continued)

Table 1 (continued)

Workpieces	Electrolytes	Input Parameters with range	Output Parameters	Major Findings	Ref.
Nickel-based superalloy	$NaNO_3$	Pulse Voltage (80 V) Pulse Duration (12 μs) Pulse Interval (20 μs) Peak current (8 A) Flushing Pressure (8 MPa) Tool Electrode Diameter (Φ 300 μm)	Tool Wear	Machining phenomena and corrugations were observed and, wrought products and recast layers were analysed. Finally, 0.5 mm deep holes are produced with high efficiency, low tool wear and better surface finish	42

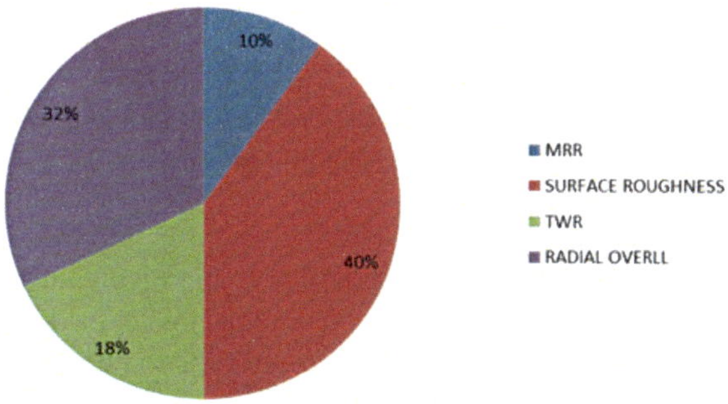

Fig. 4 Pie-chart of percentages of different output parameters of drilling-ECDM

suitable for the carbon fibre reinforced polymer (CFRP) followed by Glass Fibre Reinforced Polymer (GFRP) which is a composite material consisting of a fibre-reinforced polymer matrix. The fibres are usually glass, carbon, aramid or basalt. Metal matrix composites are composites with fibres or fibres dispersed in a metal matrix, such as copper, aluminium or steel. Ceramic matrix composites are a subset of composites and a subset of ceramics. They consist of ceramic fibres embedded in a ceramic matrix.

The line diagrams in the above series show about the works by various authors which are an effort to look into the research trends in the drilling-ECDM process. In the future, the examination might be expanded to include multiple combinations

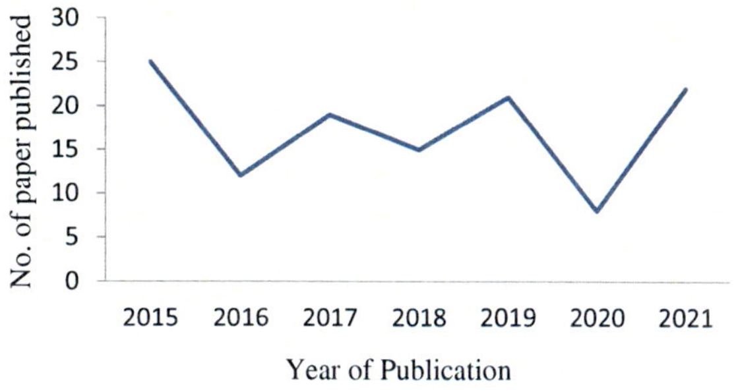

Fig. 5 Number of publications year wise

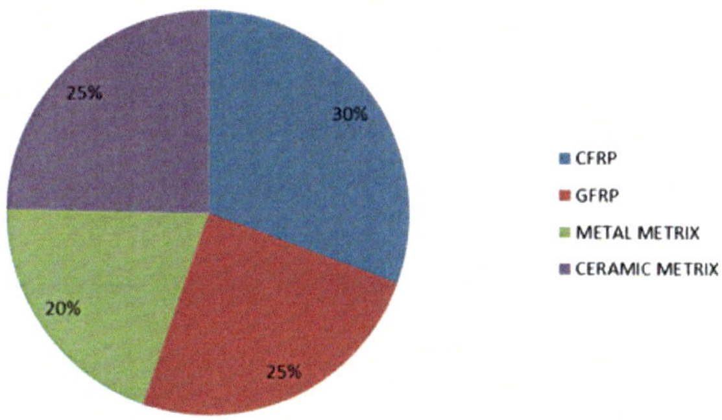

Fig. 6 Percentage of publication on composites

of device materials by converting various electrolytes with various combinations of optimization tools. The artwork may also be prolonged by using various experimental setups, and the following research gaps are addressed based entirely on a key literature review. The line diagram of the drilling-ECDM process mentioned above describe that it is a very promising method for machining electrically non-conductive materials, some areas, such as fueling movie balancing, powerful MRR with greater floor end, preheating of electrolytes, and the combined awareness of distinct electrolytes, new workpiece substances, triplex hybridization, etc. The drilling-ECDM process is widely used for making of hole on electrically non-conductive materials such as ceramics, composites, cements, and glasses that have many applications beyond the fields of automobile, aerospace, and defense sectors etc.

6 Conclusions

The present review article included more than 60 research papers and the literature review was carried out on drilling parameters of electrically non-conductive workpieces with the drilling-ECDM process. The following points have been summarised from this study:

(1) The research divulges that 40 and 32% of effort has been done for R_a and overcut, respectively, but less effort has been made like 10% and 18% for studying the MRR and TWR, respectively.
(2) Drilling-ECDM is the most efficient, appropriate and practical drilling process employed in industries for drilling of electrically non-conductive composite materials. Available literature depicts that researchers have attempted the efforts after 2015 in this area.
(3) It was concluded from meta-analysis that maximum of the work was carried out on electrically non-conductive materials such as 30% of CFRP, 25% of GFRP and 25% ceramic matrix, and least work was done on 20% of the metal matrix.
(4) The researchers have applied less effort to address the issues such as reuse of electrolyte and recovery of debris particles after machining. However, more work is required to be done in these fields.
(5) A valuable electrode arrangement for drilling a cavity in terracotta has been justified by the employment of a disruptive tool with rotating motion because it produces an additional electrical discharge that acts as a cutting action to speed up the machining process. These techniques make the process of removing material stable, which has led to a global improvement in the quality produced.

References

1. Singh T, Dvivedi A (2020) On prolongation of discharge regime during ECDM by titrated flow of electrolyte. Int J Adv Manuf Technol 107:1819–1834
2. Singh T, Dvivedi A (2018) A pressurized feeding approach for effective control on working gap in ECDM. Mater Manuf Process 33(4):462–473
3. Matharou GS, Bhuyan BK (2020) Parametric optimization of EDM processes for aluminum hybrid metal matrix composite using GRA-PCA approach. Int J Mech Product Eng Res Develop 10(3):367–378
4. Arya RK, Dvivedi A (2019) Investigations on quantification and replenishment of Vaporized electrolyte during deep micro-holes drilling using pressurized flow-ECDM process. J Mater Process Technol 266:217–229
5. Matharou GS, Bhuyan BK (2021) Modelling and combined effect analysis of electric discharge machining process using response surface methodology. Mater Today Proc 46(15):6638–6643
6. Jain VK (2005) Advanced machining processes. Allied Publishers Private Limited, New Delhi, pp 215–222
7. Bhuyan BK, Yadava V (2012) Experimental investigations of traveling wire electro-chemical spark machining (TW-ECSM) of borosilicate glass. Asian Rev Mech Eng 1(2):24–29
8. Huang SF, Liu Y, Hu HX, Sun LY (2014) Electrochemical discharge machining micro-hole in stainless steel with tool electrode high-speed rotating. Mater Manuf Process 29(5):634–637
9. Saini G, Manna A, Seth AS (2018) An experimental investigation during electro chemical discharge machining of e-glass fibre reinforced polymer composite. Int J Appl Eng Res 13(11):8899–8906
10. Dhanvijay MR, Kulkarni VA, Doke A (2019) Experimental investigation and analysis of electrochemical discharge on fibreglass reinforced plastic. J Inst Eng (India): C 100:763–769
11. Antil P, Singh S, Manna A (2018) Electrochemical discharge drilling of sic reinforced polymer matrix composite using Taguchi's grey relational analysis. Arab J Sci Eng 43:1257–1266
12. Chak SK, Rao PV (2008) The drilling of Al_2O_3 using a pulsed DC supply with a rotary abrasive electrode by the electrochemical discharge process. Int J Adv Manuf Technol 39:633–641
13. Laio YS, Wu LC, Peng WY (2013) A study to improve drilling quality of electrochemical discharge machining (ECDM) process. Proceed CIRP 6:609–614
14. Singh T, Arya RK, Dvivedi A (2020) Experimental investigations into rotary mode electrochemical discharge drilling (RM-ECDD) of metal matrix composites. Mach Sci Technol 24(2):195–226
15. Bhuyan BK, Tarun J, Saxena R (2021) Study of the effect of process variables on output response in electro chemical discharge drilling of borosilicate glass. Mater Today Proc 43(1):335–340
16. Singh T, Arab J, Dixit P (2022) A review on microholes formation in glass-based substrates by electrochemical discharge drilling for MEMS applications. Mach Sci Technol 26(2):276–337
17. Elhami S, Razfar MR (2020) Application of nano electrolyte in the electrochemical discharge machining process. Precis Eng 64:34–44
18. Kurafuji H, Suda K (1968) Electrical discharge drilling of glass. CIRP Ann 16:415–419
19. Razfar MR, Behroozfar A, Ni J (2014) Study of the effects of tool longitudinal oscillation on the machining speed of electrochemical discharge drilling of glass. Precis Eng 38(4):885–892
20. Arab J, Dixit P (2020) Influence of tool electrode feed rate in the electrochemical discharge drilling of a glass substrate. Mater Manuf Process 35(15):1749–1760
21. Kang X, Tang W (2018) Micro-drilling in ceramic-coated Ni-superalloy by electrochemical discharge machining. J Mater Process Technol 255:656–664
22. Gupta PK, Dvivedi A, Kumar P (2016) Effect of pulse duration on quality characteristics of blind hole drilled in glass by ECDM. Mater Manuf Process 31(13):1740–1748
23. Zhang Y, Xu Z, Zhu D, Xing J (2015) Tube electrode high-speed electrochemical discharge drilling using low-conductivity salt solution. Int J Mach Tools Manuf 92:10–18
24. Geng T, Xu Z, Lu J, Wang W, Zhang C (2022) A novel backflushing method for improving machining quality in electrochemical discharge drilling. J Manuf Process 82:516–529

25. Zhang C, Xu Z, Lu J, Geng T (2021) An electrochemical discharge drilling method utilising a compound flow field of different fluids. J Mater Process Technol 298:1–17
26. Arab J, Mishra DK, Dixit P (2021) Measurement and analysis of the geometric characteristics of microholes and tool wear for varying tool-workpiece gaps in electrochemical discharge drilling. Measure 168:1–8
27. Coteata M, Schulze HP, Slatineanu L (2011) Drilling of difficult-to-cut steel by electrochemical discharge machining. Mater Manuf Process 26(12):1466–1472
28. Pawar P, Kumar A, Ballav R (2019) Parametric analysis of electrochemical discharge drilling on soda-lime glass material using Taguchi L_{27} orthogonal array method. Stroj. Časop.-J. Mech. Eng. 69(4):115–132
29. Vemula VV, Chak SK (2020) Effect of loose abrasive mixed electrolyte on the aspect ratio of blind holes drilled in soda-lime glass by ECDM process. Manuf Rev 7:1–10
30. Yi X, Chen J, Jiang B, Ni J (2018) Investigation of micro-drilling using electrochemical discharge machining with counter resistant feeding. J Mater Process Technol 257:141–147
31. Kim DJ, Ahn Y, Lee SH, Kim YK (2006) Voltage pulse frequency and duty ratio effects in an electrochemical discharge microdrilling process of Pyrex glass. Int J Mach Tools Manuf 46(10):1064–1067
32. Zhang C, Xu Z, Hang Y, Xing J (2019) Effect of solution conductivity on tool electrode wear in electrochemical discharge drilling of nickel-based alloy. Int J Adv Manuf Technol 103(1–4):743–756
33. Coteata M, Pop N, Schulze HP, Slatineanu L, Dodun O (2016) Investigation on hybrid electrochemical discharge drilling process using passivating electrolyte. Proc CIRP. 42:778–782
34. Chavoshi S, Behagh AM (2014) A note on influential control parameters for drilling of hard-to-machine steel by electrochemical discharge machining. Int J Adv Manuf Technol 71:1883–1887
35. Chak SK, Rao PV (2007) Trepanning of Al_2O_3 by electro-chemical discharge machining (ECDM) process using abrasive electrode with pulsed DC supply. Int J Mach Tools Manuf 47(14):2061–2070
36. Pawar P, Kumar A, Ballav R (2020) Experimental study on the MRR, machined depth and hole diameter for soda-lime glass by electrochemical discharge machining process with copper tool. Int J Mod Manuf Technol XII 1:131–143
37. Dong S, Wang Z, Wang Y (2017) High-speed electrochemical discharge drilling (HSECDD) for micro-holes on C17200 beryllium copper alloy in deionized water. Int J Adv Manuf Technol 88:827–835
38. Huang SF, Liu Y, Li J, Hu HX, Sun LY (2014) Electrochemical discharge machining micro-hole in stainless steel with tool electrode high-speed rotating. Mater Manuf Process 29(5):634–637
39. Zheng ZP, Su HC, Huang FY, Yan BH (2007) The tool geometrical shape and pulse-off time of pulse voltage effects in a Pyrex glass electrochemical discharge microdrilling process. J Micromech Microeng 17:265–272
40. Singh M, Antil P, Singh S, Katal N, Bakshi D K, Alkesh (2022) RA-ECDM of silicon wafers using Taguchi's methodology and machine learning algorithms. Silicon 14:1–16. https://doi.org/10.1007/s12633-022-02128-1
41. Singh M, Singh S, Kumar S (2020) Experimental investigation for generation of micro-holes on silicon wafer using electrochemical discharge machining process. SILICON 12:1683–1689
42. Xu ZY, Zhang Y, Ding F, Wang F (2018) An electrochemical discharge drilling Method of small deep holes. Int J Adv Manuf Technol 95:3037–3044
43. Wuthrich R, Fascio V (2005) Machining of non-conducting materials using electrochemical discharge phenomenon: an overview. Int J Mach Tools Manuf 45(9):1095–1108
44. Bravo JL, Sanchez PC, Rodriguez AH, Alba GGH (2020) Electrochemical discharge machining modeling through different soft computing approaches. Int J Adv Manuf Technol 106:3587–3596
45. Singh T, Dvivedi A (2016) Developments in electrochemical discharge machining: a review on electrochemical discharge machining, process variants and their hybrid methods. Int J Mach Tools Manuf 105:1–13

46. Goud M, Sharma AK, Jawalkar C (2016) A review on material removal mechanism in electrochemical discharge machining (ECDM) and possibilities to enhance the material removal rate. Precis Eng 45:1–17
47. Yadav RN (2019) Electro-chemical spark machining based hybrid machining processes: research trends and opportunities. Proc Inst Mech Eng Pt B J Eng Manufact 233(4):1037–1061
48. Basak I, Ghosh A (1997) Mechanism of material removal in electrochemical discharge machining: a theoretical model and experimental verification. J Mater Process Technol 71(3):350–359
49. Bhattacharyya B, Doloi BN, Sorkhel SK (1999) Experimental investigations into electrochemical discharge machining (ECDM) of non-conductive ceramic material. J Mater Process Technol 95(1–3):145–154
50. Kulkarni AV (2007) Electrochemical discharge machining process. Def Sci J 57(5):765–770
51. Rajput V, Goud M, Suri NM (2019) Study on effective process parameters toward the better comprehension of ECDM process. Int J Mod Manuf Technol XI 2:105–118
52. Jawalkar CS, Sharma AK, Kumar P (2012) Micromachining with ECDM: research potentials and experimental investigations. Int J Mech Mechatron Eng 6(1):340–345
53. Verma M, Manna A (2017) Experimental investigations on developed electrochemical spark machining set up during machining of alumina ceramic component. Int J Creat Res Thoughts 5(4):1004–1012
54. Bhuyan BK, Yadava V (2014) Experimental study of traveling wire electrochemical spark machining of borosilicate glass. Mater Manuf Process 29(3):298–303
55. Mediliyegedara TKKR, De Silva AKM, Harrison DK, McGeough JA (2005) New developments in the process control of the hybrid electrochemical discharge machining (ECDM) process. J Mater Process Technol 167(2–3):338–343
56. Sarkar BR, Doloi B, Bhattacharyya B (2008) Experimental investigation into electrochemical discharge microdrilling on advanced ceramics. Int J Manuf Technol Manag 13(2–4):214–225
57. Mallick B, Sarkar BR, Doloi B, Bhattacharyya B (2018) Analysis on the effect of ECDM Process parameters during micro-machining of glass using genetic algorithm. J Mech Eng Sci 12(3):3942–3960
58. Cheng CP, Wu KL, Mai CC, Yang CK, Hsu YS, Yan BH (2010) Study of gas film quality in electrochemical discharge machining. Int J Mach Tools Manuf 50(8):689–697
59. Liu JW, Yue TM, Guo ZN (2010) An analysis of the discharge mechanism in electrochemical discharge machining of particulate reinforced metal matrix composites. Int J Mach Tools Manuf 50(1):86–96
60. Charak A, Jawalkar CS (2022) Experimental investigation and analysis on borosilicate glass using electrochemical discharge machining process. SILICON 14:1823–1829
61. Yusen H, Zhengyang X, Feng W, Chenxiang Z (2020) Electrochemical discharge drilling of inclined micro holes with step feeding method. Int J Electrochem Sci 15:2713–2726
62. Harugade M, Waigaonkar S, Mane N, Hargude N (2019) Experimental investigation of high speed tool rotation on heat affected zone and overcut in ECDM. Mater Today: Proc 18(3):1472–1478
63. Jain VK, Dixit PM, Pandey PM (1999) On the analysis of the electro-chemical spark machining process. Int J Mach Tools Manuf 39(1):165–186
64. Kulkarni A, Sharan R, Lal GK (2002) An experimental study of discharge mechanism in electrochemical discharge machining. Int J Mach Tools Manuf 42(10):1121–1127
65. Bhuyan BK, Gupta L, Garg C (2020) Design and development of tabletop electrochemical grinding setup. Mater Today: Proc 21(3):1479–1482
66. Wuthrich R, Bleuler H (2004) A model for electrode effects using percolation theory. Electrochim Acta 49(9–10):1547–1554
67. Arab J, Dixit P (2022) Formation of macro-sized through-holes in glass using notch-shaped tubular electrodes in electrochemical discharge machining. J Manuf Process 78:92–106

Enhancement of Stiffness of Light Duty Truck Cabin Against Frontal Impact Crush Load as Per Ais 029 Using Advanced Finite Element Approach

R. S. Chhatrawat and Dharmendra Kumar Dubey

Abstract The research is based on the cabin crash of commercial vehicles and the safety of the occupant. Several norms were implemented by the government for the safety of the occupant in commercial vehicle, these norms focus on structural strength of the cabin and its energy absorption capacity to ensure minimum damage to the occupant. The light-duty trucks also known as economic wheels of India are widely used for transportation purposes through roads. These trucks are very risky during the frontal collisions when running at high speed. The energy produced during the collision is high due to the burden or load carried by the vehicle. These collisions always raise question regarding the safety of the driver under frontal collisions. For the improvement of the safety prospective, we can investigate for more superior and advance designs. Using advanced simulation tools removal of material from the lower stress region and adding material at a higher stressed region will enhance the safety parameter and reduce the overall weight also. The risk of injury can be reduced in several frontal collisions with the help of crash technologies in the frontal structure of the vehicle or coach. A successful simulation was achieved for the AIS029 front impact test. In the CAE simulation, proper model details and material properties were used. CAE result was evaluated according to the regulations. All the options for modification were analyzed and some modifications in the design of the cabin were made to increase the safety of the occupant. The optimized design meets the regulation comfortably. That means, in the real-life scenario of an accident the physical occupant will be safe under the front impacts.

Keywords FEA · Optistruct · AIS 029 · Front impact test · Occupant Safety

---

R. S. Chhatrawat (✉) · D. K. Dubey
Mechanical Engineering Department, Bhagwant University, Ajmer, India
e-mail: rscbk1@gmail.com

H. Kumar et al. (eds.), *Recent Advances in Intelligent Manufacturing*,
Lecture Notes in Mechanical Engineering,
https://doi.org/10.1007/978-981-99-1308-4_18

1 Introduction

A vehicle which runs with the help of motor is generally used for the purpose of transportation of goods and passengers. Any motorised vehicle of any type, designed for roads and capable of transportation for payment or not is termed as "commercial motor vehicle" by the European Union [1–3].

In the development of India, the commercial vehicle plays an important role of transporting goods or passenger. It is recorded that while transporting these goods, many frontal impact and collision occurs that leads to the death of the occupant and driver. To overcome this death casualty, safety regulations are required [4–6].

The safety of the occupant was ensured by checking the survival space in the cabin for the occupant in different collision scenarios under AIS029 regulations. Some researchers categorized the types of truck collision and based their observations based on the occupant survival space [7]. Some researchers studied that the rollover of the vehicle is equally responsible as the frontal impact for the injuries to the occupant [8, 9].

During collision, the impact energy was usually absorbed by the structure of the cabin. To improve the strength and support to the cabin, cabin mounts are used [10]. The mounts are very useful for the absorption of energy from the cabin to chassis. The design considered for the cabin was taken from a large number of CAE iterations results [11–13]. The vehicle cassis and suspension systems are deliberated and considered for the detailed modelling of the front cabin. To check the real time failure of material, the properties of the materials were considered in simulation and the effects on the pattern deformed was analysed [14].

For the physical test resemblance, all the boundary conditions for the cabin were analysed on the model in CAE. The results were recorded with the help of a post-processing software in terms of space for survival and intrusion and compared with a physical test [15].

For the safety of the occupant, 3 cases safety standard are established on the basis of survival under a crucial safety standard AIS-029 as shown in Fig. 1 [16].

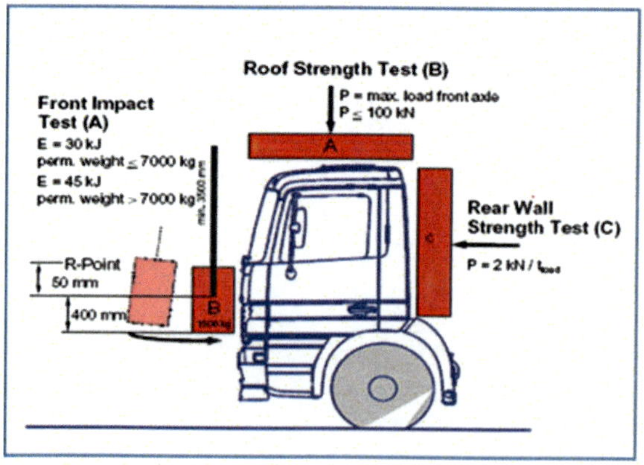

Fig. 1 AIS 029

A. Front impact test [TEST A].
B. Roof strength test [TEST B].
C. Rear wall strength test [TEST C].

A. **Front Impact Test**

The Front-impact test also known as the Swing Bob impact test is a method to analyse and calculate the cabin's strength during frontal impact crash. To satisfy the needs of requirement, a pendulum in motion is subjected towards the cabin with energy of 30 or 45 kJ for the frontal impact crash. The vehicle's GVW is responsible for the impact of energy. For the successful result in this testing, the cabin should not get detached from the chassis and also the manikin gets enough space for survival in the cabin.

B. **Rear Wall Strength Test**

The strength or the rear wall can be analysed with the help of the rear wall test for the cabin. The rear wall must be capable of bearing a static load of 2kN per tonnes of permissible useful load. To apply load on the rear wall, a rigid barrier is used. The rigid barrier should be positioned perpendicular to the vehicle's longitudinal medium axis. Also, the rear wall area of the cabin must be covered.

C. **Roof Strength Test**

The strength of the cab roof was analysed by roof strength testing the study the impact bearing of roof under rollover during crash. The roof of the vehicle cab must capable to bear the static load of about 10 tonnes equal to the maximum mass that can be applied to the front axles of vehicle. To get satisfied results from the testing, the manikin placed in the cabin should have enough space for survival when the load is applied on the roof of the cabin. The violation of this test is due to the failure of roof under impact load conditions and buckling of columns.

2 Methodology

The solid works CAD modelling software was used initially to design the cabin's CAD model. After the initial modelling, a FE model was prepared with the help of an initial model through HYPERMESH. After several analyses and study, the main power train components of the vehicle including axles and engine were designed with rigid materials with the help of 3D solid elements. Shell elements are used to design the components of the sheet metals. Several other components of the vehicles such as suspension, steering system and dashboard were designed and properly connected with the previously designed model after detailed study about mounting and interaction of cabin [17, 18]. For the simulation of the front impact test, Optistruct was used. All the stages of simulation were shown in Fig. 2 [19] (Fig. 3).

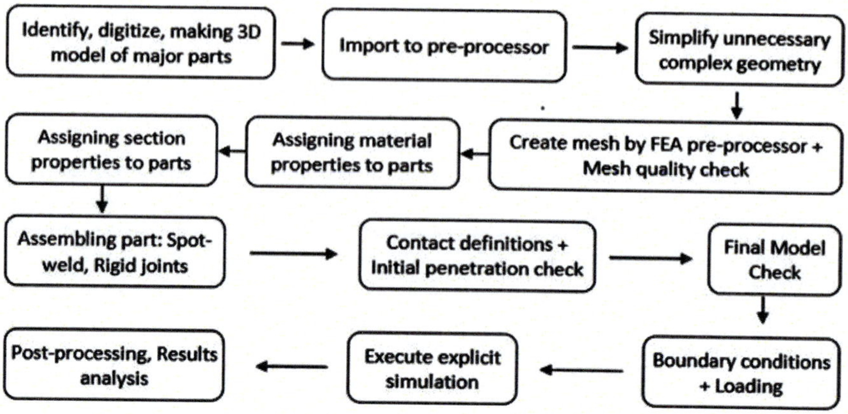

Fig. 2 Stages of simulation

Fig. 3 CAB model

3 Technical and Safety Requirements of the Truck Cabin

1. Truck Cab Structural Strength

The truck cab must fulfill the requirements of survival space as per AIS-029 as amended from time to time. If there should arise an occurrence of non-OE assembled cabs, local weight lifters should utilize type endorsed cab designs or potentially the parts only.

2. **Cab Mounting Arrangement and Strength Requirements**

The mounting arrangement for the cab shall ensure adequate rigidity of the cab chassis. For mountings, use must be made of circular openings and reasonable jolts with the least possible play, in any event at the premier and rearmost connection points. The mountings shall ensure adequate resistance against lateral/ transverse movement of the cab in case of braking or sudden acceleration or while driving on slopes. The cab mountings shall be of shock and vibration dampening type. The joint between the cab and suspension frame will be furnished with elastic pads or any appropriate arrangement.

3. **Protection of the Occupants in the Event of a Lateral Collision**

The protection of the occupants in the event of any lateral collision shall be done by the following parts:

Doors: the doors shall be hinged to the A-pillar and the locking mechanism shall be provided by authorized agencies. The structure of the door must have the option to withstand mutilation because of force incited during its operation and the operability of door parts basic to the expulsion of the driver and co-driver after an accident.

Vehicle seats and strength requirements

Cab construction material: wood must not be permitted to be used in the cab construction till no other suitable material as a substitute is available. A composite material that meets the functional requirement may be used

Exterior and interior panels: the exterior and interior panels of the cab shall be protected against corrosion. If a suitable material is unavailable then the surfaces must be coated with a galvanizing paint.

Roof luggage carrier: no luggage shall be permitted to be carried on the roof of the cab. If any provision is to be made, the mounting of the same shall not be provided on the cab.

4. **Cab Mountings for Tilting Type Cabs**

The mounting arrangement for the cab shall provide adequate rigidity to the cab with the chassis. For mountings, use must be made of circular openings and reasonable jolts with minimal play, at any rate, the principal and rearmost connection points. The mountings shall ensure adequate resistance against lateral/ transverse movement of the cab in case of breaking or sudden acceleration or while driving on slopes. The cab mountings shall be of shock and vibration damping types/ the attachment between the cab and the chassis should be provided with rubber cushions or any suitable arrangement. Self-locking fasteners may be used to prevent loosening due to vibrations. The fasteners used should be of suitable strength and must ensure the safety of the cab against vibrations because of road irregularities.

A. **Material Details**

The bolt components are attributed for the Material properties. The Elastic material (MALT1) is used for designing of the components which shows higher deformation with reference of their strain versus stress rate.

B. **The Boundary Conditions**

A rigid rectangle shaped pendulum of 800 mm height, 2500 mm width and 1500 ± 250 kg weight, is used to generate the 45 kJ of impact energy. To meet the requirement of the test, the centre of gravity of the pendulum must come 50 ± 5/0 mm below the seating reference point when in rest position.

C. **Acceptance Criterion**

- At the end of simulation, intrusion must not more than 50% on manikin by any deformed component of the cabin.
- Door latches should not exceed the target limit.
- chassis must be prevented from complete damage or breakdown and the Cabin must be attached with the chassis (Fig. 4).

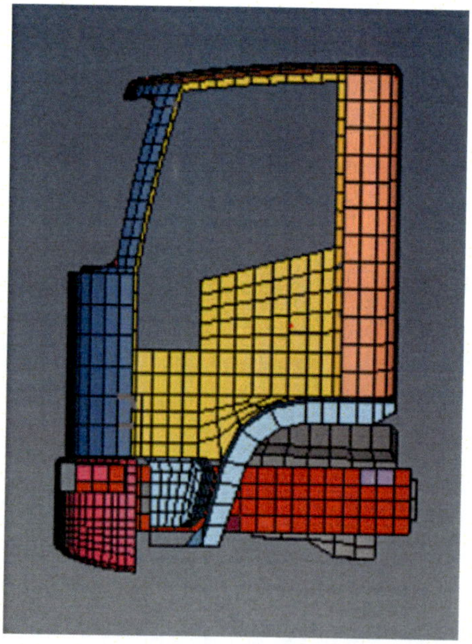

Fig. 4 Complete CAE model

4 Results

The post processor hyper view is used for the calculation of total structural damage, manikin clearance and survival space with the help of front impact test's simulation as shown in Fig. 5.

The Pendulum is developed in Hypermesh as per the government norms. It is developed to behave as per its actual working. The Pendulum is located at a distance as per the said rules and is hinged at the top as shown in Fig. 5.

A. **Design Modification**

- **Iteration 1—Dummy placed at position as per the norms and simulated**

The dummy inside the model needs to be least the harm and should not get effected by any intrusion to make simulation and result as per our desire (Fig. 6). But, in the given iteration it can be seen that the steering wheel gets intruded inside the abdomen of the dummy. This kind of intrusion can cause heavy loss to the upper body. It can break multiple ribs and may lead to internal bleeding. Figures 7 and 8 shows that the intrusion is 79 mm inside the abdomen part which fails the current model.

The impacted distance of 79 mm also determines that the truck body is not capable of absorb the force and energy by the impact in the after manner by which the occupant doesn't get hurt by any impact. The design can be changed to make model more optimized and save the occupant from the impact. Changing the geometry, dimension or adding different features can be a helpful option to make more optimized model.

- **Iteration 2—By adding cross member behind radiator**

During iteration 1, the impact is highly recognized at the radiator component. Where if, by adding any support or change in design can help to prevent the iteration to be a failure. By adding a cross member behind the radiator can prevent its rotation and direct change as a result of intrusion with the steering wheel.

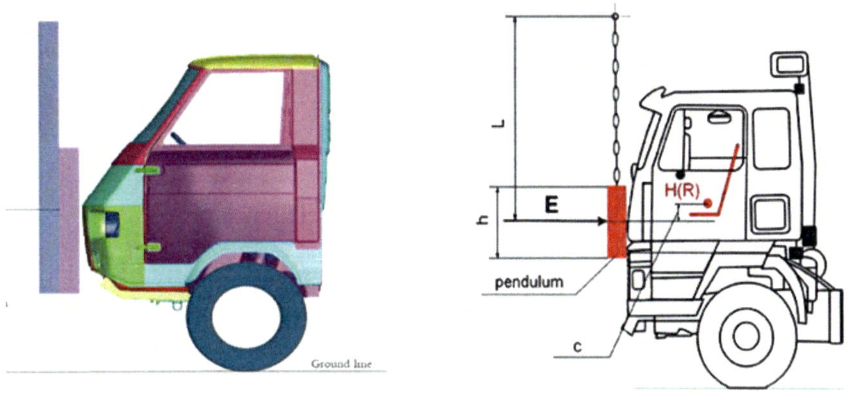

Fig. 5 Norm virtual setup side

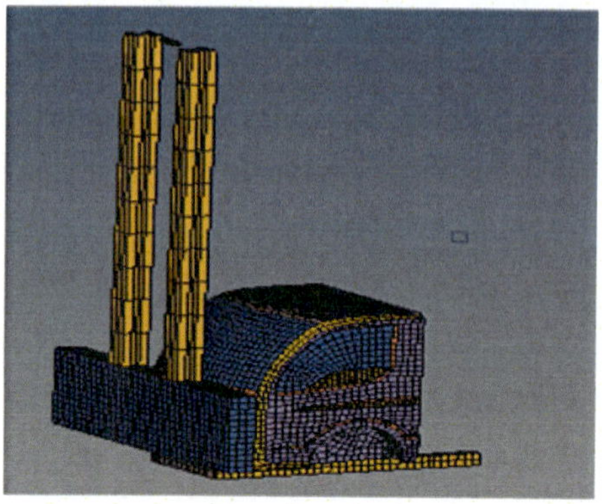

Fig. 6 Setup of meshed model to the pendulum

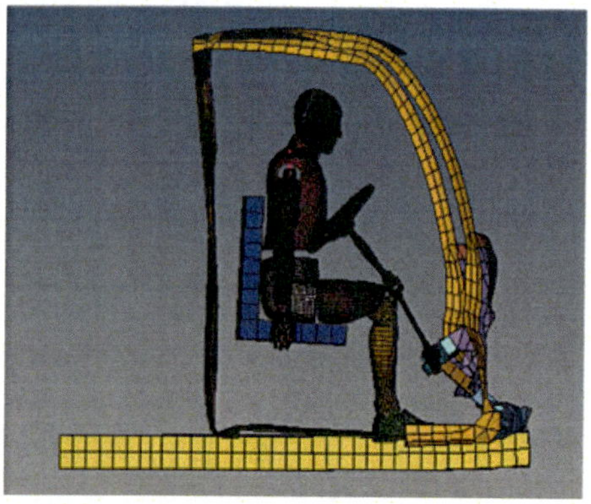

Fig. 7 Intruded dummy in intrusion

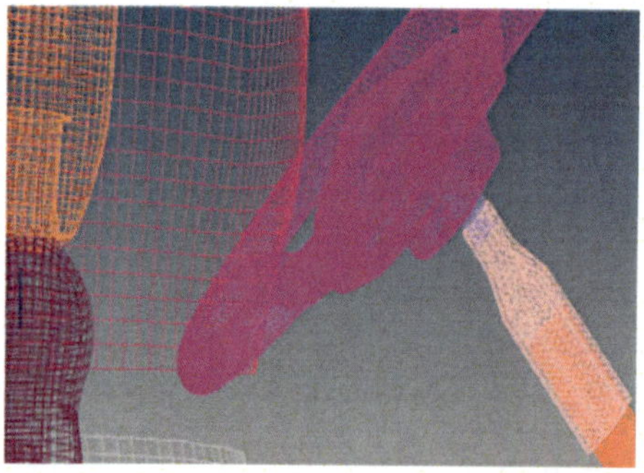

Fig. 8 79 mm of intrusion detection into the dummy, which can be a fatal damage

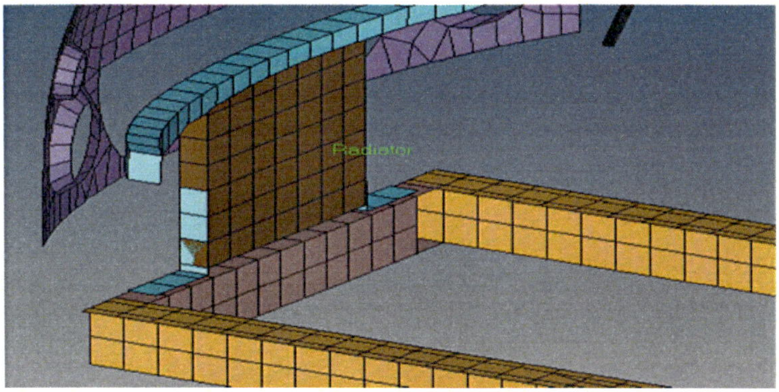

Fig. 9 Observing the point, where changes can be done

This is all about, when the cross members are not added to the model. It can be said that the model needs support to prevent rotation from this portion to prevent intrusion and it can be one of the non-failure case (Fig. 9).

The cross member can be seen on the model for the second iteration. The member here will try to support directly to decrease or minimize the intrusion. Figure 10 shows the L-shaped Cross members added to the radiator and results after iteration for 34 mm of Intrusion after adding the cross member have been shown in Fig. 11.

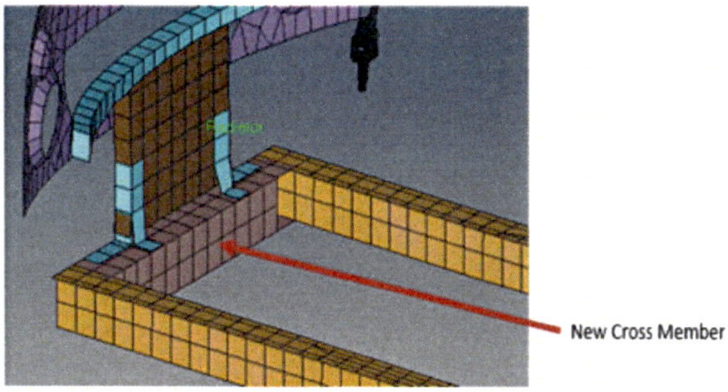

Fig. 10 L-shaped cross members added to the radiator

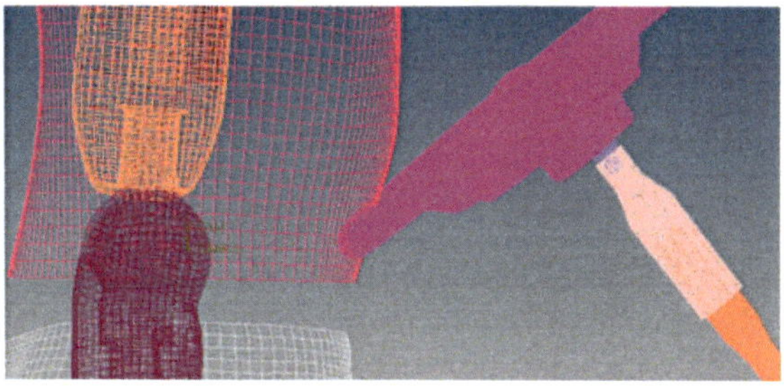

Fig. 11 34 mm of Intrusion after adding cross member

- **Iteration 3—by Adding Gussets Behind the Radiator**

Thus, in last iteration the L cross members are not successful. The L cross members need more attention by which the geometry needs to change or add more members. The change take place in this iteration by adding gussets behind the radiator that can be seen in Fig. 12. The position is recognized by the previous to add gusset, so that the model can be better in terms of decrement in intrusion. The gussets proved in the previous research are more effective than any other solution for better support. Table 1 briefly delineates the effect of design modifications on the LCV (Fig. 13).

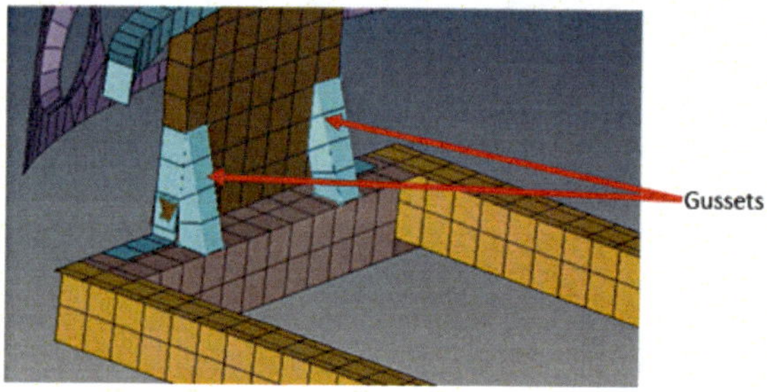

Fig. 12 Gussets added at radiator

Table 1 Changes in design to perform different Iteration

Design changes	Current model	Addition of cross member	Addition of gussets
Intrusion of Steering wheel into dummy model	79 mm	34 mm	1 mm

Fig. 13 Result after the final iteration with no intrusion

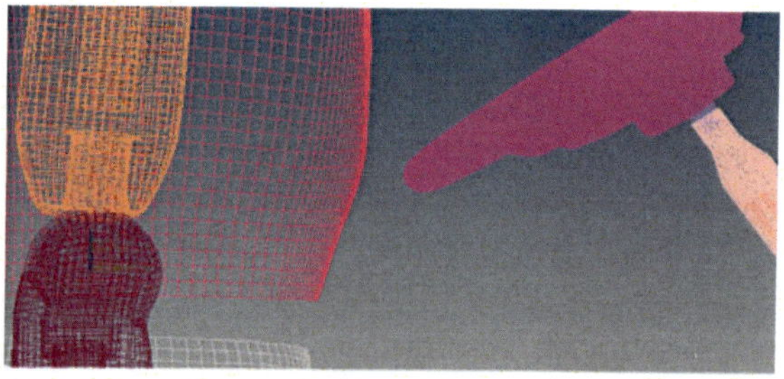

5 Conclusion

The AIS 029 simulation was successfully evaluated with the help of FEA software for the frontal impact test. Several conclusions were carried out after the evaluation of the detailed model with properties of the material. The evaluation of the results was carried out according to the regulations.

1. The CEA simulation concluded that the safety of the driver can be increased at the time of crash and testing of the standard equipment can be done easily.as shown by the CEA.
2. This method also concluded that the prototyping method can be eliminated which is economical for production.
3. It was observed that there are many options for modification and some modifications were made in the design to increase the safety of the driver.
4. Adding L shape cross members and gussets behind the radiator shows higher protection from crash, and gussets offer help to make the model pass the criteria.
5. The clearance between the occupant and steering is appropriate to pass the norms of the government and the model safer to the occupant too.
6. The optimized design meets the regulation comfortably in the real-life scenario of an accident where the physical occupant will be safe under the front impacts.

References

1. Burns LR (ed) (2014) India's healthcare industry: innovation in delivery, financing, and manufacturing. Cambridge University Press
2. Cheng L, Werner S, Khatua T, Ray R, Lau E (1996) Heavy truck crashworthiness-case studies of heavy truck accidents involving truck occupant fatality. In: Proceedings: international technical conference on the enhanced safety of vehicles, vol 1996. National Highway Traffic Safety Administration, pp 1880–1888
3. Dhole A, Shrivastava R, Raval C (2019) Study of energy absorbing front cabin mount, its stiffness balance with chassis and test correlation in frontal impact of commercial vehicle cabin (No. 2019-01-1034). SAE Technical Paper
4. Shirsath MSM, Student PG, Yeola SNDCOER, Londhe BC, Yeola SNDCOER (2018) Design & weight optimization of the front cab mounting bracket of truck. Resincap J Sci Eng 2(8):2456–9976
5. Krzikalla D, Mesicek J, Petru J, Sliva A, Smiraus J (2019) Analysis of torsional stiffness of the frame of a formula student vehicle. Appl Mech Eng 7:315
6. Gupta M, Goswami C, Khandelwal N (2021) Design and structural analysis of 3-sided and 4-sided columns of a CNC machine using finite element analysis. Mater Today: Proc
7. Hallquist JO (1998) LS-DYNA theoretical manual-nonlinear dynamic analysis of structures. Livermore, California: Livermore Software Technology Corporation, Livermore, California
8. Nieberle T, Kumar SR, Patnaik A, Goswami C (2021) Composite materials for armour application. Adv Eng Des 239–248
9. Standard AI (2000) Automotive vehicles-seats, their anchorages and head restraints for category M1-specifications
10. Mungale S, Leo S (2013) Evaluation of truck driver safety in various crash scenarios (No. 2013-26-0029). SAE Technical Paper

11. Mehara M, Goswami C, Kumar SR, Singh G, Wagdre MK (2021) Performance evaluation of advanced armor materials. Mater Today: Proc
12. Mirzaamiri R, Esfahanian M, Ziaeirad S (2012) Crash test simulation and structure improvement of IKCO 2624 truck according to ECE-R29 regulation
13. Zhang H, Zhang Y, Khattak A (2012). Analysis of large-scale incidents on urban freeways. Transp Res Record: J Transp Res Board 2278:74–84
14. Seiff H (1985) Heavy truck safety-what we know. SAE Technical Paper 851191. https://doi.org/10.4271/851191
15. Stan C, Iozsa D, Oprea R. Study concerning the optimization of the mounting system of the truck cab. ISBN: 978-960-474-383-4, pp 272–276
16. Choudhury P, Krishna CH (2014) Mathematical modelling and shape optimisation of front damper mount of Ashok Leyland 1612 truck using 3d finite element method. Int J Eng Sci (IJES) 3(4):70–76
17. Ambroz R, Kollhammer D, Palousek D (2007) Optimalization of cabin mounting. Conference of diploma thesis 2007 institute of machine design, institute of solid mechanics, mechatronics and biomechanics, faculty of mechanical engineering, Brno University of Technology June 5–6, 2007, Brno, Czech Republic
18. Murali Krishna D, Madhu S (2014) Design and analysis of tiltable truck cabin floor. Int J Res Stud Sci Eng Technol 2(7):69–73
19. Fui TH, Rahman RA (2007) Statics and dynamics structural analysis of a 4.5 ton truck chassis. Jurnal Mekanikal 24:56–67

A State-Of-Art: Review on Ultrasonic Welding Process (UW)

Sachinkumar Madhukar Wani, V. G. Arajpure, Rajkumar S. Sirsam, and Avinash D. Bagul

Abstract The Ultra-sonic Welding (UW) processes are evolving day to day for good strength of welded joints and optimum weld parameters for similar as well as dissimilar materials. Therefore, it is necessary to have a basic knowledge of the Ultra-sonic welding process and its parameters. The study presented here is the theoretical view in concept of Ultra-sonic welding and its applications. Several literatures were studied that correlated with UW and presented for developments in the UW process. In this paper, the applications of the UW process are discussed. Certain advantages and limitations are discussed with the UW process.

Keywords Ultra-sonic welding · Welded joints · CFRP

1 Introduction

In the 19th Century, it was demonstrated that the high frequency oscillations could break the similar of dissimilar materials apart from them. From that conception investigators were trying to produce the approach which join the components by an analogous high frequency. This results in the development of Ultra-sonic Welding operation which join two parts with help of high frequency application [1]. The Ultra-sonic welding is a welding technology in which Ultra-sonic frequency oscillations

S. M. Wani (✉)
Department of Mechanical Engineering, Government College of Engineering, Jalgaon, India
e-mail: gcoesmw@gmail.com

V. G. Arajpure
Department of Mechanical Engineering, Suryodaya College of Engineering & Technology, Nagpur, India

R. S. Sirsam
Department of Chemical Engineering, UICT, KBC N. M. University, Jalgaon, India

A. D. Bagul
Department of Mechanical Engineering, NES's Gangamai College of Engineering, Nagaon, India

H. Kumar et al. (eds.), *Recent Advances in Intelligent Manufacturing*,
Lecture Notes in Mechanical Engineering,
https://doi.org/10.1007/978-981-99-1308-4_19

were applied to objects that kept together by some holding pressure applied to produce the weld. This methodology is normally applied to plastics and specifically to two or further different materials. This methodology doesn't demand any ultra-expensive weld bolts and any list materials. Thus, this approach acquires the genuinely useful operations from medical equipment to shoes to motor vehicles [2].

1.1 Historical Development

Practical Operation of Ultra-sonic welding was executed for unplasticized plastics during the decade of 1960. Now unplasticized plastics may be weld in most of the Cases. Robert Soloff and Seymour Linsley received the patent in 1965 for the Ultra-sonic system for welding hard thermoplastics. At Branson Instruments, where Ultra-sonic tests were used to weld thin plastic flicks into bags and tubes by Soloff, the author of Sonics & Accoutrements Inc., served as the lab director. Accidentally moving the query to a plastic video dispenser's tip caused the dispenser's pieces to fuse together. He understood that the probe did not need to be physically moved around the component because hard plastics could be penetrated by the Ultra-sonic energy, which could also be used to weld an entire bond. He later invented the first Ultra-sonic press. The toy industry was where this new technology initially found its applications. In 1969, Ultra-sonic welding was used to create the first car made entirely of plastic. Although plastic buses were not popular, Ultra-sonic welding was. Since the 1980s, it has been adopted frequently by the automotive sector. Today, it is employed in a variety of procedures [5, 6].

2 Literature Review

Or et al. and Zhao et al. studies point out that Ultra-sonic welding has the advantage of being more rapid than traditional adhesives or bonding. The parts do not have to be left in a container for extended periods of time while waiting for the joint to dry or cure because the drying process is rapid. Automation of the welding process makes for clean and accurate joints. The weld location is also exceedingly clean and hardly ever needs to be touched up. A smaller number of accessories can be joined together due to the reduced thermal impact on the components supplied [1, 3].

Zhao et al. and Chen et al. study pointed over that the Ultra-sonic welding equipment is frequently easily adjusted to fit the exact specifications of the components being welded when combining complex injection-molded thermoplastic components. A 20 kHz low-amplitude auditory vibration is released from the components when they are trapped between a strong and instantly shaped nest (anvil) and a sonotrode (horn) attached to a transducer. The two surfaces interface is specifically created to concentrate the melting action when welding plastics [3, 10].

Musa et al. concluded as some of the materials frequently have an energy director that resembles a spike and connects the plastic component. A joint is formed when the purpose contact between the components is fused by Ultra-sonic energy. This process may be a sensible automated option for screwing or snapping together designs. Although it is often used on components as large as a small automotive instrument cluster, it is generally used with little components (such as mobile phones, consumer devices, temporary hospital equipment, toys, etc.). However, Ultra-sonic welding is often only utilized for small welds of thin, moldable metals such as aluminum, copper, and nickel. Due to the installation rankings required, Ultra-sonic would not be used to fuse the chassis of a car or the parts of a motorcycle together [4].

Al-Sarraf et al. explained as the concentration of vibrational energy along the joint-to-be-welded induces localized melting of the thermoplastic during Ultra-sonic welding. Due to the natural mobility of the materials and the high-pressure dispersion of surface oxides, welding can take place in matter. Although there is heating, it is not sufficient to melt the materials at the bottom. For both raw materials, hardened and soft plastics, such as semi-crystal plastics, Ultra-sonic welding is often utilized. With research and experimentation, understanding of Ultra-sonic welding has improved. The development of more advanced materials that are also affordable, along with rising demand for plastic and electronic components, have expanded the understanding of the basic process. Despite this, several elements of Ultra-sonic welding remain unexplored, such as the relationship between weld quality and reuse parameters. The field of Ultra-sonic welding is still growing rapidly [5].

Cai et al. focused on a paper that was funded from the German Research Foundation, where researchers from the Institute of Materials Science and Engineering at the University of Kaiserslautern have demonstrated that carbon-fiber reinforced polymer (CFRP) sheets and light essence can form largely durable bonds using ultra-sonic welding processes [6].

Bergmann et al., described the principles of Ultra-sonic welding especially for the metal joining process in the chapter. The authors have been investigated on the process and product parameters of Ultra-sonic welding in details. Due to the clean nature and safer processes of joining, it is mostly adopted in surgical as well as production industries [11].

Xie et al., investigated that selection of proper parameters of Ultra-sonic welding machine will assist to obtain the strong joint between two different kinds of alloys. Also, the brittle fractures origin in the intralayer of the Al_3Ti which will affect the tensile strength of the bond or joint are discussed [12].

de Leon et al., reviewed that Ultra-sonic welding is one of the most promising welding technologies now a days. For joining the components in electric vehicle batteries like the UW method are used. The study mostly focused on rare-earth barium copper oxide (REBCO)-based superconductors contained batteries which have constrained to join because of different stabilizers added but Ultra-sonic welding is most suitable for such kind of joining applications [13].

3 Parts of Ultra-Sonic Welding Process

The same fundamental components are found in all Ultra-sonic welding systems, and they are illustrated in Fig. 1. [courtesy by: https://blog.xfurth.com/blog/how-do-Ultra-sonic-welding-machines-generate-heat].

- A press to join different materials under load, typically using a pneumatic as well as an electric motor.
- A nest, anvil, or other specially created fixture where the materials are arranged and the high frequency oscillations can be directed to the interfaces
- A piezoelectric transducer, an elective promoter, and a Sonotrode make up an Ultra-sonic stack (US Horn). The stack's three essential components have been specially calibrated to reverberate at an exact Ultra-sonic frequency (normally 15, 20, 30, 35 or 40 kHz). Using the piezoelectric principle, the transducer converts the electrical pulse into a mechanical oscillation.
- The oscillation's amplitude is mechanically altered by the booster. In analogous systems, this is used to secure the stack in the press. Horn or Sonotrode grasp the form of the part, correspondingly modifies the amplitude mechanically and applies the mechanical vibration to the materials to be welded.
- A high-power electronic signal sent by an electronic Ultra-sonic generator (US Power force), whose frequency matches the resonance frequency of the stack.
- A regulator that manages the press's movement and the production of Ultra-sonic energy [7].

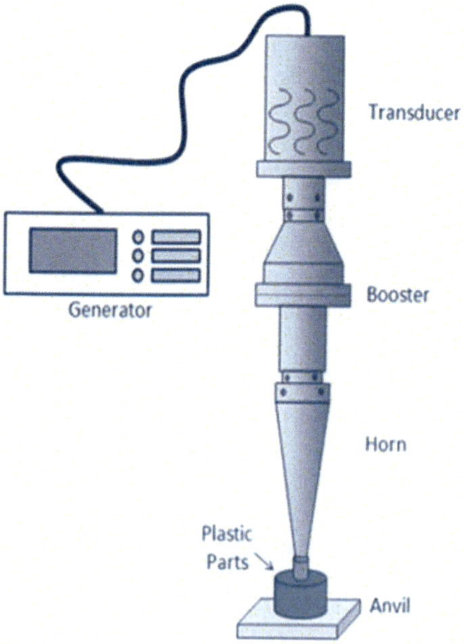

Fig. 1 General schematic view of ultra-sonic welding machine

4 Ultra-Sonic Welding Applications

(A) Electronics and Computers Sectors

To connect wired junctions and create connections in tiny, fragile circuits, Ultra-sonic welding is broadly applied in the electrical and computer industries. Ultra-sonic welding is extensively used to join the junctions of cable harnesses. Large bunches of cables called cable harnesses are used to distribute the electrical power and signals. Ultra-sonic welding can also be used to assemble electric motors, field coils, mills, and capacitors. Due to the high-volume requirements, it is also widely chosen in the assembly of storage media such as memory sticks and computer drives [7].

(B) Aerospace and Automobiles Sectors

Large plastic and electrical components like instrument circuits, door circuits, lighting, air pipes, steering components, upholstery, and machine components are frequently joined together in motor vehicles by Ultra-sonic welding. The joining of plastic components together and their closeness has suddenly become a major problem as plastics have continued to replace other components in the design and production of motor vehicles. Low capital costs, automation, negligible process times, and workability are just a few advantages of Ultra-sonic welding [8].

(C) Medical Equipment

Ultra-sonic welding is often used in health organizations since it doesn't introduce contaminants or declination into the weld and the devices may be tailored for use in hygienic conditions. The process is also largely automatable, offers exact control over dimensional tolerances, and doesn't interfere with a material's biocompatibility. It therefore raises part quality while lowering product charges. Ultra-sonic welding can be used to create details such as artery refiners, anesthetic refiners, blood refiners, IV catheters, dialysis ducts, pipettes, cardiometry stocks, blood/air refiners, face vizards, and IV shaft/pollutants. Fabrics processing is a crucial step in the medical support for Ultra-sonic welding. Ultra-sonic welding can be used to seal and rip items like sterile clothing, masks, transdermal patches, and materials for clean dwellings. This lowers the risk of infection by preventing impurity and dirty products [6].

(D) Packaging Sector

Ultra-sonic welding is also used in the packaging of hazardous materials such as explosives, fireworks, and highly reactive chemicals. These details frequently call for hermetic sealing but cannot withstand high temperatures. The butane lighter is one of the examples. In order to hold the butane, the vessel weld needs to be leakproof, able to withstand high pressure, and load. The packaging of safety products and fuels is another illustration. To protect the customer from these issues, these packaging must be able to withstand greater pressure and weight. Safety is a primary concern while sealing hazardous components [9].

5 Safety, Precautions While Using UW

Like all mechanical apparatus, Ultra-sonic welding equipment carries some risks. These involve being exposed to extreme temperatures and voltages. To prevent damage, this equipment should be handled according to the safety instructions provided by the manufacturer. For illustration, users must in no way dispose hands or arms near the welding point while the motor is actuated. The handler should also receive a hearing protector and safety glasses. Operators should be informed of and subject to the administration department's bylaws for the equipment used for Ultra-sonic welding. Motors used for Ultra-sonic welding require regular maintenance and inspection. For feeding and care, panel doors, casing covers, and defensive guards may need to be removed. This should be serviced by the qualified professional servicing the motor while the power to the machinery is off and they are by themselves [1, 2, 5, 6].

6 Conclusion

Ultra-sonic welding, which was initially developed for the joining of plastic, has since developed into other elements like material, exterior mount component, and fabric material. The studies suggest here confirm the use of Ultra-sonic welding for other elements. Thin film foils are joined together by multi-material bonding using a technique known as high frequency oscillation at low amplitude to create solid objects. The results of the study on multiple-material components demonstrate how successfully these accessories worked together. Using modified plastic welding and a resonant flexural mode welding tip, Ultra-sonic welding to SMT factors is carried out in order to convert longitudinal oscillation in the horn into shear oscillation at the weld interface.

References

1. Or SW, Chan HLW, Lo VC, Yuen CW (1998) Ultrasonic wire-bond quality monitoring using piezoelectric sensor. Sens Actuators A 65:69–75
2. Levy A, Le Corre S, Chevaugeon N, Poitou A (2011) A level set based approach for the finite element simulation of a forming process involving multiphysics coupling: Ultrasonic welding of thermoplastic composites. Eur J Mech A/Solids 30:501–509
3. Zhao N, Li W, Cai WW, Abell JA (2014) A fatigue life study of ultrasonic ally welded lithium-ion battery tab joints based on electrical resistance. J Manuf Sci Eng 136:1–7
4. Musa MA, Wan Saidin Wan, Kasim NI, Ibrahim MI, Noor NM (2015) A review on ultrasonic welding capability: breakaway from traditional plastic. Appl Mech Mater 789–790:136–140
5. Al-Sarraf Z, Lucas M, Harkness P (2012) A numerical and experimental study of ultrasonic metal welding. IOP Conf Ser Mater Sci Eng 42:12–15
6. Cai W, Daehn G, Vivek A, Li J, Khan H, Mishra R, Komarasamy M (2018) A state of the art review on solid state metal joining. J Manuf Sci Eng 1–39

7. Rani RM, Suresh KS, Prakasan K, Rudramoorthy R (2007) A statistical study of parameters in ultrasonic welding of plastics. Exp Tech 52–58
8. Kang B, Cai W, Tan CA (2014) Dynamic stress analysis of battery tabs under ultrasonic welding. J Manuf Sci Eng 136:1–18
9. Parmar U, Pandya DH (2016) Experimental investigation of ultrasonic welding on non-metallic material. Procedia Technol 23:551–557
10. Chen Y, He Z, Wang X (2018) Experimental study of the vibration weld strength of dissimilar thermoplastic. IOP Conf Ser Mater Sci Eng 394:1–7
11. Bergmann JP, Köhler T, Pöthig P (2021) Chapter 7—ultrasonic welding. Advanced joining processes, welding, plastic deformation, and adhesion, pp 239–267. https://doi.org/10.1016/B978-0-12-820787-1.00007-3
12. Xie J, Chena Y, Yin L, Zhang T, Wang S, Wang L (2021) Microstructure and mechanical properties of ultrasonic spot welding TiNi/Ti6Al4V dissimilar materials using pure Al coating. J Manuf Process 64:473–480. https://doi.org/10.1016/j.jmapro.2021.02.009
13. LeonHyung-SeopShin M (2022) Review of the advancements in aluminum and copper Ultrasonic welding in electric vehicles and superconductor applications. J Mater Process Technol 307:117691. https://doi.org/10.1016/j.jmatprotec.2022.117691

Development of a Combined RSM-GA Approach for Improving and Optimising Soyabean Oil Bio-diesel Production

Pardeep Kumar and Aswani Kumar Dhingra

Abstract In this paper, bio diesel was produced from soybean oil using the transesterification process. Transesterification process is affected by different variable parameters and carried out at different molecular ratio, temperature, catalyst concentration and time to examine the combined effect of variable factors on process optimisation. The central composite axial full factorial design of the Response Surface Methodology (RSM) and GA was applied for optimising the biodiesel production yield. The combine effect of the variable factor was investigated and optimised. The biodiesel production yield (95.172) was optimised at the following factor value: molar ratio (6), catalyst concentration (1.3), temperature (70) and time (40.001 min) which was obtained by using Genetic Algorithm. Moreover, different properties of the bio diesel have been compared with the properties of conventional diesel.

Keywords Transesterification · Response surface methodology · Genetic algorithm

1 Introduction

Due to urbanisation and industrialisation, the world's energy consumption has been increasing consistently [1]. Different industrial Sectors such as power generation and transportation are dependent upon fossil fuels [2]. Due to this, there will be scarcity of the fossil fuel reserves [3]. However, when an IC engine is running, combustible materials are produced, which causes the environment to degrade and emit different pollutants [4]. To overcome these problems different alternative fuels have been tested by number of researchers [5, 6]. Biodiesel is the most widely used among the alternative fuels [7]. Numerous benefits of biodiesel include its ability to degrade naturally, economic operating costs, use of renewable energy, and its ability to lower emissions

P. Kumar · A. K. Dhingra (✉)
Mechanical Engineering Department, University Institute of Engineering and Technology, Maharishi Dayanand University, Rohtak, Haryana, India
e-mail: ashwani_dhingra1979@rediff.com

H. Kumar et al. (eds.), *Recent Advances in Intelligent Manufacturing*, Lecture Notes in Mechanical Engineering,
https://doi.org/10.1007/978-981-99-1308-4_20

[8]. Different feedstocks, including edible and non-edible materials and animal fats have been used to produce biodiesel [9]. Due to the increased durability, power output and combustion efficiency of the biodiesel, it can be widely acceptable substitute for the conventional diesel. Diesel engines are commonly used for the purpose of transportation, agricultural activities, and manufacturing organization [10]. Biodiesel is renewable, biodegradable, and non-toxic and almost has the similar characteristics for conventional diesel. It can therefore be blended with conventional diesel for the fueling diesel engines [11]. Alcohol has been used in the transesterification process for production of biodiesel. Due to increased viscosity of the biodiesel, alcohol can be mixed for decreasing the viscosity [12, 13]. In the transesterification process, different alcohols such as methanol, ethanol, propanol, and butanol have been used [14]. Significant reductions in THC and CO emissions can be seen with soybean biodiesel and hence without any Engine modifications, soybean biodiesel can be utilised in diesel engines [15].

For optimising the soyabean biodiesel production, a central Composite Rotatable Design (CCRD) with a quadratic model in Response Surface Methodology (RSM) and GA has been considered in the present study. The central composite rotatable design can predict responses on the basis of small number of experimental data sets where all the parameters change will be within a predetermined range. The RSM model's prime drawback is its inability to perform precisely over a wide range of input variables and process parameters [16, 17]. Due to the success in addressing challenging optimisation issues like complex networks, the genetic algorithm and other evolutionary algorithms have been widely used in different application [18]. Because of their adaptability, quick to discover targets, and strong coders, genetic algorithm has a substantial lead over other search algorithms, making the effective random search algorithms with vital applications in numerous fields. [19].

To understand how different variables will effect the biodiesel yield, the study reported in this paper optimises soyabean biodiesel yield by utilising the response surface methodology with Genetic Algorithm (GA) approach.

2 Materials and Methods

Transesterification reaction has been performed using the cylindrical beaker and magnetic stirrer with thermocouple temperature measurement. Sodium Hydroxide (NaOH), methanol (CH_3OH) and refined soyabean oil have been procured from market. The different experiments have been conducted by using a magnetic stirrer. The biodiesel yield obtained by mixing of the above said chemicals. Design Expert 13 and GA have been considered for the optimal production of the bio diesel.

2.1 Apparatus

The transesterification process uses a biodiesel reactor (5-L capacity) with several temperature sensors which are attached to the reactor for maintaining the temperature in the reactor along with a cylindrical beaker and stirrer.

2.2 Problem Formulation

The variables effecting biodiesel production and its range have been studied and screened from the existing literature [20, 21]. The key findings are related to four significant reaction variables and how the constraint levels influenced the responses. It has been found that the biodiesel yield varied with these factors. Consequently, the function of four variables can be used to express the biodiesel yield as follows:

The calculating biodiesel yield (y) is the function of {A, B, C, and D}, where A represents the molar ratio, B the catalyst concentration, C the reaction temperature in °C, and D the reaction time in minutes.

2.3 Constraint Function

The factors that influence the generation of the biodiesel have upper and lower limits which can been seen in Table 1. Therefore, different range of molar ratio, catalyst concentrations, reaction time and temperatures have been considered for optimising the biodiesel production.

2.4 Transesterification Process

The reaction among soybean oil and methanol have been carried out by the using a stirrer in the presence of the catalyst NAOH. Each experiment produced two products, biodiesel and glycerol, which are required to be separated and refined in a specific

Table 1 Process variables constraints

Serial number	Variables	Constraints
1	Molar ratio	$1.5 < A < 7.5$
2	Catalyst concentrations	$0.1 < B < 1.7$
3	Reaction temperature in °C	$32.5 < C < 82.5$
4	Reaction time in minute	$15 < D < 115$

beaker. The calculated percentags of soyabean oil, catalyst, and methanol are as follows:

Triglyceride molecular weight 920 kg/kmol. Methanol's molecular weight 32.04 g/mol.

1 g mol of soyabean oil equal to 920 g.
1 g mol of methanol equal to 32.04 g.
Quantity of soyabean oil equal to 100 g.
Sample which are used for biodiesel yield calculation:
Amount of methanol is 100 g

$$\text{for molar ratio } 3:1 = (32.04/920) * 100 * 3 = 10.447 \text{ g and}$$

$$\text{for molar ratio } 6:1 = (32.04/920) * 100 * 6 = 20.895$$

Then the biodiesel production amount is 90 g (say).
Therefore, Amount of methanol: 20.895 g (for the molar ratio 6:1).

$$\begin{aligned} \text{Yield} &= (\text{Biodiesel produced amount/amount of oil taken}) * 100\% \\ &= (90/100) * 100\% = 90\% \end{aligned}$$

2.5 *Design of Experiments*

The different experiments have been carried out using Design Expert. Design Matrix has been developed according to the output and number of variables for optimising the biodiesel production yield. It was first necessary to establish the amount of the factors influencing the biodiesel output [22–27], as seen in Table 1. Table 2 shows the coded and actual values for the considered variables. For several studies, the central composite design (CCD) method in the Response Surface Method (RSM) model was used to develop the regression model for the response with four relevant components and shown in Table 3. The central composite design has been selected for the best solution for the experimentation and the quality of the data collected [28]. Total of Thirty experiments have been formulated for optimising the soyabean bio diesel yield for which eight are xial points, six centre points, and sixteen factorial points. In order to obtain a reliable estimate of the experimental error, centre points are often repeated 4–6 times (pure error). For example, when there are two factors, the design will be by default have five centre points. As per the design matrix acquired from the RSM with the central composite design, the experiments have been carried out by using stirrer.

The combination of reaction conditions for measuring responses in the experimentation are shown in Table 3.

Table 2 Coded and actual variables values

Factor	Name	Units	Type	Type	Lower value	Upper value	Low coded value	High coded value	Mean value	Std. dev
A	Molar ratio		Numeric	Continuous	1.5	7.5	$-1 \leftrightarrow 3.00$	$+1 \leftrightarrow 6.00$	4.50	1.36
B	Catalyst	w/v %	Numeric	Continuous	0.1	1.7	$-1 \leftrightarrow 0.50$	$+1 \leftrightarrow 1.30$	0.9000	0.3639
C	Temperature	C	Numeric	Continuous	32.5	82.5	$-1 \leftrightarrow 45.00$	$+1 \leftrightarrow 70.00$	57.50	11.37
D	Time	Minute	Numeric	Continuous	15	115	$-1 \leftrightarrow 40.00$	$+1 \leftrightarrow 90.00$	65.00	22.74

Table 3 Biodiesel yield (in %) at various groupings of factors

Standard	Run	Type	Factor: A	Factor: B	Factor: C	Factor: D	Response
			Molar ratio	Catalyst concentration w/w %	Temperature	Time (min)	Yield
1	1	Factorial	3	0.5	45	40	69
2	2	Factorial	6	0.5	45	40	83
3	3	Factorial	3	1.3	45	40	84
4	4	Factorial	6	1.3	45	40	92
5	5	Factorial	3	0.5	70	40	56
6	6	Factorial	6	0.5	70	40	78
7	7	Factorial	3	1.3	70	40	79
8	8	Factorial	6	1.3	70	40	87
9	9	Factorial	3	0.5	45	90	63
10	10	Factorial	6	0.5	45	90	80
11	11	Factorial	3	1.3	45	90	81
12	12	Factorial	6	1.3	45	90	92
13	13	Factorial	3	0.5	70	90	49
14	14	Factorial	6	0.5	70	90	72
15	15	Factorial	3	1.3	70	90	67
16	16	Factorial	6	1.3	70	90	89
17	17	Axial	1.5	0.9	57.5	65	49
18	18	Axial	7.5	0.9	57.5	65	84
19	19	Axial	4.5	0.1	57.5	65	60
20	20	Axial	4.5	1.7	57.5	65	90
21	21	Axial	4.5	0.9	32.5	65	88
22	22	Axial	4.5	0.9	82.5	65	81
23	23	Axial	4.5	0.9	57.5	15	77
24	24	Axial	4.5	0.9	57.5	115	72
25	25	Center	4.5	0.9	57.5	65	76
26	25	Center	4.5	0.9	57.5	65	71
27	25	Center	4.5	0.9	57.5	65	74
28	25	Center	4.5	0.9	57.5	65	70
29	25	Center	4.5	0.9	57.5	65	69
30	25	Center	4.5	0.9	57.5	65	76

3 Results and Discussion

3.1 *Significant Testing of Variables by Using of Analysis of Variance*

After experiments, the Analysis of Variance (ANOVA) have been considered for predicting the regression equation which provide the information about the significance of the suggested model. The quadratic model was found to be significant for the specified response by comparing the F-values of several potential regression equation terms which is shown in Table 4. From the experimental data, biodiesel yield was obtained by using the mathematical regression model.

$$\begin{aligned} \text{yield} = {} & 138.693626.25625 \text{ molar ratio} + 9.72135 \text{ catalyst} - 2.97125 \text{ temperature} \\ & - 0.293125 \text{ time} - 2.8125 \text{ molar ratio} * \text{catalyst} + 0.083333 \text{ molar ratio} \\ & * \text{temperature} + 0.035 \text{ molar ratio} * \text{time} + 0.1625 \text{ catalyst} * \text{temperature} \\ & + 0.05625 \text{ catalyst} * \text{time} - 0.0022 \text{ temperature} * \text{time} - 0.597222 \text{ molar ratio}^2 \\ & + 4.88281 \text{ catalyst}^2 + 0.0202 \text{ temperature}^2 + 0.00105 \text{ time}^2 \end{aligned} \quad (1)$$

The result of the ANOVA test has been presented in the Table 5. The f-value of the model is 33.20. The model is significant as per model f-value. There is only a 0.01% chance that change in F-value could occur because of noise. P-values of the model are less than 0.0500, indicating that the model is significant. In this model, the value of A, B, C, D, AB, AC, A^2, and C^2 is significant in terms of model. The model terms are not significant if their values are greater than 0.1000. In this model B^2, AD, BC, BD, CD, D^2 are not significant. If there are many insignificant values available, then the model is improved by reduction.

The F-value for lack of fit is 0.80, which shows that the lack of fit is insignificant compared to the pure error. A significant Lack of Fit F-value has a 64.28% chance of being caused by noise. A little lack of fit is advantageous to the model's fitness.

Table 4 Various sequential models for sum of squares

Model	Sequential p-value	p-value for lack of fit	R^2 for adjusted value	R^2 for predicted value	
Linear	<0.0001	0.1064	0.8132	0.7594	
2FI	0.5422	0.0912	0.8067	0.7541	
Quadratic	0.0002	0.6428	0.9396	0.8718	Suggested
Cubic	0.6392	0.4453	0.9312	0.3221	Aliased

Table 5 ANOVA for RSM quadratic model

	Value of sum of squares	df	Value of mean square	F-value	p-value	
	3818.62	14	272.76	33.2	<0.0001	Significant
A-molar ratio	1584.38	1	1584.38	192.82	<0.0001	
B-catalyst	1365.04	1	1365.04	166.13	<0.0001	
C-temperature	273.38	1	273.38	33.27	<0.0001	
D-time	84.38	1	84.38	10.27	0.0059	
AB	45.56	1	45.56	5.55	0.0326	
AC	39.06	1	39.06	4.75	0.0456	
AD	27.56	1	27.56	3.35	0.087	
BC	10.56	1	10.56	1.29	0.2747	
BD	5.06	1	5.06	0.6161	0.4447	
CD	7.56	1	7.56	0.9204	0.3526	
A^2	49.53	1	49.53	6.03	0.0268	
B^2	16.74	1	16.74	2.04	0.174	
C^2	273.24	1	273.24	33.25	<0.0001	
D^2	11.81	1	11.81	1.44	0.2491	
Residual	123.25	15	8.22			
Lack of fit	75.92	10	7.59	0.8019	0.6428	Not significant
Pure error	47.33	5	9.47			
Total	3941.87	29				

The confidence interval indicates a range of possible values within which it may accurately locate the population's statistical measure. The margins of error contribute to determining the confidence interval. The interval margins of error and corresponding range also vary depending on our confidence level. The confidence interval used in this study is 95%, which generally has been the most common confidence interval for almost all statistical measures [29].

For different values of the reaction parameters of the biodiesel yield, the contours of the response surface can be determine using the mathematical model Equation (1). The production of the biodiesel will be impacted by changes in molar ratio and reaction time which is shown in Fig. 1. With the combination of the variables , the yield find out to be constant. The contour plot revealed that the biodiesel output increased as one moved farther from the origin and maximised at a specific place. With a time of 90 min and a molar ratio of 5.996, the production of soyabean biodiesel will be at its optimised value among the combinations of variables at the optimal design point which is shown in Fig. 1.

Figure 2 shows how temperature and molar ratio effect biodiesel yield. The yield will be constant at a molar ratio of 5.996 and a temperature of 45 °C. The yield

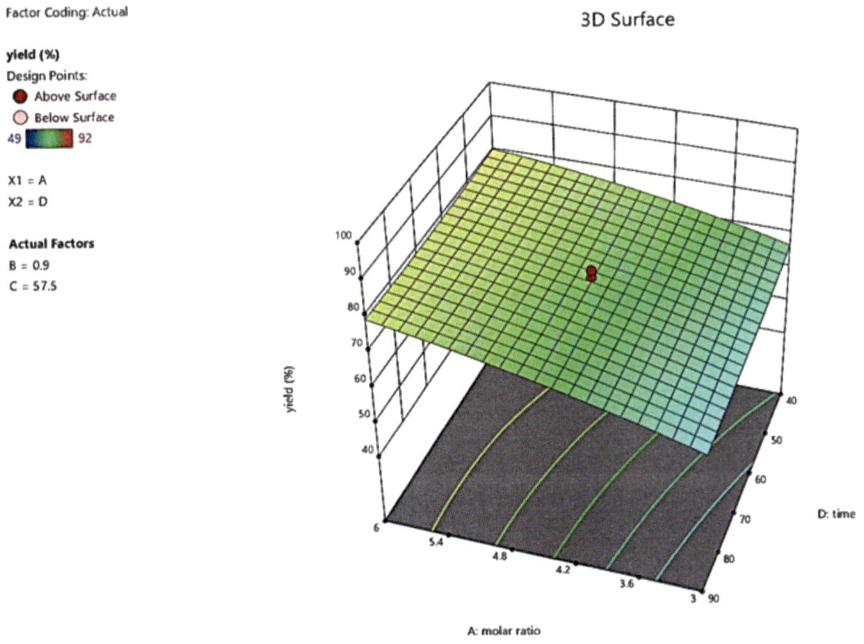

Fig. 1 Graphical representations of soyabean biodiesel yield with reaction time (min) and molar ratio

initially increased as the molar ratio and temperature increased, and at a certain point, it achieves its optimality and therafter deceased in the value of yield till 91.467.

Figure 3 shows the effect of molar ratio and catalyst concentrations on the production of the biodiesel at a reaction temperature of 45 °C with a reaction period of 90 min. The pattern in this case will be similar as shown in Fig. 2. The yield will be constant and increased initially, then decreased as the molar ratio increased from 3 to 6 and the catalyst concentration increased from 0.5 to 1.3. The optimum yield has been obtained during these modifications, serving as the design point for various factor combinations. The yields of actual and predicted biodiesel are shown in Fig. 4. Through various combinations of factors, an experimental result varies from 49 and 92. The existence of noise in the trials led to substantial residual differences, as seen in Fig. 5, which compares residuals with experimental runs. The regression equation shows that every run had noise, which caused the actual yield to vary from the predicted yield.

3.2 *Genetic Algorithm (GA)*

The Genetic Algorithm (GA) has been used to optimise different process factors for maximising the biodiesel yield using the regression model of RSM (Response

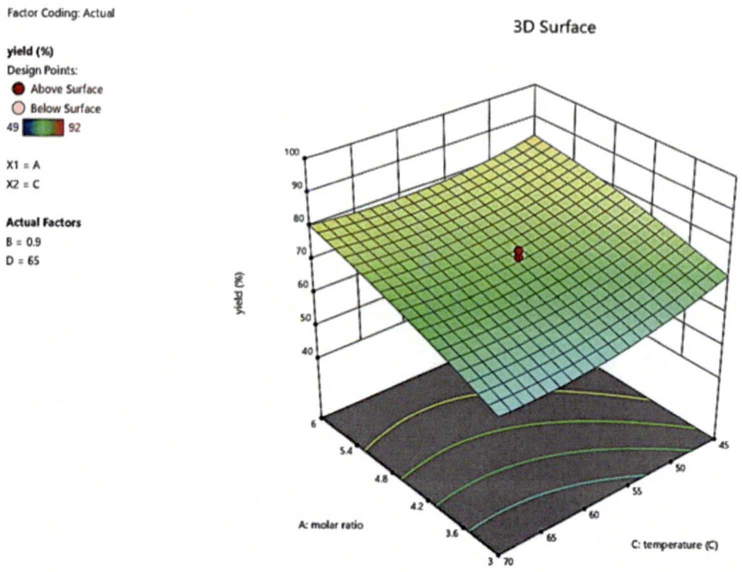

Fig. 2 Graphical representations of yield of soyabean biodiesel with reaction temperature (°C) and molar ratio

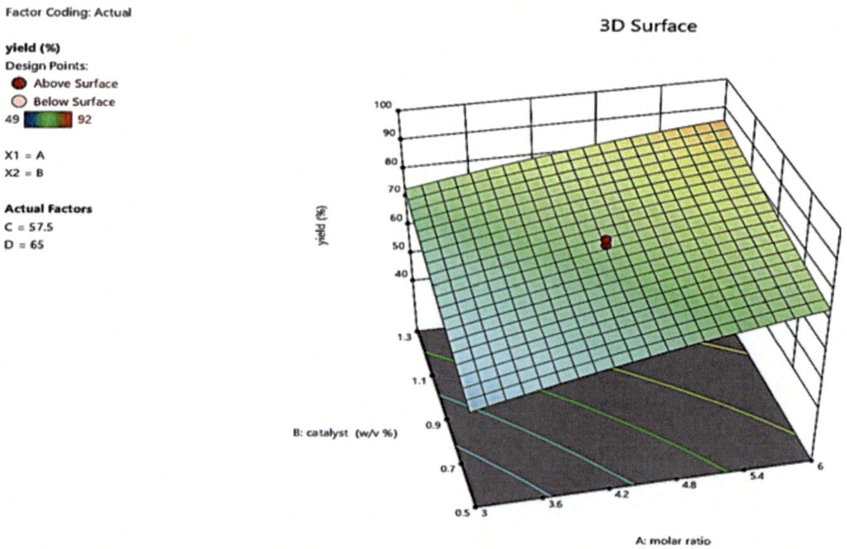

Fig. 3 Graphical representations between soyabean biodiesel yield with catalyst concentration (%) and molar ratio

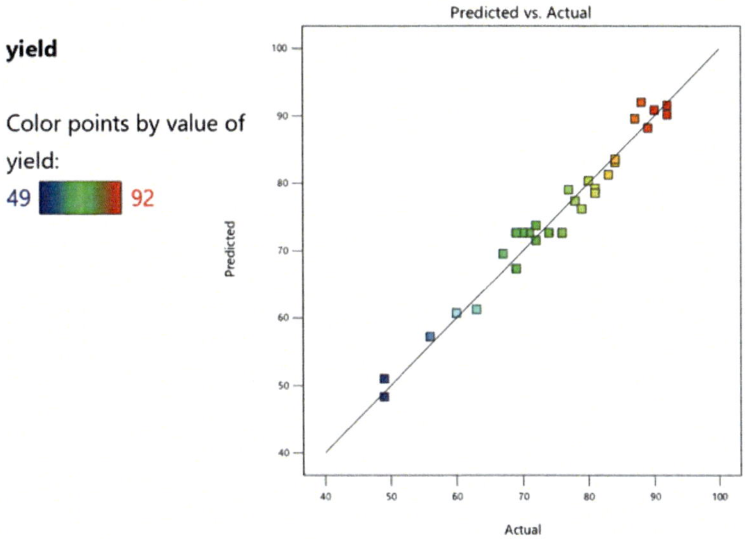

Fig. 4 Predicted biodiesel versus actual yield

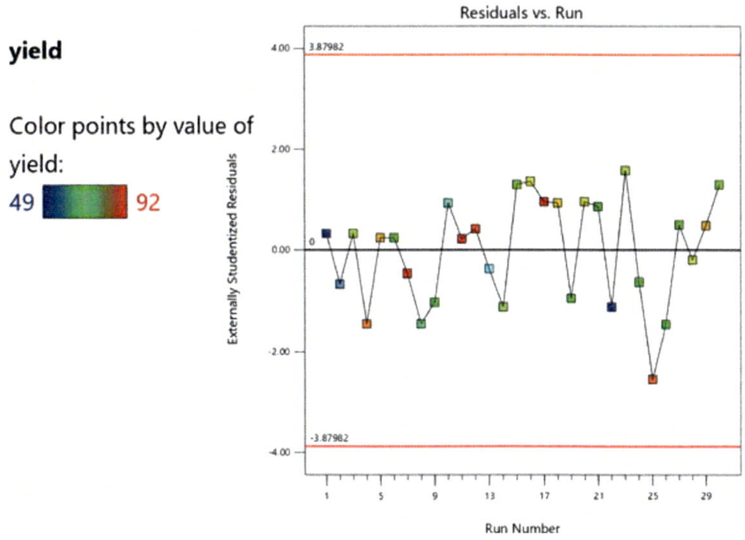

Fig. 5 2-Dimensional presentation of residual differences between actual yield and predicted yield of experimental runs in response surface methodology (RSM)

Table 6 Various parameters used for genetic algorithm

Parameter name	Value
Size of population	50
Population type	Double vector
Elite count	0.05
Cross over fraction value	0.8
Crossover function	Constraint dependent
Migration fraction value	0.2
Migration interval	20

Surface Method). It provides details on selection by probabilistic basis for creating the population solutions for the problem. An initial population has been considered to fix a specific mutation value, after which further generations obtained through predetermined breeding. The initial population has been generated automatically by the GA in the MAT Lab in accordance with the constraints shown in Table 6. The regression equation from the RSM has been considered for executing the GA parameters until the optimisation. Population size, elite count, crossover function, crossover function, migration fraction, and generations are some of the varied GA parameters. Generally, the chosen elite solution is the best. Figure 6 shows the general outline steps of the GA. The fitness function has been generated using a regression equation and the range of various parameters which is shown in Table 6. Each iteration requires the creation and testing of the fitness function using MAT Lab. The generation of the population known as the parents is defined by the computating the best fit for each group in the population. The computing of best fitness for the elite, crossover, mutation known as reproduction. The genetic algorithm used crossover and mutation since it was possible to alter their values to get the optimum yield. This process will be going until the stopping condition's limit will be reached. As shown in Fig. 7, the mean and best solution from the genetic algorithm (GA) is 95.1724% which can obtained by repeatedly running of similar parameters which are related to the Genetic Algorithm.

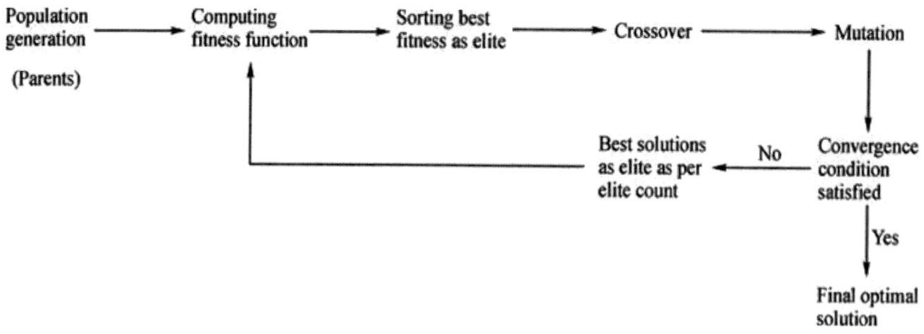

Fig. 6 Various steps for genetic algorithm to achieve optimum solution

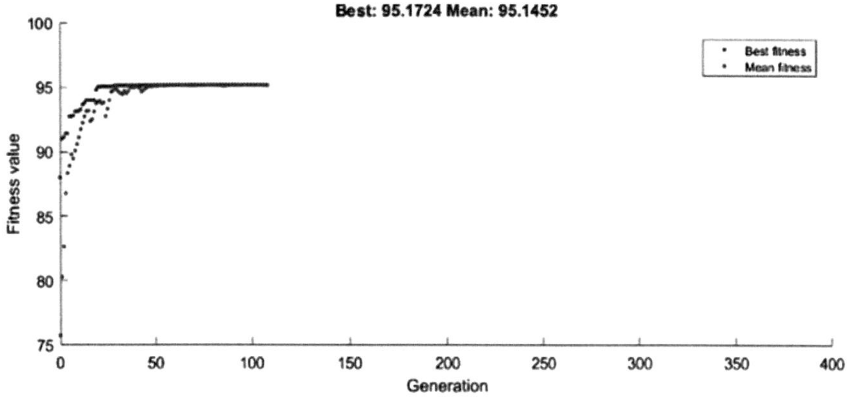

Fig. 7 Best and mean solution from genetic algorithm

4 Validation of Experiments for the Optimum Conditions Which Are Obtained from Genetic Algorithm (GA) and Response Surface Method (RSM) Approach

A minimum of six tests have been carried out under ideal reaction conditions for evaluating the accuracy of the predicted model by using two approaches. The studies reported an average yield of 85%, as indicated by the RSM model, which is lower than the expected yield of 91.467%. The output was low since another GA approach was used because of the response required to optimise it. The regression model obtained via the RSM was used in the GA's fitness function. The yield has been predicted to be 95.1724% for the reaction conditions which is shown in Table 7, and the following parameter variations: population size 50, crossover function is Constraint dependent, Migration Fraction ins 0.2, elite count 0.05, crossover fraction 0.8, Migration Interval is 20 and Population type is a double vector. Similar experiments have been conducted under the GA's optimum reaction conditions to verify the suggested output. An average yield of 92% has been obtained.

5 Improvement in Yield of Soyabean Biodiesel and Comparison of Its Properties with Diesel

A yield of 85% was observed on average when various experiments were conducted using the RSM to determine the best reaction conditions. Still, a yield of 92% was observed when the GA's optimised variables were used. The analysis results are provided in Table 7, indicating that the GA produced an optimal biodiesel yield that was 8.23% higher than the yield obtained using the RSM method. The parameters of the biodiesel produced under the GA's optimum reaction conditions were

Table 7 Comparison of genetic algorithm and response surface method (RSM) optimal design

	Molar ratio	Catalyst concentration	Temperature	Time	Bio diesel (predicted)%	Yield actual	Percentage improvement
Recommendation using GA	6	1.3	70	40.0001	95.1724	92	8.23
Response surface methodology	5.986	1.295	45	90	91.467	85	

Table 8 Comparison between soyabean biodiesel and conventional diesel

Properties	Soyabean biodiesel	Diesel
Density (g/L)	0.869	0.836
Flash point	154	64
Cetane index	48	53
Distillation range (90%)	352	338
Sulphur content value (% mass)	0.015	0.048
Kinematic viscosity (mm^2/s)	5.19	3.073
Cloud point	2	-7
Acidic value (mg NAOH/g)	3.758	6.032
Lower heating value	38.81	42.50

quite similar to the values expected. Compared to commercial diesel, this biodiesel's performance in engines did not show significant changes in power output. Table 8 shows the comparison between soyabean bio diesel and conventional diesel.

6 Conclusions

Based on the findings of the work, the specific conclusions are as follows:

(1) The biodiesel yield can be optimised using GA with four input variables. A molar ratio of 6 with a catalyst concentration of 1.3%, a reaction temperature of 45 °C, and a reaction duration of 40.0001 min resulted in an average yield of 92% of the obtained soyabean oil.
(2) When comparing the Genetic Algorithm approach to the RSM with the same reaction variables, it has been found that the average biodiesel yield increased by 8.23%.
(3) The biodiesel's fuel characteristics finds to be similar to the conventional diesel.

Therefore, it is concluded that the soyabean biodiesel can be used as an alternative fuel by blending it with regular diesel to run the compression ignition engines.

References

1. Ulusoy Y, Arslan R, Tekin Y, Surmen A, Bolat A, Sahin R (2018) Investigation of performance and emission characteristics of waste cooking oil as biodiesel in a diesel engine. Petrol Sci 15(2):396–404. https://doi.org/10.1007/s12182-018-0225-2
2. Chuah LF, Mohd Salleh NH, Osnin NA, Alcaide JI, Abdul Majid MH, Abdullah AA, Bokhari A, Jalil A, EE, Klemes JJ (2021) Profiling Malaysian ship registration and seafarers for streamlining future Malaysian shipping governance. Aust J Marit Ocean Aff 13(4):225–261. https://doi.org/10.1080//18366503.2021.1878981

3. Yadav AK, Khan ME, Pal A, Dubey AM (2017) Experimental investigations of performance and emissions characteristics of Kusum (Schleichera oleosa) biodiesel in a multi-cylinder transportation diesel engine. Waste Biomass Valor 8(4):1331–1341. https://doi.org/10.1007/s12649-016-9658-2
4. Chuah LF, Aziz ARA, Yusup S, Kleměs JJ, Bokhari A (2016) Waste cooking oil biodiesel via hydrodynamic cavitation on a diesel engine performance and greenhouse gas footprint reduction. July Chem Eng Trans 50:301–306. https://doi.org/10.3303/CET1650051
5. Papagiannakis RG, Krishnan SR, Rakopoulos DC, Srinivasan KK, Rakopoulos CD (2017) A combined experimental and theoretical study of diesel fuel injection timing and gaseous fuel/diesel mass ratio effects on the performance and emissions of natural gas-diesel HDDI engine operating at various loads. Fuel 202:675–687. https://doi.org/10.1016/j.fuel.2017.05.012
6. Geng L, Xiao Y, Li S, Chen H, Chen X (2020) Effects of injection timing and rail pressure on particulate size-number distribution of a common rail DI engine fueled with Fischer-Tropsch diesel synthesized from coal. J Energy Inst. https://doi.org/10.1016/j.joei.2020.08.008
7. Aydın S (2020) Comprehensive analysis of combustion, performance and emissions of power generator diesel engine fueled with different source of biodiesel blends
8. Subramaniam M, Solomon JM, Nadanakumar V, Anaimuthu S, Sathyamurthy R (2020) Experimental investigation on performance, combustion and emission characteristics of DI diesel engine using algae as a biodiesel. Energy Rep 6:1382–1392. https://doi.org/10.1016/j.egyr.2020.05.022
9. Elkelawy M et al (2022) A comprehensive review on the effects of diesel/biofuel blends with nanofluid additives on compression ignition engine by response surface methodology. Energy Convers Manag X 14:100177
10. Pradelle F, Leal S, Rosa A, De Aguiar F, Turkovics F, Nohra R, Pradelle C (2019) Performance and combustion characteristics of a compression ignition engine running on diesel-biodiesel-ethanol (DBE) blends e Potential as diesel fuel substitute on an Euro III engine. Renew Energy 136:586–598. https://doi.org/10.1016/j.renene.2019.01.025
11. Yilmaz N, Atmanli A (2017a) Experimental evaluation of a diesel engine running on the blends of diesel and pentanol as a next generation higher alcohol. Fuel 210:75–82. https://doi.org/10.1016/j.fuel.2017.08.051
12. Ma Q et al (2021) The performance and emissions characteristics of diesel/biodiesel/alcohol blends in a diesel engine. Energy Rep 7:1016–1024
13. Atmanli A (2016) Effects of a cetane improver on fuel properties and engine characteristics of a diesel engine fueled with the blends of diesel, hazelnut oil and higher carbon alcohol. Fuel 172:209–217. https://doi.org/10.1016//j.fuel.2016.01.013
14. Ma FR, Clements LD, Hanna MA (1998) Biodiesel fuel from animal fat. Ancillary studies on transesterification of beef tallow. Ind Eng Chem Res 37:3768–3771
15. Ozener O, Yüksek L, Ergenç AT, Özkan M (2014) Effects of soybean biodiesel on a DI diesel engine performance, emission and combustion characteristics. Fuel 115:875–883. ISSN 0016-2361. https://doi.org/10.1016/j.fuel.2012.10.081
16. Jena PC, Raheman H, Kumar GP et al (2010) Biodiesel production from mixture of mahua and simarouba oils with high free fatty acids. Biomass Bioenerg 34(8):1108–1116
17. Shanmugaprakash M, Kirthika J, Ragupathy J et al (2014) Statistical based media optimization and production of naringinase using Aspergillus brasiliensis 1344. Int J Biol Macromol 64:443–452
18. Liu D, Chang Z, Yang G, Chen E (2022) Hiding ourselves from community detection through genetic algorithms. Inf Sci 614:123–137. https://doi.org/10.1016/j.ins.2022.10.027
19. Wang H (2022) Research on the application of genetic algorithm in physical education. J Math 8p. Article ID 8477945. https://doi.org/10.1155/2022/8477945
20. Cao XL, Wang SD, Li X, Jiang ZX, Ren BZ (2017) Optimal synthesis of soybean oil methyl ester by response surface methodology. Soybean Sci 36(1):124–130 ref. 21
21. Bala VSS, Thiruvengadaravi KV, Kumar PS, Premkumar MP, Kumar VV, Sankar SS, Kumar MH, Sivanesan S (2012) Removal of free fatty acid in Pongamia Pinnata (Karanja) oil using divinylbenzenestyrene copolymer resins for biodiesel production. Biomass Bioenerg 37:335–341

22. Khayoon MS, Olutoye MA, Hameed BH (2012) Utilization of crudekaranja (Pongamia pinnata) oil as a potential feedstock for the synthesis of fatty acid methyl esters. Biores Technol 111:175–179
23. Kamath HV, Regupathi I, Saidutta MB (2011) Optimization of two stepkaranja biodiesel synthesis under microwave irradiation. Fuel Process Technol 92(1):100–105
24. Maheshwari N, Balaji C, Ramesh A (2011) A nonlinear regression basedmulti-objective optimization of parameters based on experimental data from an IC engine fuelled with biodiesel blends. Biomass Bioenerg 35(5):2171–2183
25. Sahoo PK, Das LM (2009) Process optimization for biodiesel production from Jatropha, Karanja and Polanga oils. Fuel 88(9):1588–1594
26. Jena PC, Raheman H, Prasanna GV, Machavaram R (2010) Biodiesel production from mixture of mahua and simarouba oils with high free fatty acids. Biomass Bioenerg 34(8):1108–1116
27. Rajendra M, Jena PC, Raheman H (2009) Prediction of optimized process parameters for biodiesel production using ANN and GA. Fuel 88(5):868–875
28. Rakic T, Vujanović I, Jovanović M, Jancic Stojanovic B, Ivanović D (2014) Comparison of full factorial design, central composite design, and Box-Behnken design in chromatographic method development for the determination of fluconazole and its impurities. Anal Lett 47. https://doi.org/10.1080/00032719.2013.867503
29. Simundic A-M (2008) Confidence interval. Biochemia Medica 18:154–161. https://doi.org/10.11613/BM.2008.015

Improvement and Optimisation of Castor Oil Bio-diesel Using RSM and ANN-GA

Ashish Chhikara, Aswani Kumar Dhingra, and Pardeep Kumar

Abstract The Present paper studies the castor oil biodiesel production and the effect of input variables by using the trans-esterification process using response surface methodology (RSM) and artificial neural network (ANN) joined with genetic algorithm (GA). The Optimization of reaction parameters has been considered for maximizing the castor biodiesel production yield using RSM based central composite rotational design with a quadratic regression model. A multilayer back propagation feeds forward neural network model has been developed by a mathematical quadratic regression model with three inputs, two hidden and one output layer. The output-generated data of ANN is further used in GA to optimize variable parameters (concentration of catalyst, molar ratio, and reaction time). The proposed value of castor biodiesel yield by GA is 92.469% at variables combination of 11.576 molar ratio, 1.59% w/w catalyst and 94.825 min of reaction time. The experimentation validation showed an average castor biodiesel yield of 89.70% under these variable optimization conditions.

Keywords Castor oil · RSM · ANN-GA · Transesterification

1 Introduction

Biodiesel is termed as a fuel mostly made of mono-alkyl esters generated from sustainable vegetable oils or animal fats that satisfy the requirements of ASTM D6751 as per American Society for Testing and Materials. Because its disposable greenhouse gas emissions are lower as compared to fossil fuels and hence biodiesel is called as a green biofuel [1]. Castor beans or Ricinus communis are high-potential feeds stock that might provide around 60% of the non-edible oil required to make biodiesel [1]. Ricinoleic acid (C_{18}-H_{34}-O_3), which makes up about 80–90% of castor

A. Chhikara · A. K. Dhingra (✉) · P. Kumar
Mechanical Engineering Department, University Institute of Engineering and Technology, Maharishi Dayanand University, Rohtak, Haryana, India
e-mail: ashwani_dhingra1979@rediff.com

H. Kumar et al. (eds.), *Recent Advances in Intelligent Manufacturing*, Lecture Notes in Mechanical Engineering,
https://doi.org/10.1007/978-981-99-1308-4_21

oil's total fatty acid content, is its primary fatty acid. Due to its high miscibility, high viscosity, low freezing point and low iodine content it is a suitable material for the production of biodiesel [2, 3].

The chemical industry has commercially utilized castor oil for many years as a highly renewable resource and superior raw material in terms of cost and quality. India is the top worldwide for castor crop production, area, and productivity. India occupied 870 thousand hectares (69.22%) of the world's land in 2016 while producing 1554 thousand tones (87.42%) of it. Regarding global castor crop productivity in 2016, India placed in second position with 1786.2 kg/ha. Castor beans from India contain 48% oil, of which 42% may be extracted.

The area under cultivation of castor seeds increased upto 1.3 thousand ha. during the year 2016–17. The average acreage under castor cultivation was 965 thousand ha. in the country during the time period 2006–07 to 2016–17 [4]. In this nation, the annual production of castor oil varies typically between 250,000 and 350,000 tonnes. Gujarat produces the majority of India's castor seeds—about 86%—followed by Rajasthan and Andhra Pradesh [5].

Castor oil seeds have been conventionally in situ transesterified. Ethanol and Methanol are the most commonly used alcohols for the transesterification reaction method for making biodiesels. Because the product contains many free fatty acids and contaminants, choosing a suitable catalyst is particularly important. Most researchers employ a primary homogeneous catalyst like potassium hydroxide (KOH) or sodium hydroxide (NaOH) as the specific mechanism for producing biodiesel [6]. Blends of castor oil methyl ester and ethyl ester have been examined for their ability to reduce viscosity [7]. This process kinetics has been studied by varying the molar ratio temperature and time of alcohol to oil [8]. Numerous reports have been studied on castor oil transesterification reaction optimization to make biodiesel. However, there are few references in the literature to transesterification reaction optimization using ANN modelling. From the perspective of kinetic parameter estimates, it can be said that these processes parameter have not been studied in the presence of an acid catalyst. Due to its time-saving benefits and the ability to optimize the biodiesel production on a pilot scale utilizing experimental data refined by ANN, ANN modelling can be a viable method for optimizing acid-catalyzed reactions for manufacturing biodiesel. For use with an alkaline catalyst, castor oil's FFA concentration needs to be less than 1% [9].

In this work, which focuses on the castor oil transesterification in the presence of an acid catalyst, genetic algorithms considered for maximising the production of castor oil-based biodiesel. The experimental data has been determined by RSM central composite design (CCD) for predicting the fractional production profile of FAME under optimal conditions.

2 Materials and Methods

Analytical-grade chemicals like catalyst KOH in pallet form and ethanol (99.5%) has been considered for the base catalyst for a process. The caster oil has been procured from the local market and the investigations have been carried out using a magnetic stirrer.

2.1 *Apparatus*

The reaction was carried by using the laboratory reactor. The reactor was a double-jacketed, 5-L borosilicate-glass vessel with a 4-L working capacity. On the system's display, the temperature was continuously tracked. Using the reflux system, ethanol from the reaction system was kept from vaporising. The temperature sensor managed the reaction system's temperature. Mechanical stirring was employed to mix the reaction mixture in the reactor properly.

2.2 *Problem Formulation*

The factors influencing biodiesel yield were identified from literature before starting the design, and their boundaries were also examined [10, 11]. The major observations concerned three significant reaction variables and their constraint values influencing the response. The biodiesel yield was analysed to differ with these set-ups of factors.

The effect of the transesterification process' operating variables, such as the oil: ethanol ratio, KOH content, and reaction time, on yield and ester content, was investigated using a factorial design. The response surface methodology and residual analysis were used to examine the results. For a total of 20 experiments, the second-order central composite (CCD) design—which includes eight face points (+1 and +1), six centre points, and six axial points (−1.68 and +1.68)—was employed. The operational constraints of the experimental setup and experiments, as shown in Table 1, informed the selection of factor levels. The system determined the oil: ethanol ratio speed values ranged from 1:9.65 to 1:12.35. Depending on the weight of the oil, catalyst concentrations ranged from 1.33% to 2.17%. The range of reaction times was 39.55 to 140.45 min. The average temperature throughout all experiments was roughly 55 °C.

Table 1 Value of various variable which are used in Response Surface Methodology in design expert

Quadratic model in response surface methodology with experiments							
Factor	Name	Units	Type	−α	+α	Coded low	Coded high
A	Molar Raito	%v/v	Numeric	9.65	12.35	−1 ↔ 10.20	+1 ↔ 11.80
B	Catalyst	%w/w	Numeric	1.33	2.17	−1 ↔ 1.50	+1 ↔ 2.00
C	Time	Minute	Numeric	39.55	140.45	−1 ↔ 60.00	+1 ↔ 120.00

2.3 Development of RSM Model

To evaluate the influences of the three-process parameter for the optimization the yield of castor oil (FAEE) fatty acid ethyl ester, the central composite design was selected as the method for experiment design. To identify the uncontrolled variables in these studies, they were randomized. The mathematical model between yield and various experimental variable are showing the first, second and interaction terms of the following polynomial Eq. 1.

$$y = \beta + \sum_{j=1}^{k} \beta_j x_j + \sum_{i<j}\sum \beta_{ij} x_i x_j + \sum_{j=1}^{k} \beta_{jj} x_j^2 + \varepsilon \quad (1)$$

2.4 ANN Model Development for Best Correlation Coefficient

Data gathering, network construction, configuration, initialization of weights and biases, network validation, network training and network use are the main processes in this technique. As depicted in Fig. 1, the multidimensional mapping issues were fitted by 2-level sigmoid hidden neurons with feed-forward network. The multilayer input–output function provides adequate information about neurons and consistent data in its hidden layer. The three layers express the ANN technique to choose the best model fitness with neurons, inputs and outputs, [12].

To build the network model using MATLAB R2016a software, the biodiesel yields obtained through the experimental study were employed for network training. The Lavenberg-Marquardt backpropagation algorithm was used in the training phase because it can produce the highest R2 and the lowest standard deviation and mean square errors compared to other algorithms. Training, validation, and testing data were the three subsets randomly selected from the data for ANN modelling. In the initial training phase, 70% of the experimental data set's data were randomly selected, and the weighted parameters of the connections were clarified until the mean squared error (MSE) between the experimental and predicted biodiesel yield reached a minimum value. The MSE value is defined in Eq. 2 where P is the total

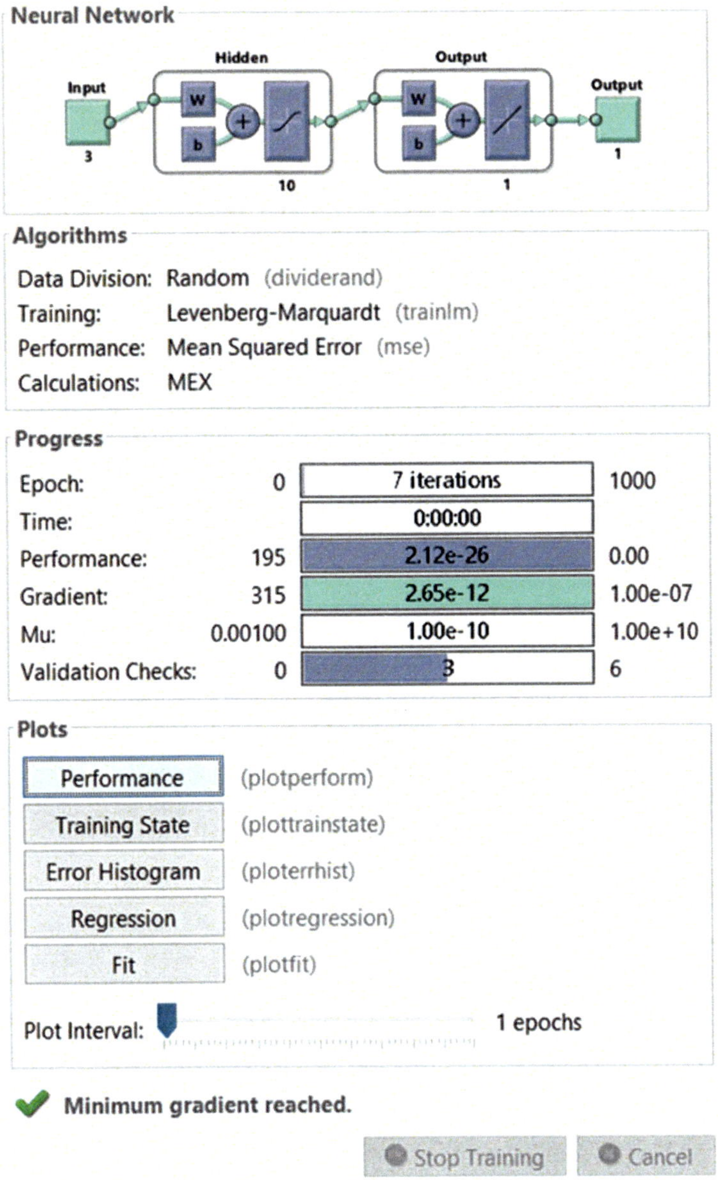

Fig. 1 Artificial Neural Network between the input and output function of bio diesel

Table 2 ANN training parameters for estimation of biodiesel yields

Number of output neurons	1
Minimized error function	Mean squared error (MSE)
Number of input neurons	3
Algorithm	Lavenberg-Marquardt back propagation
Number of hidden neurons	10
Number of output neurons	1

number of training data, dp is the desired output, and op is the computed output. In the second stage, testing, 15% of the data were randomly chosen to test the "trained" ANN, which employed the weighted parameters established in the previous step. In the validation stage, 15% of the remaining data subsets were used to confirm the accuracy of the ANN created in the first two stages' final estimation of biodiesel yield. The coefficient of determination (R) and mean squared error were used to verify the ANN's performance (MSE). Table 2 presents the training parameters for the ANN used to estimate biodiesel yields [13].

$$MSE = 1/P \sum_{P=1}^{P} (dp - op)^2 \tag{2}$$

The design matrix which are based on Response Surface Method with CCD was used to obtain the output and input data. There were three variables for each input. The fitness function was run with training functions (trainlm) is used to get the optimum correlation coefficient. Output, input, net file, and fitness functions were the parameters obtained from ANN. These were employed in the Genetic Algorithm to predict the optimum condition of the reaction parameters.

2.5 *Transesterification Process Variable Optimization Using ANN-GA*

The reaction parameters were optimized for optimal biodiesel (FAEE) production using a GA programme built on the MATLAB 2016 platform. To begin with, the problem definition was used to make the GA with random population. The subsequent steps involved creating fitness, finding encoding solutions, and determining each generation's genetic parameters (crossover, reproduction, and mutation). In order to develop the ANN for estimating the biodiesel production, the population generation has been selected for given castor oil through Genetic Algorithm in the binary form and was first transformed to actual values. By analysing population strings, the fitness function was proposed. Among the top elite chromosomes of size "N," the yield of castor biodiesel higher as compared to expected by the response surface technique approach which was chosen through a tournament. A particular portion of strings were switched between the pairs of two chromosomes in the "N" elite chromosomal pairing. The probability or mutation percentage is the proportion of chromosomes that have undergone mutation since crossing. The chromosome's binary digit form switched from 0 to 1 or vice versa. One generation is the collective name for the process (crossover, reproduction and mutation) [14].

The following GA parameters must be changed to obtain the best values for reaction variables: population, rate of crossover, mutation, and the generation number. The fitness functions are assessed for each trial while changing one of these factors while keeping the others fixed [15, 16]. The final solution obtained by the GA-suggested optimal parameters. They may constitute the ideal reaction variables for the particular castor oil biodiesel. The optimal combination has also been validated for verifying the predicted ANN-GA model.

2.6 *Validation of ANN Model*

To evaluate the model's suitability, an appropriate variables combination is chosen (for a maximum response value). To validate the projected model as recommended by ANN-Genetic Algorithm, experiments were done using the essential variables. The reaction was conducted as per GA's recommended ideal solution. A glycerol layer and two layers of biodiesel were created. The biodiesel content was calculated once these were separated in the flask. The recommended properties were verified using the recommended methods, including density, cetane number, kinematic viscosity, flash point and calorific value, etc. Important fuel characteristics of produced castor oil biodiesel were tested using the recommended techniques and compared with available standards of biodiesel.

3 Results and Discussion

3.1 *Significanace of Model verified through ANOVA in Response Surface Methodology*

The catalyst concentration (KOH) (% w/v), the molar ratio (% v/v), and the reaction time (minute) were chosen as the independent variables in the current study. These variables' ranges were discovered in several research articles, and they were subsequently included in Table 1 utilizing RSM techniques based on Central Composition Design. To improve the prediction of optimal parameters, the software used automatically extends the range of these process variables. According to Table 3, various experiments were carried out, and the resulting biodiesel yield was calculated. The measured biodiesel yield was between 67.01 and 86.32 (wt%). The quadratic model fitness indicates that there is only one set of response variables where biodiesel production is optimized. Equation 3 was generated from the Design Expert to examine the model. Table 4 displays the results of the F and p test, which was used to calculating the significance of each term in the model. It denotes if a term's coefficient in a regression Eq. 3 is significant and if it has a greater impact on the outcome than other terms. The model was significant, as indicated by the model's F value of 41.73 and p-value of 0. 0500. The predicted mathematical regression equation obtained is Second order

$$\begin{aligned} &\text{YIELD} = -506.7725084 + 852285.0494855911455 * \text{A} + 150.53432368299 * \\ &\text{B} - 0.21742552255527 * \text{C} - 3.4308013995583 * \text{A2} - 42.457032584573 * \\ &\text{B2} - 0.0038990708241905 * \text{C2} - 4.9812499999989 * \text{A} * \text{B} + 0.021927083333345 * \\ &\text{A} * \text{C} + 0.40383333333337 * \text{B} * \text{C} \end{aligned} \tag{3}$$

Table 3 ANN and RSM matrix for the optimisation of castor oil bio diesel

STD				RSM			ANN	
	Molar ratio	Catalyst	Time	Actual	Predict	Error	Predict	Error
1	10.2	1.5	60	81.67	80.5339	1.136104	81.70076	−0.03076
2	11.8	1.5	60	85.75	85.99886	−0.24886	85.578	0.171998
3	10.2	2	60	67.01	68.21188	−1.20188	67.10028	−0.09028
4	11.8	2	60	68.95	69.69184	−0.74184	69.29837	−0.34837
5	10.2	1.5	120	76	75.14277	0.857225	72.76115	3.238851
6	11.8	1.5	120	84.03	82.71274	1.317257	83.96701	0.062988
7	10.2	2	120	75.3	74.93575	0.364245	75.3897	−0.0897
8	11.8	2	120	77.5	78.52072	−1.02072	77.69192	−0.19192
9	9.65457	1.75	90	74.68	75.31156	−0.63156	71.7042	2.975796
10	12.3454	1.75	90	83.39	82.92162	0.468384	82.96955	0.420448
11	11	1.32955	90	83	84.76489	−1.76489	83.04812	−0.04812
12	11	2.17045	90	72.48	70.87829	1.601712	72.41674	0.063258
13	11	1.75	39.5462	74.64	73.95619	0.683812	74.57249	0.06751
14	11	1.75	140.454	76	76.84699	−0.84699	76.05085	−0.05085
15	11	1.75	90	86.32	85.327	0.992999	86.42871	−0.10871
16	11	1.75	90	86.32	85.327	0.992999	86.42871	−0.10871
17	11	1.75	90	85.29	85.327	−0.037	86.42871	−1.13871
18	11	1.75	90	85.06	85.327	−0.267	86.42871	−1.36871
19	11	1.75	90	84	85.327	−1.327	86.42871	−2.42871
20	11	1.75	90	85	85.327	−0.327	86.42871	−1.42871

Table 4 Response surface ANOVA for quadratic model of produced biodiesel

Quadratic model for ANOVA						
Response: yield						
Source	Value for summation of Squares	df	Value for Mean Square	Value for F	Value for p	
	690.54	9	76.73	41.73	<0.0001	Significant
A-molar ratio	69.91	1	69.91	38.02	0.0001	
B-catalyst	232.78	1	232.78	126.59	<0.0001	
C-time	10.09	1	10.09	5.49	0.0412	
AB	7.94	1	7.94	4.32	0.0644	
AC	2.22	1	2.22	1.2	0.2981	
BC	73.39	1	73.39	39.91	<0.0001	
A^2	69.48	1	69.48	37.78	0.0001	
B^2	101.48	1	101.48	55.19	<0.0001	
C^2	177.46	1	177.46	96.51	<0.0001	
Residual	18.39	10	1.84			
Lack of fit	14.48	5	2.9	3.7	0.0887	Not significant
Pure error	3.91	5	0.7825			
Cor total	708.93	19				

Figure 2 shows the variation of surface in three-dimensional standard error of castor biodiesel production concerning ratio of molar with catalyst concentration. The graph show that the standard error is initially increased as the corresponding reaction variables decreased, reaching a maximum value of molar ratio of 11.75 (% v/v) and catalyst of 1.58% as per their weight. This increase in the standard error was preceded by a decrease in the molar ratio and catalyst. It was discovered that the usual minimal mistake (between the actual and anticipated amount of biodiesel yield) was obtained for a specific set of factors. The multiple response approach is desirable. The desirability function, or D (X), converts the response into d (X), a free value known as desirability. It is between 0 and 1. Figure 3 displays three-dimensional plots for the desirability. These graphs show how attractiveness first rose with falling concentration of catalyst and molar ratio, peaked at a molar ratio of 11.75 (% v/v) and a concentration of catalyst is 1.58 (% w/v), and then began to decrease. The best model forecast is what is most desirable (Fig. 4).

The bottom of Fig. 4 displays a graph plot for desirability together with the variation in the reaction parameter. Therefore, as the molar ratio and concentration of catalyst decreased, desirability increased. At a specific point, it turned into "1" (most desired value). The design point is the point where the standard error is at its lowest, and the level of desirability is at its highest. Figure 4 displays the actual and expected biodiesel yield for 20 tests using the combination of three variables. The model's significance can be seen by the closer values obtained between the biodiesel's actual

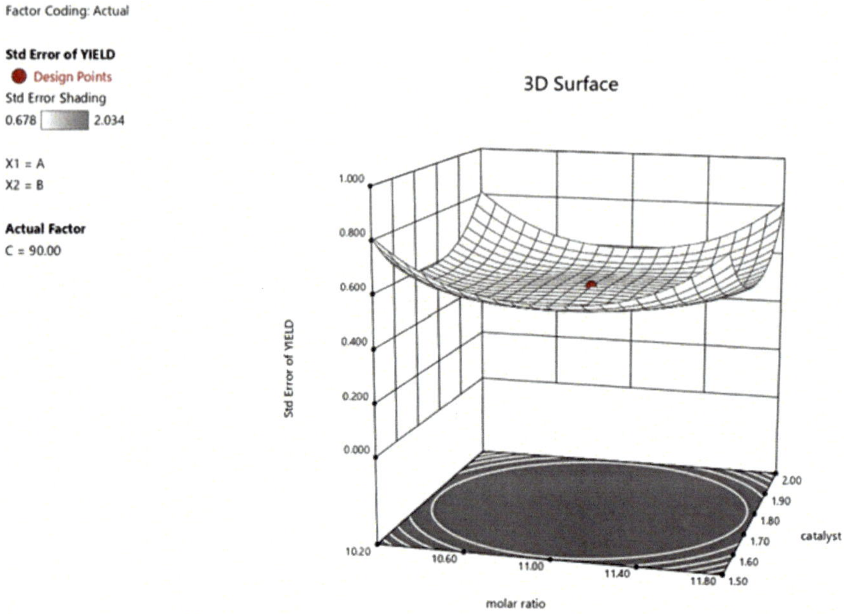

Fig. 2 3-D surface graph of standard error for castor biodiesel yield

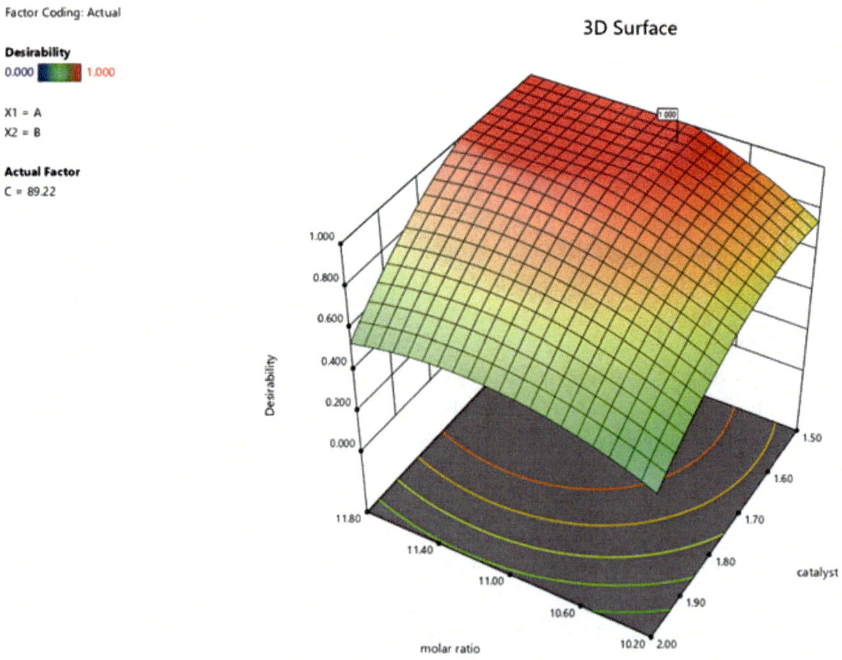

Fig. 3 3-D Surface graph between Biodiesel yield and Desirability

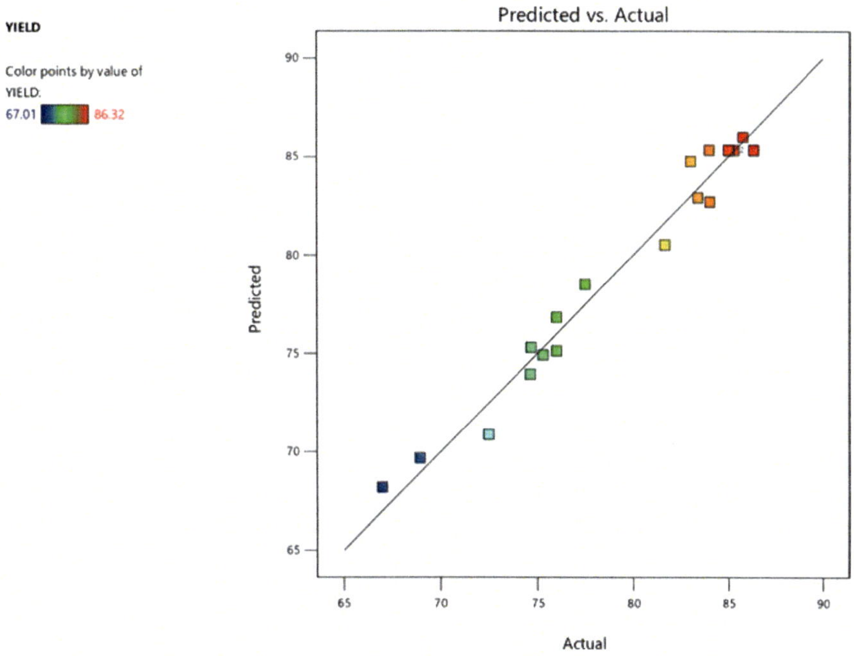

Fig. 4 Graph between predicted and actual yield

and expected values. To maximize the objective function, optimization was used for various ranges of reaction variables. As indicated in Table 5, multiple combinations of variables led to varied solutions. Since the output was inadequate, a different strategy was used to determine the ideal. To maximize the objective function, mathematical optimization was used for various ranges of reaction parameters.

Table 5 Various parameter combination for optimization in transesterification process for maximum production of biodiesel

Number	Molar ratio	Catalyst	time	YIELD	Desirability	
1	11.75	1.65	77.03	86.4178	1	Selected
2	11.39	1.65	99.13	86.9859	1	
3	11.15	1.65	87.80	86.9742	1	
4	11.32	1.58	95.71	87.4051	1	
5	11.01	1.59	98.39	86.3495	1	
6	11.10	1.53	86.64	87.4049	1	
7	11.47	1.61	73.18	86.8293	1	
8	11.06	1.61	82.74	86.9719	1	
9	11.22	1.51	100.44	86.473	1	
10	11.37	1.61	93.65	87.5088	1	

Table 6 Validation of experiments under Response Surface Method

Number	Value of molar ratio	Catalyst	Time	Yield
1	11.75	1.65	77.03	84
2	11.75	1.65	77.03	83.5
3	11.75	1.65	77.03	83.8

3.2 Training and Development of Fitness Function Using ANN

Twenty input sets (A, B, and C) and output have been used to train the predicted model from ANN (biodiesel yield). The matrix's rows and columns provided the neural fitting tool with the input of [10.2, 1.5, 60] and output data [11.8, 2, 120]. There were 20 experiments conducted, 16 of which were for training and four each for validation and testing. While validation was used to gauge network generalization and stop training when it stopped out, training was used to represent the network during training and the network was modified per its mistake. Testing had no bearing on training hence it offered a neutral indicator of network performance both during training and after training. Fitness Function is a MAT LAB has been created. It is regularly performed with various neurons and a different fraction of training data. As shown in Table 7, the training functions were chosen one at a time and added to the fitness function until we received the best value of R (correlation coefficient) was achieved. Regression values (actual) examined the correlation between actual and predicted results. R's value of 1 denotes a close link, whereas R's value of 0 denotes a random relationship. The hidden layer size was anticipated to be 10, the training R was 0.99967, the validation R was 0.99613, and the testing R was 0.99976, resulting in an overall correlation value of 0.98389 (Fig. 5). It was discovered that the sequential weight with learning function "trainlm" was the training function with the best correlation coefficient. A duration of 12 s and an epoch of 1,000 iterations were obtained using the fitness function from the abovementioned parameters. The mean squared error was another tool used to evaluate the effectiveness of the ANN model. The average squared difference between the goal (actual) value of biodiesel yield and the outputs (predicted) value is what this term means. Lower values are preferable, while 0 indicates no error. The information listed below was saved for usage in Genetic Algorithm: MATLAB networks (net), performance and data set information to outputs to MATLAB matrix (output), MATLAB (info), and errors to MATLAB matrix (error). An optimization software was made to predict the biodiesel's superior value using these characteristics to ensure that the output predicted is accurate (biodiesel yield).

Table 7 Training function to develop the ANN fitness function

Ra	Training	Validation	Test	Overall
Trainlm	0.99967	0.99613	0.99976	0.98389

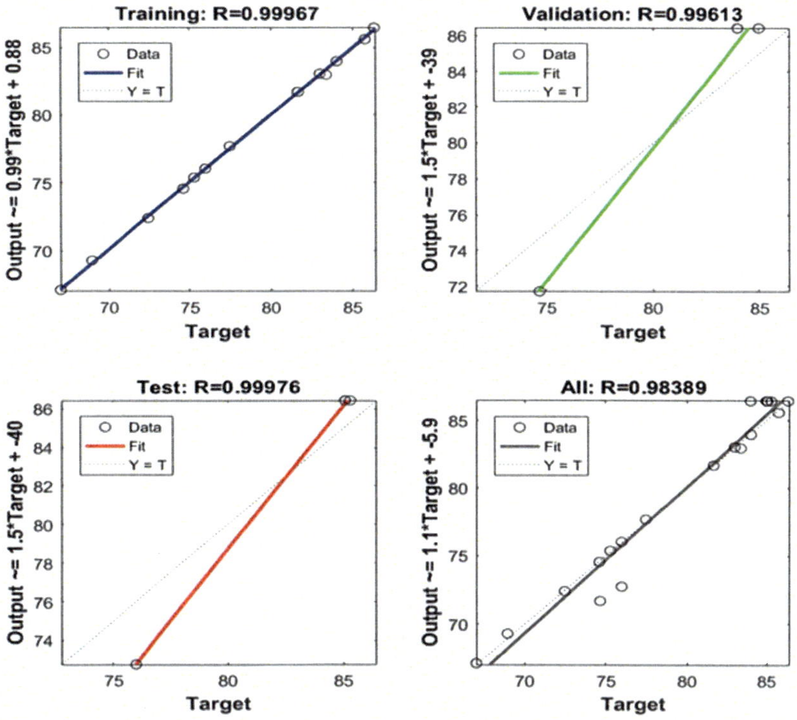

Fig. 5 Coefficients of correlation (R) between actual (target) and predicted value of castor oil biodiesel

3.3 Variables Optimization by ANN-GA

Using data from a neural model, a genetic algorithm programme was created to predict the optimization set of parameters for the maximum possible yield of biodiesel. The design matrix of the RSM was used to obtain the lower bounds and upper bounds for the reaction parameters used in the GA optimization tool. The MAT LAB-GA software (Fig. 6) was repeatedly run to find the best solutions while changing the GA settings (Table 8). For the maximum response, the best optimal solution among these was chosen (biodiesel yield). GA parameters included a population of double vector with a size of 50, 2 elite count, a constraint dependent mutation function, a crossover function scattered, a mutation rate value of 0.05, 50 generations, and value of stall generations is 50. This resulted in the best and meant the

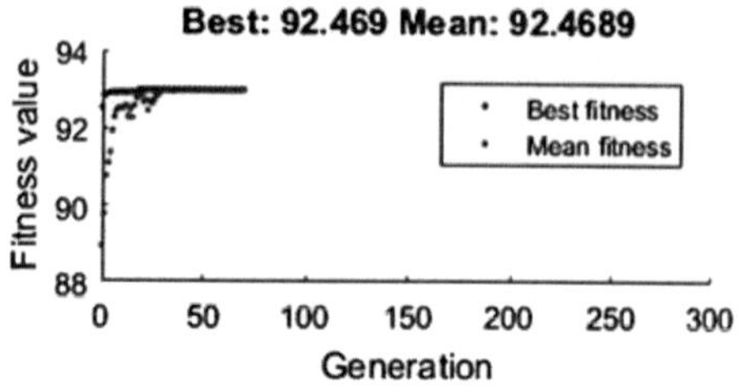

Fig. 6 Development of ANN-Genetic Algorithm for optimizing transesterification process variables

Table 8 Various parameter for training of ANN-GA

S. No	Various parameters	ANN-GA
1	Types of population	Double vector
2	Size of population	100
3	Mutation function	Constraint dependent
4	Value of crossover fraction	0.8
5	Crossover function	Constraint dependent
6	No. of generation	50
7	Value of elite count	0.05
8	Stall generation	50

optimized value of 92.469% by weight (as shown in Fig. 6). For the best biodiesel yield, the following reaction parameters were suggested: molar ratio 11.576 (% v/v), catalyst concentration 1.589% by weight, and reaction time 94.825 min.

3.4 *ANN-GA Model Validation*

The proposed variables combination for castor biodiesel has been used to validate the developed ANN-GA model (Table 9). The transesterification process is carried out using the optimum combinations of the reaction parameters obtained from ANN GA model. To confirm the predicted model, a minimum of three tests has been carried out at the best critical values of the three reaction parameters. The model's acceptability confirmed by an average biodiesel output of 89.70% by weight, close to the projected figure of 92.475%. Table 10 contains measurements of the significant fuel properties of the produced castor biodiesel which is within the ASTM biodiesel standard range.

Table 9 Experimentation validation of ANN-GA

Number	Molar ratio	Catalyst	Time	Yield
1	11.576	1.589	94.825	89.70

Table 10 Comparison of crude castor oil, castor oil ethyl ester (COEE), conventional diesel according to biodiesel standards

Property	Castor crude Oil	Castor biodiesel	Conventional diesel
Kinematic viscosity at 40C (cSt)	231.23	15.45	2.42
Calorific value (MJ/kg)	37.20	38.34	44
Cetane number	–	43.8	51
Density at 15C (g/c m^3)	0.9621	0.9465	0.8389
Cloud point (C)	–	−18	7
Pour point (C)	2.7	−30	3
Flash point (C)	229	192	69
Moisture content (wt%)	0.34	0.12	0.02

4 Conclusions

It has been concluded that the high correlation coefficient (0.98389) for the specified castor oil indicates that a back-propagation neural network with three inputs, ten hidden layers, and one output layer produces a better-proposed model with the following observations.

1. The combined technique of RSM, ANN coupled with GA is capable of predicting the optimum reaction parameters for yield of castor biodiesel of 92.469% by weight compared to 86.681% from RSM alone (A = 11.576% v/v, B = 1.589% w/v and C = 94.825 min,). In contrast to the suggested value, the experimentation validation showed an average yield of 89.70%.
2. Castor biodiesel produced using optimised reaction variables almost satisfies American biodiesel criteria for fuel characteristics.
3. Due to the same properties as commercial diesel, biodiesel can be used as an alternate fuel for compression ignition engines by mixing it with commercial diesel.

References

1. Osorio-González C, Gomez Falcón N, Sandoval-Salas F, Saini R, Brar S, Ramírez A (2020) Production of biodiesel from castor oil: a review. Energies 13:2467. https://doi.org/10.3390/en13102467
2. Bauddh K, Singh K, Singh B, Singh RP (2015) Ricinus communis: a robust plant for bio-energy and phytoremediation of toxic metals from contaminated soil. Ecol Eng 84:640–652
3. McKeon TA (2016) Castor (Ricinus communis L.). In: Industrial oil crops. Amsterdam, The Netherlands, Elsevier, pp 75–112. ISBN 978-1-893997-98-1
4. Pawar N, Kumar N, Kundu KK, Malik DP (2020) Introduction growth and economic profitability of castor cultivation insemi-arid zone of Haryana. Green Farm (2 & 3):258–262

5. Shrirame H, Panwar N, Bamniya B (2011) Bio diesel from castor oil—a green energy option. Low Carbon Econ 2:1–6
6. Hundie KB, Shumi LD, Bullo TA (2022) Investigation of biodiesel production parameters by transesterification of watermelon waste oil using definitive screening design and produced biodiesel characterization. South Afr J Chem Eng 41:140–149. ISSN 1026-9185
7. Hincapie H, Mondragon F, Lopez D (2011) Conventional and in situ transesterification of castor seed oil for biodiesel production. Fuel 90:1618–1623
8. Meneghetti SMP, Meneghetti MR, Wolf CR, Silva EC, LimaGES CMDA, Soletti JI, Carvalho SHV (2006) Ethanolysis of castor and cottonseed oil: a systematic study using classical catalysts. JAOCS 83:819–822
9. Shukri MR, R. Mustafizur, D. Ramasamy, Kadirgama K (2015) Artificial neural network optimization modeling on engine performance of diesel engine using biodiesel fuel. Int J Autom Mech Eng 11:2332–2347. https://doi.org/10.15282/ijame.11.2015.15.0196
10. Optimization of transesterification of castor oil with ethanol using a central composite rotatable design (CCRD). Fuel 89:1172–1176
11. Madankar CS, Pradhan S, Naik SN (2013) Parametric study ofreactive extraction of castor seed (Ricinus communis L.) for methyl ester production and its potential use as bio lubricant. Ind Crops Prod 43:283–290
12. Garhenson. Artificial neural networks for beginners' Carlos Stephen JC (2000) MATLAB, Programming for Engineers. Cole Publishing Company
13. Farobie O, Hasanah N, Matsumura Y (2015) Artificial neural network modeling to predict biodiesel production in supercritical methanol and ethanol using spiral reactor. Procedia Environ Sci 28:214–223. ISSN 1878-0296
14. Izadifar M (2005) Neural network modelling of trans isomerformation and unsaturated fatty acid changes during vegetable oil hydrogenation. J Food Eng 66(2):227–232
15. Hashemi N, Clark N (2007) Artificial neural network as a predictive tool for emissions from heavy duty diesel vehicles in Southern California. Int J Engine Res 8(4):321–336
16. Ghadge SV, Raheman H (2006) Process optimization for biodiesel production from Mahua (Madhuca indica) oil using response surface methodology. BioresourTechnol 97(3):379–384

Reducing the Maintenance Cost and the Customer Issue to Improve the Customer Satisfaction as Well as the Quality of Crane by Using the 8D Problem Solving Tool

Suraj Kumar, Munna Verma, and Dharmendra Kr Dubey

Abstract 8D's tool is a problem-solving approach, to find out the main reason of issues. 8D's is a remarkable first step to enhance the standard and reliability when there is any product issue or not satisfying the consumer. To attain this, the 8D solving strategy was used to solve the problem by the help of Pareto analysis. Cause and effect diagram was constructed as well as the 5 why survey was carried out after the brainstorming session. First the data collected then an investigation was done to identified the matter which influencing the consumer objection management system in crane industry. This tool is used not only as a problem solving tool, and it also play an important role in reducing the cost, to improve the quality along with improving the customer satisfaction. During the whole case study, the problem was defined, the variables were measured and the causes for failure were analysed and to improve over these failures were planned and a standard performance level was achieved. By integrating and managing the quality of record the reorganization of tasks and methods were improved. In order to achieve this 8D's tool was used to allocate and eliminate the issues arises in which Pareto analysis plays an important role. 5why analysis carried out after the brainstorming session.

Keywords 8D process · Functional team · 5 why · Root causes · Prevent recurrence · Permanent solution

S. Kumar (✉) · M. Verma
Department of Mechanical Engineering, Bhagwant University, Ajmer 305004, India
e-mail: gmpsuraj@gmail.com

D. K. Dubey
Department of Mechanical Engineering Chhatrapati, Shivaji Maharaj Institute of Technology, Navi Mumbai, India

H. Kumar et al. (eds.), *Recent Advances in Intelligent Manufacturing*,
Lecture Notes in Mechanical Engineering,
https://doi.org/10.1007/978-981-99-1308-4_22

1 Introduction

This paper depends on two goal is objective; first one is look out the 8D tool approach. In which eight-steps consists which is followed by team of quality upgradation for problem and for product and system improvement. Second objective is to analyze its effectiveness after applying the 8D methodology. Product quality plays an important role to survive in market along with the competitor and internal and external. Consumer Loyalty play key role for an organization in the global market as well as it also helps to increase the production rate as per supply chain management [1]. 8D Methodology is 8 disciplinary approaches to solve the problem. 8 step is used to improve the quality by the analyzing team of quality along with correlated team member. In this research, first the 8D's tool and their technique is studied along with effectiveness by the help of a case study. The 8D's is nothing it's a group of those who is working together to remove the problems by facts along with proper data, rather than the opinions. Table 1 shows the eight disciplines in problem solving.

8D procedure are as follows:

Y1—(1 day) A1;

Y2—(2 days) A2 along with A3;

Y14—(14 days) A4 and A5;

Y60—(60 days) Everything should be finished.

As per example, after getting the issues from the consumer, from A1 to A3 should be finalized within two days second feedback of consumer should be received.

Table 1 Eight disciplines in problem solving

Discipline number definition	Discipline details
D1	To form the team
D2	Describe the problem
D3	Instant action
D4	Find out the root causes
D5	Curative steps
D6	Look out their effectiveness
D7	Extension
D8	Boost up the teams

2 Literature Review

Following will be the analysis of literature behalf of 8D method and its steps to solve the problem. Wan Ahmad [2] supervise the case study on implementation of 8D technique for defective product in automotive sector. His research proves that 8D approaches fulfill all the assurance by reducing the issues in the part of assembly line. By Marjanca [3] find out 8D approach is important key to prevent the issues arises again and again. Cheng [4] uses 8D approach as well as Kanos tool in his study and with the help of them he resolved the issues deduce the error percentage from 28 to 0.5%. As per Rambaud [5] 8D tool can be in to improve the military standards. To decrease the refusal Yadav [6] used QC tools, as pareto analysis, cause and effect diagram along with why analysis.

3 Introduction to 8D Methodology

D0: Make Plan—Learn the basic principle and make the plans for resolving the Problem

D1: Make the tram—Make the specialist team

D2: Draw & define the problem—To resolve the problem by the help of 5W2H

terms that is (who, what, where, when, why, how, and how many) (Suraj Kumar, Munna Vrma 2021).

D3: Containment Action—Find out the issues and implement containment actions i.e. it is an action where after detecting the problem supplied of material stopped after allocating he root causes.

D4: To identify the root cause—Locate all the applicable issues that can explain why obstacle has passed and identify where they took place.

D5: Confirm Corrective action- Verify that whether that corrective action can solve the problem.

D6: Validate Corrective Action—Verify that corrective action is beneficial.

D7: TO prevent the reoccurring issues—regroup the system either man Planning, operating or processing to prevent the reoccur same problem.

D8: Congratulate your team—. It's necessary the officially thanks.

Figure 1 shows the 8D system architecture.

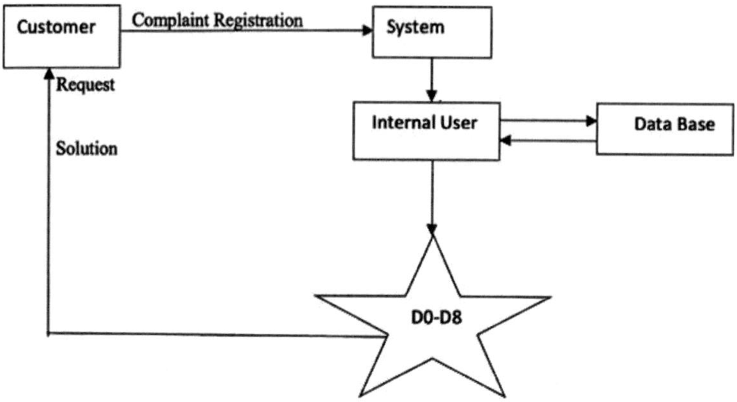

Fig. 1 8D system architecture

4 Case Study

For this research work, the customer complaint record/data was collected because the complaint was reoccurred many times.

4.1 Data Collection and Data Analysis

For pre phase test, previous three months rejection sample was collected along with the Pareto analysis done. Aim of this process was to find out the major issue and eliminate them. Table 2 shows the rejection data. Figure 2 shows the Issue versus running hour plot.

Rejection data table.

4.2 Main Aim of Implementation of 8D Tool is to Reduce the Issue as Well as Maintenance Cost to Improve the Customer Satisfaction

After collecting the data, 8D technique implemented to allocate the issues and eliminate the problem. Activities of 8D systems are as follows.

Step 1: Forming the Cross-Functional Team (D1)

To eliminate the issues first team should be formed. All the expert of concern department (Production, Quality, Maintenance along with R&D) along with the consumer side person.

Table 2 Rejection data

S. no	Capacity (T)	Quantity	Test hour	Running hour	No. of issue
1	12	8	12–13	80	2
2	12	4	12–13	80	2
3	12	6	12–13	82	3
4	12	3	12–13	110	4
5	12	12	12–13	112	4
6	12	4	12–13	122	5
7	12	24	14–15	395.5	8
8	14	8	14–15	92	3
9	14	5	14–15	80	2
10.‘	15	6	14–15	102	3
11	15	18	14–15	158	5

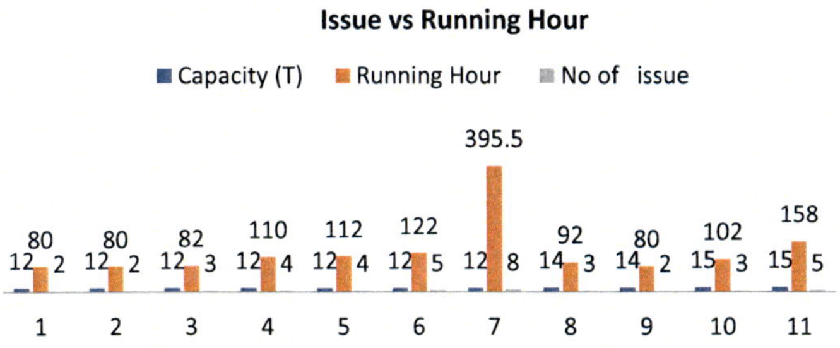

Fig. 2 Issue versus running hour plot

Step 2: Description of the Problem (D2)

Table 3 shows the description of the problem.

Step 3: Containing the Problem (D3)

All the issues were allocated and pinpointed. Lots already on the line were rechecked.

Step 4: Identify the Root Causes (D4)

To identify the root causes brainstorming technique and why analysis implemented and the main issues were allocated for the occurrence of problem:

- No use of service detail record

Table 3 Description of the problem

Issue arises	Crane (customer sight)
At what place defect is observed?	Consumer End after 80 h
No of defects?	41

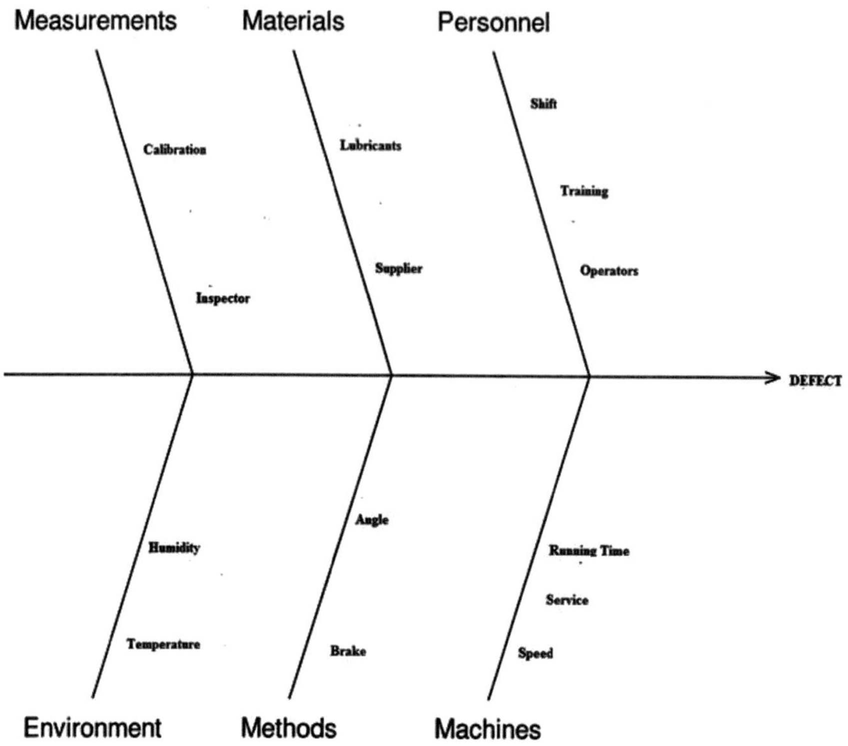

Fig. 3 Fish bone diagram

- Wrong sampling inspection (one part checked in one lot)
- Less skill of the operator
- Negligence for customer review

Figure 3 shows the fish bone diagram.

Step 5: Inscribe the Corrective Actions (D5)

Below mentioned corrective actions were approved in order to avoid the occurrence of issue.

- Mentioned documented record.
- Use proper checking the customer complaint
- Set alarm at schedule time for servicing (Instead of one
- Keep the tool history card

Step 6: Apply the Corrective Actions (D6)

To maintain the distance from the issues corrective actions were implemented that is at after 80–90 h running one service should be provided on behalf of company.

Step 7: Precaution for the Reoccurring the Issues (D7)

Some protective steps were suggested and applied in order to stop the regularity of such issue. A reminder popup set up as a reminder to service provider, maintenance and customer coordinator system which again and again remind them to take review from customer. Table 4 shows some test and trail results.

2: Improved data report after implementation of 8D tool.

Step 8: Praise the Team (D8)

This one is the last step of the 8 D'S toll. After resolving the issues, all team member appreciated for the handsome work done by 8D team for solving the issue. After that 8D steps a closured report was prepared and submitted.

5 Results

After implementation of lasting corrective and preventive actions solution of problem is provided by the 8D tool by which customer satisfaction increase as well as product demand increased and reduces the traveling cost decreases for maintenance.

6 Conclusions

The 8D methodology used as a magnificent tool to resolve the issues for improving the customer satisfaction and reducing the cost and this conclusion is supported by the results of this work. Result of this research show that the 8d's tool is more effective rather than others. This provides the well organizes guidelines for all. After implementing the 8D steps the issue reduces to zero percent after that company declared that the 8D steps had been closed for such issues.

Table 4 Some test and trail result

S. no	Capacity (T)	Quantity	Test hour	Running hour	1st service	Issue	Running hour	2nd service	No of issue
1	12	2	12–13	80	C	0	165	D	0
2	12	2	12–13	80	C	1	170	D	0
3	12	4	12–13	82	C	0	180	P	0
5	12	5	12–13	80	C	0	170	D	0
6	12	4	12–13	80	C	0	170	P	0
7	12	3	14–15	80	C	0	170	D	0
8	14	3	14–15	80	C	1	163	D	0
9	14	2	14–15	80	C	0	162	P	0
10.'	15	3	14–15	80	C	0	165	D	0
11	15	3	14–15	80	C	0	165	P	0

References

1. Awaj YM, Singh AP, Amedie WY (2013) Quality improvement using statistical process control tools in glass bottles manufacturing company. Int J Qual Res 7(1):107–126
2. Najmuddin WA, Saidin W, Ibrahim AM, Azir MZ, Ngah H, Noor NM, Norhidayah MH (2014) 8D methodology: an effective approach for problem solving in automotive assembly line. Aust J Basic Appl Sci 8(22):31–34
3. Krajnc M (2012) With 8D method to excellent quality. J Univ Excellence 118–129
4. Chen HR, Cheng BW (2010) A case study in solving customer complaints based on the 8Ds method and Kano model. J Chin Inst Indus Eng 27(5):339–350
5. Rambaud L (2006) 8D structured problem solving: a guide to creating high quality 8d reports. PHRED Solutions, Breckenridge, Co, p 147
6. Yadav VV, Shaha SJ (2016) Quality analysis of automotive be casting for productivity improvement by minimizing rejection. Int J Mech Prod Eng 4(6):1–8
7. Kumar S et al (2023) Reducing the defect and improving the quality of manufacturing product (CT Wheel/Crain part) using 8d problem solving tool. Ind J Sci Res ISSN 2583–2913 3(1)
8. Kumar S et al Improving the profit by finding the root cause and quality of manufacturing product (CT Wheel/Crain part) by the help of 8D problem solving tool ISSN: 2582–8150 IJSMRT 8(2):4

Analysis of Submerged Arc Welding (SAW) Surface Defects Using Convolutional Neural Network (CNN)

Mirza FarhatullaBaig, Khadersab Adamsab, and Dharmendra Dubey

Abstract The industrial applications of submerged arc welding are widely used in recent times. To produce reliable weld quality, it is very important to predict and monitor welding parameters during the welding process. For any product to attain high material toughness, hardness and strength depends on the quality of weld. Evaluation of the weld quality can be characterized based on various parameter such as weld bead, hardness of weld, deposition rate and weld geometry. By changing the welding speed, current, electrode stick out and arc voltage, the characteristics of weld can be controlled. In this paper, submerged-arc welding surface defect identification has carried out using Convolutional Neural Network (CNN). Using a design convolutional neural network, feature extraction is performed automatically. Following feature extraction, the random forest method is used to perform step classification. Weld surface and defect images were segmented and enhanced using image thresholds. To assess the performance of CNN, the database of weld surface images was compared to experiments. With combined application of CNN and random forest, accuracy of 0.96% has achieved.

Keywords Submerged-Arc welding · Weld surface defects · CNN

1 Introduction

Modernization and application of welding has been growing in Aircraft, Automobiles, Shipping, Electronics industries and even in Construction. With this extensive application, welding industry is turning to a multibillion-dollar sector. Advancement of welding sector has increased the involvement of process parameters, which

M. FarhatullaBaig (✉) · D. Dubey
Department of Mechanical Engineering, Bhagwant University, Ajmer, Rajasthan, India
e-mail: jawamiulkalim9@gmail.com

K. Adamsab
IEA Consultants, Bangalore, India
e-mail: abdulkhader851@gmail.com

H. Kumar et al. (eds.), *Recent Advances in Intelligent Manufacturing*,
Lecture Notes in Mechanical Engineering,
https://doi.org/10.1007/978-981-99-1308-4_23

required the coupled field knowledge of engineering and basic science. For metals and alloys, a common type of joining method is welding, which can be done by varying the parameters to achieve various levels of strength. Welding has been a significant part of the manufacturing and construction industry in recent years. In order to attain required standard of welding quality different parameters are involved like bead porosity and geometry. Development of weld defect may lead to induced stress in the weld joints. Depending upon the stress intensity induced due to weld defect may led to failure of weld. Identifying and correcting the weld defect may help in avoiding any catastrophic failure of Weld Joints. Monitoring of weld is done by changing the process parameters. The variation of quality of welding depends upon these welding parameters directly and indirectly during the arc welding process. To determine how the bead geometry affected by welding parameters is one of the important tasks to develop a model for predicting welding quality. For any welding process to obtain high quality welds, selection and close monitoring of parameters are quite important steps. Estimating the effect of welding parameters to obtain optimal design of bead geometry many attempts need to be carried out. Welding process automation has increased the complexity to manage the data generated and to establish relationship between variables and parameters. These parameters and welding variables include supply of heat input high or low, instability in welding arc, errors in the welding joint and respective distortion, formation of porosity and undercut in weld, lack of penetration, welding fusion development of irregularities in bead geometry. Deterioration in microstructure and crack in weld also lead to loss of mechanical properties [1–4]. To get high production and precision many fabrication industries preferring submerged arc welding (SAW) in recent times. Figure 1 shows schematic diagram of submerged arc welding (SAW) process. To achieve high quality weld bead in SAW process, it is very important to establish relationship between weld bead and parameters of welding process. Use of SAW for metal joining process has increased tremendously in industries. As the flux in SAW, process covers weld pool ensures high penetration in weld pool with good surface appearance. Also, it eliminates formation of smoke and arc flash during the welding process. Key parameters such as welding arc voltage, speed and current, polarity will define the quality of the weld joint in SAW. Factors that have direct effect on the submerged welding process is shown in Fig. 1.

Arc development during submerged arc welding process take place between electrode and workpiece. Lag and protective gas are produced by the flux generated during the SAW process, which help in the deposition of metal. Deposition of metal allows control in composition during the formation of weld pool. During the SAW process, work piece surface coated with flux powder. Movements of arc is along the joint of weld in submerged condition. Heat loss is minimum and high thermal efficiency during SAW process arc is covered flux layer. Factors that have direct effect on the submerged arc welding process is shown in Fig. 2.

Components and systems with critical weld joints involved with high stresses need the weld monitoring. For monitoring defects in welds classification and identification of defect is required. Weld monitoring is done by evaluating the welding variables

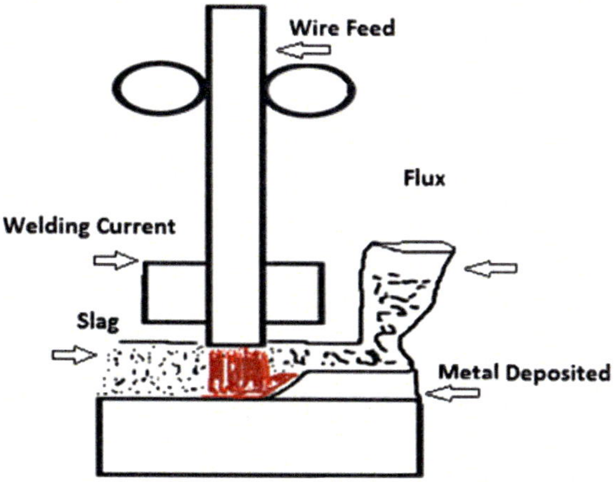

Fig. 1 Schematic diagram of SAW process

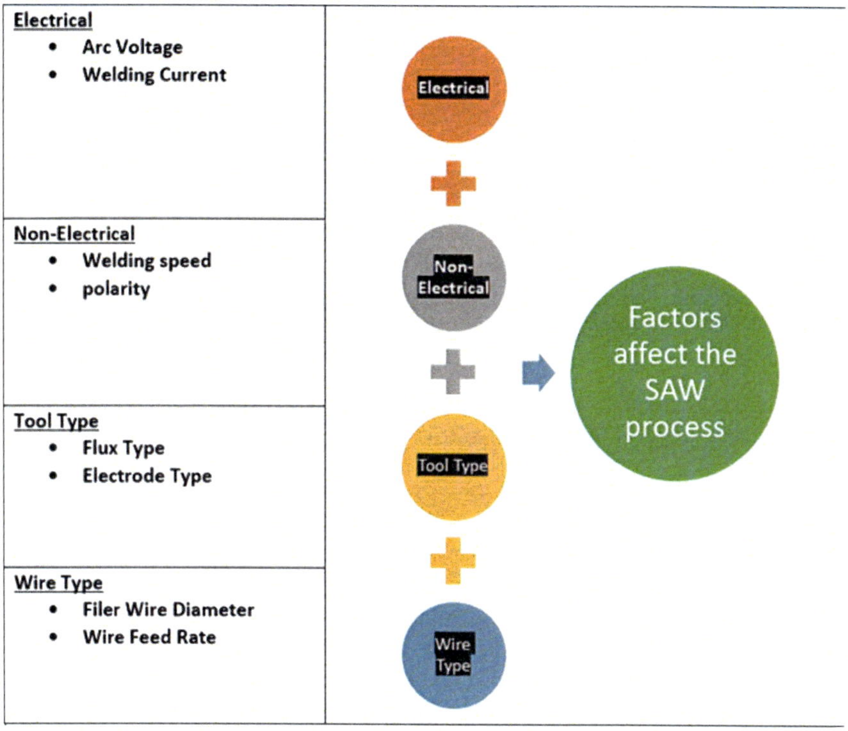

Fig. 2 Factors affecting the SAW process

in accordance with standard of welding manual. Many of the non-destructive techniques such as Ultrasonic test is used in identification of defect in weld. Data based assessment were carried out by comparing the samples data from the defect free weld and with defected weld joints. Designing of critical construction and components with advancement of new manufacturing methods has increased considerably in many divisions of Engineering. In recent time, the use of additive technologies has easy transformation process of use high-end equipment. Factors such as high cost, use of new materials and lack of method led to hinder additive technology. To overcome this drawback associated with use of additive technologies application of artificial intelligence is required. Initial methods of empirical, software based numerical theoretical studies were developed and future advancement has led to application of artificial intelligence technology. With application of artificial intelligence has increased the capacity of analysis of the weld quality [5–7]. Advancement of manufacturing process has directly affected the welding process in many ways. In order to establish interrelationships between the various variables of welding such as input and output need to be address quite efficiently. Neural networks application in modelling the welding variables has given greater advantage in recent times. Neural networks played very important role in managing the various variables and reducing the effect of complication in better way. Regardless the number of welding process conditions to find best weld bead geometry application of neural networks able to cover all the parameters with different set of conditions [8, 9]. Many studies can be categorized in terms of weld pool shape as function of welding variables is carried out using the different types of neural network techniques [10, 11]. By varying welding arc process parameters such as voltage and current, the arc length is analyzed using the neural network method to obtained optimal bead geometry. Using the neural network model and techniques the welding process is analyzed to obtained optimize the parameters of the process [12, 13].

2 Methodology and Experiment Setup

Some of the variables of SAW such as arc voltage, welding speed and current, electrode extension and diameter, flux type and flux layer, current type and polarity are considered for study. Welding arc voltage speed and current has direct effect on weld size, bead geometry and appearance and other hand on penetration rate. In addition, the weld bead geometry and depth penetration are affected by electrode diameter. Rate of deposition and penetration will change with respect to electrode stick out. To get the necessary weld quality very important to select right welding process parameters such as welding speed, current and arc voltage. These parameters may have direct and indirect effect on the weld quality in submerged arc welding. Electrode stick out also affects the weld bead geometry. These parameters were kept constant in the current investigation in order to obtain and optimize the weld. Images of weld surface are collected under the good condition and with defected conditions. Weld Specimen with dimensions of 200 $\times$ 50 $\times$ 5 mm for experiment. The SAW machine generates constant voltage of 20 V and rectifier type power supply. Coated type electrodes are used with wire rate of 4 m/min. Basic model of electrode granular flux with welding speed of 40 cm/min and gas shield of Co2. Images with defects such as pore spatter, overlap, and with constant parameters are taken into consideration. In addition, standard resolution is set for obtaining various postures and appearances. Moreover, group is made for different postures and appearances for same resolution. Performance of welding process is regularly affected by process-based factors such as welding parameters and environment and human engineering.

Images obtained for defected weld surface contain very large amount of redundant information. For removing redundant information and extract defect feature in convolutional neural network (CNN) images are pre-processed.

Sample images of original and with defect weld surface are shown in Fig. 3. Images with defect are pre-processed with segmentation enhancement and filtering techniques at different stages. Initial, to remove noise from original images of weld surface is done using median filter. Future images are enhanced at weld area by processing to gradation technique this enhance the value of weld area. Images of weld surface with defect and without are subjected to image thresholding intensity by using OTSU method. During this process, categorization is performed between the defect free images and with defected images. After setting the threshold intensity for weld, surface images are transfer for feature extraction.

3 Results and Discussion

Extraction of features for weld surface and defected surface images is done using the Convolutional Neural Network (CNN). Architecture of CNN has created using multiple layers of convolutional network. In convolutional neural network, layer consists of convolutional, pooling, one fully connected and SoftMax layers. Total three and two convolutional and pooling layers are used to train the data set more effectively. Images with specific size are validated and trained in the network to obtained optimal difference value. Feature extraction of images carried out using optimizer technique. Data set trained as whole over the time (epoch). Stochastic and batch gradient descent used to perform feature extraction. Gradient descent of the entire batch of training data at a time. Moreover, in stochastic gradient descent the single data trained per time.

Adaptive value is set for learning rate were train data with value of 0.001 and rate of decays are set to constant value 1×10^{-8}. Momentum value is varied from 0.5 to 0.9 to get optimal value. Database of weld surface with defects images are pre-processed using CNN. Figure 4 shows the data set training process were iteration increase with training data set is shown in Fig. 4a. And validation of data set is trained with 95% accuracy. Loss function of data set gradually convergent with time. Features extraction for geometrical and texture for weld surface defect images are modelled artificially. In order to increase the feature extraction, speed the images of weld surface defect are modelled in accordance with matrix in terms of vectorize.

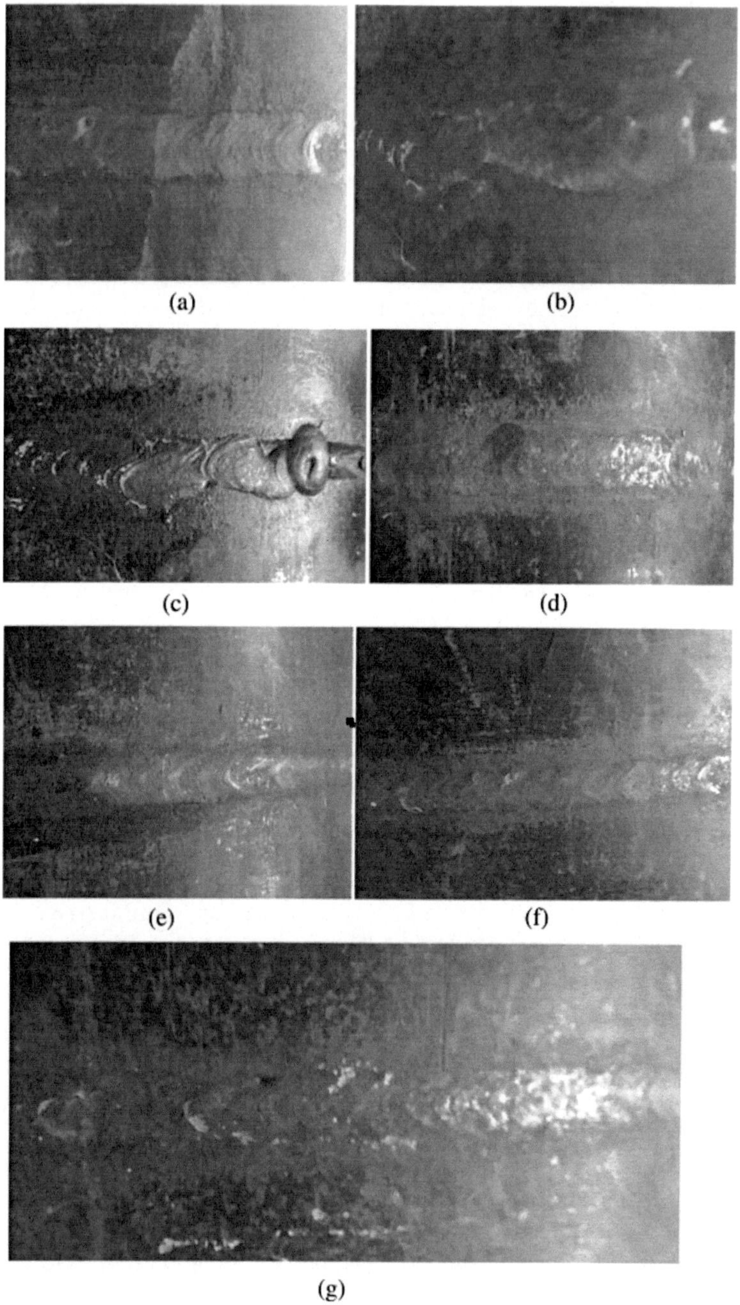

Fig. 3 Sample images of original and with defected weld surface

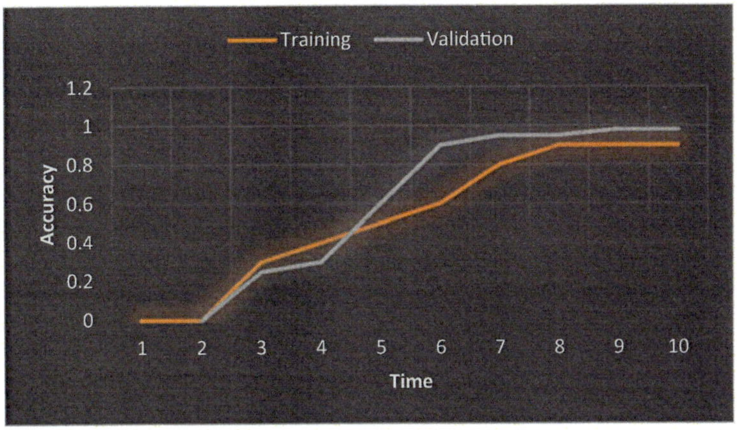

(a)

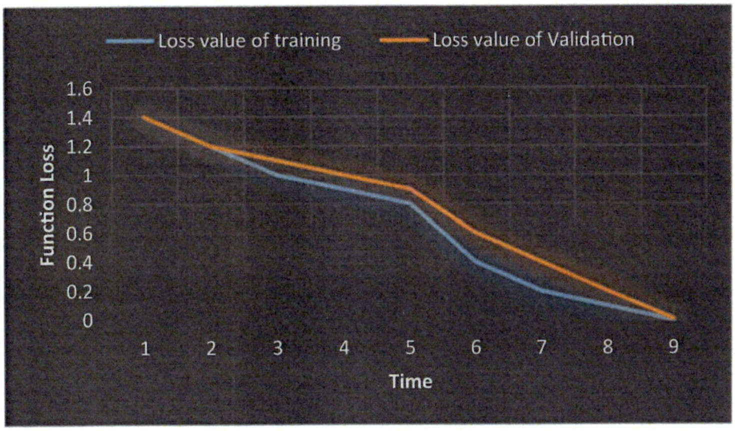

(b)

Fig. 4 Data set training process with accuracy (**a**) and function loss (**b**)

Modelled CNN network will perform the feature extraction automatically. Images are associated with classic feature extraction methods to increase effectiveness. The images extracted via the traditional technique in the feature extraction procedure are shown in Fig. 5a-e.

The CNN feature extraction module studies advanced features from the weld surface defects images. Established image-processing methods are check with our module of feature extraction in order to evaluate the effectiveness. Our method extracted high-level features in better way as compared with traditional method of feature extraction as shown in Fig. 5.

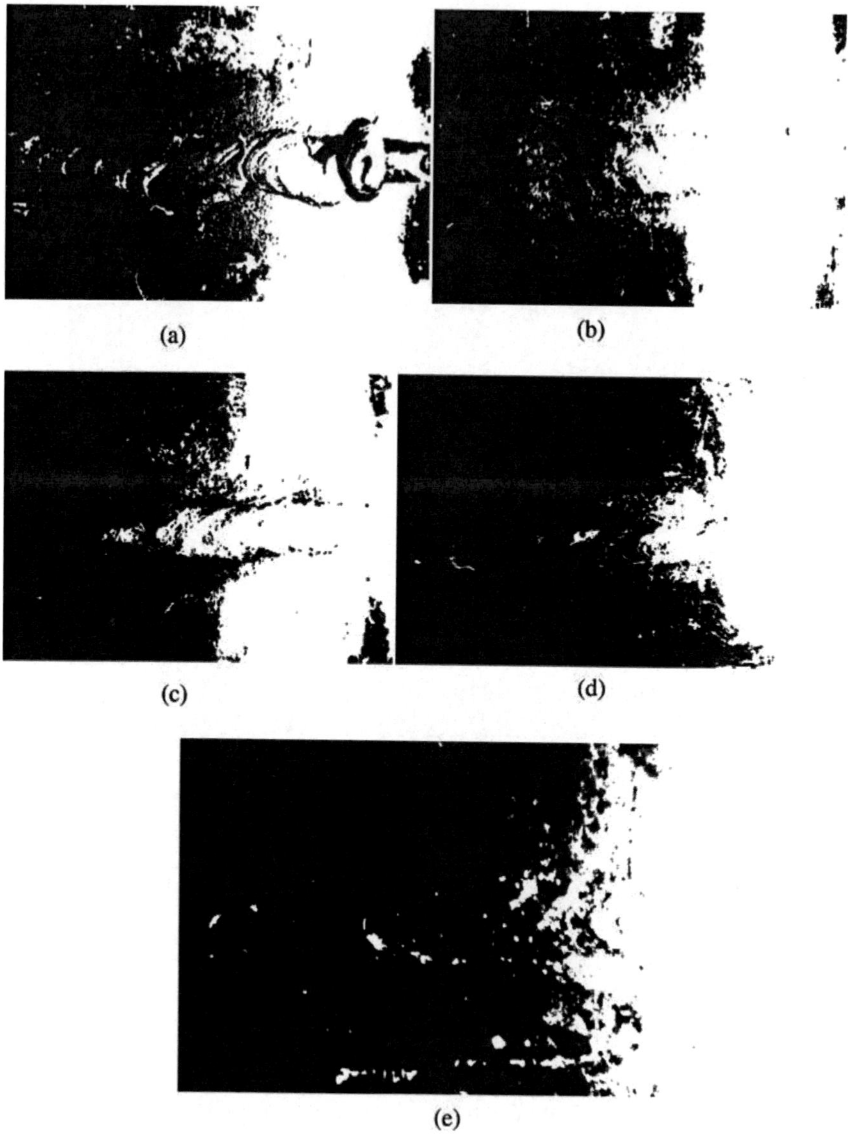

Fig. 5 Sample images after feature extraction technique

4 Conclusions

This research provides a Deep Learning-based approach for detecting weld flaws. The CNN extracts the high-level characteristics of the weld surface defects images, completely utilizing the CNN's capabilities to extract image features. The random forest classification algorithm, which is of a higher generalization ability and robustness, used to identify the classifications of network model. Experiments are modelled accordingly with weld surface images database. The findings demonstrate that the suggested method's highest accuracy is 0.96%, which will identify weld surface defects with better classification of standards and feature extraction to improve overall production process. In the offline method, the model is used to detect weld surface defects and extract features.

References

1. Gunaraj V, Murugn N (1999) Application of response surface methodology for predicting weld bead quality in submerged arc welding of pipes. J Mater Process Technol 88:266–275
2. Yang LJ, Bibby MJ, Chandel R (1993) Linear regression equations for modelling the submerged-arc welding process. J Mater Process Technol 39:33–42
3. Tao X, Zhang D, Ma W, Liu X, Xu D (2018) Automatic metallic surface defect detection and recognition with convolutional neural networks. Appl Sci 8(9):1575
4. Nowacki J, Rybicki RP (2005) The influence of welding heat input on submerged arc welded duplex steel joints imperfections. J Mat Proc Technol 164:1082–1088
5. Sumesh A, Rameshkumar K, Mohandas K, Shyam Babu R (2015) Use of machine learning algorithms for weld quality monitoring using acoustic signature. Proc Comp Sci 50, 316–322, Elsevier
6. Khadersab Adamsab (2020) Machine-learning algorithms for rotating machinery bearing fault diagnostics. Mater Today Proc 44(6), 4931–4933. Elsevier
7. Yan Li, Miao Hu, Taiyong Wang (2018) Weld image recognition algorithm based on deep learning. Int J Patt Rec Artif Intell. https://doi.org/10.1142/S0218001420520047
8. El Ouafi, Bélanger R, Méthot JF (2011) Artificial neural network-based resistance spot welding quality assessment system 108, 343–355. EDP Sciences
9. Won-Bin Oh, Tae-Jong Yun (2017) A study on intelligent algorithms to control welding parameters for lap-joint. Proc Manuf 00, 000–00, Elsevier
10. Wenhui Hou, Ye Wei, Jie Guo, Yi Jin, Chang'an Zhu (2018) Automatic detection of welding defects using deep neural network. IOP Conf Series J Phys Conf Series 933, 012006
11. Sang-jinOh , Min-jae Jung , Chaeog Lim, Sung-chul Shin (2020) Automatic detection of welding defects using faster R-CNN. Appl Sci 10, 8629. https://doi.org/10.3390/app10238629
12. John Stavridisa, AlexiosPapacharalampopoulosa, Panagiotis Stavropoulos (2018) A cognitive approach for quality assessment in laser welding. Procedia CIRP 72, 1542–1547, Elsevier
13. Warren Liao T (2008) Classification of weld flaws with imbalanced class data. Exp Syst Appl 35, 1041–1052, Elsevier

The Study of Sensors in Soil-Less Farming Techniques for Modern Agriculture

Gaganjot Kaur, Prashant Upadhayaya, and Paras Chawla

Abstract Agriculture plays an important role in the economic sector that contributes to a nation's overall economic development. According to the United Nations FAO report, there will be a tremendous food demand globally, which is expected to rise by 70% by 2050. The key parameters such as declining resources, global population, and climate change are becoming challenging for conventional agriculture. In today's scenario, modern agricultural practices such as soil-less agriculture techniques are helping farmers to get rid of the above challenges. Soil-less agriculture techniques like hydroponics and aeroponics are helping farmers to increase yield per unit with no soil and limited use of land and water. These modern agriculture techniques include use of various sensors which help the farmer provide information about moisture, humidity, temperature, etc., which are easily accessible to farmers with the help of smartphones. This paper presents a review study of the modern agricultural technologies that are integrated with sensors to help the farmer analyze the performance of agriculture growth.

Keywords Soil-less farming · Hydroponics · Sensors · IoT · Vertical farming

1 Introduction

India is the sixth largest economy with a GDP of 3.17 trillion as per data available from the world bank [1]. In this economy, the agriculture sector plays an important role and contributes roughly 14% of the Indian GDP. On the other hand, India is also becoming the most populated country (currently at the second position) and is likely to surpass China and will become the world's most populated country by 2028 as per recent UN

G. Kaur (✉) · P. Upadhayaya
Department of Electronics and Communication Engineering, Chandigarh University, Mohali, India
e-mail: gaganbaidwan93@gmail.com

P. Chawla
iNurture Education Solutions Pvt Ltd, Bangalore, India

H. Kumar et al. (eds.), *Recent Advances in Intelligent Manufacturing*,
Lecture Notes in Mechanical Engineering,
https://doi.org/10.1007/978-981-99-1308-4_24

reports. The current world population is estimated to be 7.2 billion and is expected to reach 9.6 billion by 2050 [2]. This tremendous increase in the population has created the biggest challenge in front of the country to achieve Sustainable Development Goal (SDG) two i.e., SDG-2, which talks about zero hunger. Also, as per the report, food production needs to be doubled by 2050 in order to meet the demand of the growing population [3]. As GDP plays an important role in the development of the country it measures the economic health of the country. In India's economy, agriculture is one of the biggest industries, and therefore, the most common word used in India is the Agro-Economy. However, as the population is increasing, the demand for food supply within India is also increasing. As per the report of IMF (International Monetary Fund), the GDP per capita of India is 2,357.643 (in US Dollars). Also, with the increase in the population, there is a heavy depletion of natural resources such as cultivated land, freshwater, etc.

Agriculture resources such as clean water and arable land are declining [4, 5]. The variation in the climate conditions and reduction in bio-diversity are other major concerns in the agriculture field. Further, the lack of rainwater often results in a declined supply of groundwater, lakes, or rivers, leading to a decrease in groundwater levels and the scarcity of natural water resources. Increasing demand for agricultural land has led to the destruction of tropical rainforests around the world, leading to high CO_2 emissions. Crude oil is still the main source of energy in the food, fertilizer, and transport sectors and therefore has a significant impact on climate change [6]. Also, during the present scenario, especially during the phase of the COVID pandemic, there were large-scale disruptions in the food supply, which posed a detrimental impact on the rural Indian economy [7]. Therefore, to sustainably feed the growing population modern methods of farming such as soil-less farming techniques viz. hydroponics, aeroponics, aquaponics, and vertical farming are the latest innovation techniques that may provide some cure to the food supply demand, especially in the urban areas helping to deal with such alarming situation [8, 9]. Thus, aiming at feeding the growing population, different agricultural revolutions in India took place such as the green revolution, the gray revolution, the Proteins revolution, and the Evergreen revolution [10, 11]. These involve the use of hybrid crop species, new tools, high-yielding crop varieties, pest management, and greenhouse technology. Therefore, in this paper, we present the study of the various IoT-enabled soil-less techniques which are used for smart agriculture practices. The organization of the paper is as follows: Sect. 2 explains the various soil-less agriculture techniques. In Sect. 3, we talk about the role of IoT in soil-less farming along with the various soil-less monitoring systems of literature. Section 4 deals with the results and discussion, followed by a conclusion in Sect. 5.

2 Soil-Less Agriculture Techniques

In general, soil-less farming involves growing plants in nutrient-rich solution without the use of soil preferably in an indoor environment whereas vertical farming is all about stacking the various plant layers in order to increase the yield per unit area [12–14]. As these systems do not need soil to be effective, they require far less space and water in comparison to conventional agriculture. Generally, soil-less farming techniques can be categorized mainly into three parts viz., Hydroponics, Aquaponics, and Aeroponics. Hydroponics is the process of growing crops in the absence of soil. In this process, the plants are grown using a nutrient-based solution [13, 14]. The plants are planted with their roots exposed to the mineral solution and are physically supported by using some inert medium such as perlite, pebbles, or other fractions of the soil. One of the key features of the hydroponics system is that it requires the minimum amount of water for plant production i.e., it uses only 10% water compared to conventional agriculture with more than 75% reduction in land requirement [15]. These types of agriculture techniques are very useful in areas where water is declining rapidly and are also useful in urban areas. The hydroponic system is also gaining popularity, due to the non-usage of pesticides and the better plant growth it provides as compared to conventional agriculture as plants receive the perfect amount of nutrition [16]. Aeroponics is also similar to a hydroponic system, in which the roots are exposed continuously in an area filled with fine droplets (fog or aerosol) of the nutrient-rich solution under controlled conditions [17] which do not require any substrate. It involves the growth of plants in a growing chamber with their roots suspended freely in deep air with a fine mist of nutrients. A fine spray mist of nutrients is created with the use of nozzles that provides maximum aeration which is one of the key requirements of aeroponics [18]. Aeroponic systems have lesser water usage by 98%, and lower labor costs with a 45–75% increase in yield when compared with hydroponic systems [19]. Aeroponic systems are commercially very successful for potato production seed germination, tomato production, leafy crops, and young vegetables [20, 21]. On the other hand, Aquaponics is the fusion technique of aquaculture and hydroponics; aquaculture stands for growing fish or other aquatic animals whereas hydroponics refers to soil-less agriculture techniques. The fusion technique uses a combination such that plants are fed with the waste of aquatic animals [22–24]. In this technique, the plant absorbs the required nutrient from the given solution which is required for growth and discards the remaining solution back to the fish reservoir.

Amongst the above-mentioned techniques, hydroponic is a popular technique used by researchers and farmers [25–27]. The hydroponic systems use three main components that include, growing media, air pump, and net pots as shown in Fig. 1. The growing media is an inert media that provides support to the roots of plants. It is a substitute for soil but has no nutritional role in plant growth. This is supposed to retain moisture and nutrients from nutrient solution and hence needs to be porous and have good CEC (Cation Exchange Capability) and WHC (Water Holding Capacity). The available materials are perlite, sand, clay balls, vermiculite, coco-coir, Rockwool, etc.

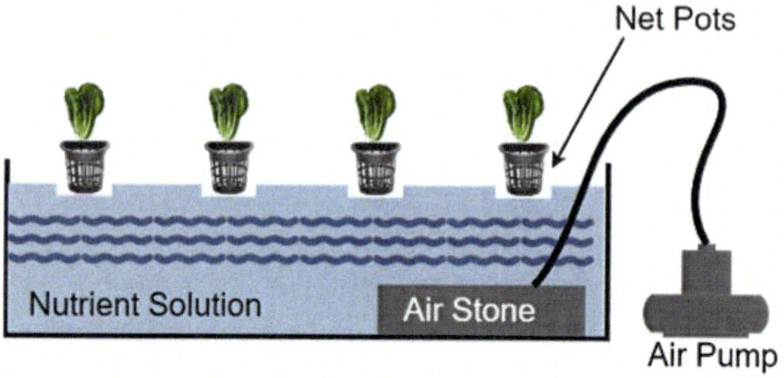

Fig. 1 Basic hydroponic system

[28]. Amongst all these, coco-coir is organic of all and offers good pest management too [28]. Air pumps are used to disperse oxygen in the nutrient solution in the form of air bubbles. The bubbles also help in evenly distributing the nutrients in the solution. They don't generate any oxygen on their own but need to be attached to the air pump and net pots are mesh planters that hold the roots of plants [29].

The hydroponic system can be further classified based on the uptake of the nutrient solution for plant growth as shown in Fig. 2. For example, in Deep Water Culture (DWC), the plants are held in a Styrofoam sheet with net pots over a deep reservoir of nutrient solution. So, nutrients could be easily absorbed by plants [30]. This is one of the simplest systems and plants with this method grow quickly. In the Wick system, the plant is held in an absorbent medium like vermiculite or perlite and then the wick is placed around to carry nutrient solution from the reservoir to the roots. These techniques are suitable for plants that require less amount of water for their growth [31]. One of the advantages of this system is that it does not require a pump for nutrient circulation as a result save electricity. Another technique used in the hydroponic system is Nutrient Film Technology (NFT) [32]. It is the simplest one as it scales well to a variety of applications. In this system, there is a nutrient solution in a large reservoir, and from there it is pumped into the system and then collected back at the reservoir. Net pots are recommended to be used in this system. In the EBB and flow technique [33], plant roots are flooded with nutrient solution and then drained after some time. It uses timers for this purpose to switch off the pump after a certain time. It is also known as the flood and drain method. Large plants are not recommended to be grown in this system. Finally, the Drip System techniques use drip lines to feed the nutrient solution straight to the plant base. Drip emitters are present at end of each tube to control the quantity of solution releases. It could be a circulating or non-circulating system. Circulating is a constant dripping system. As size and flow rate are variable, any plant could be grown in this type of system [34].

Another emerging agriculture technique is Vertical farming. This modern method of vertical gardening was promoted by Dickson Despommier in 1999, who was a Professor at Columbia University in the Department of Public Health and Environmental Sciences [35]. In vertical farming, plants are grown using stacked layers in

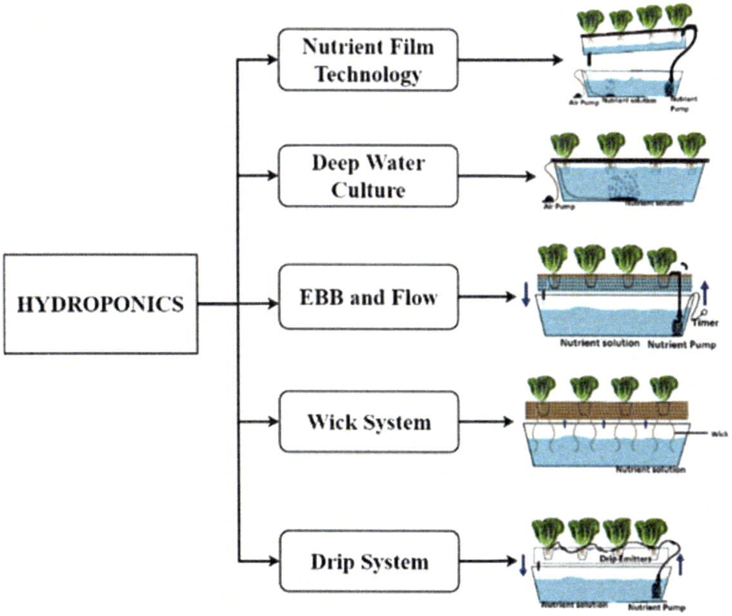

Fig. 2 Different types of hydroponic techniques

the presence or absence of soil [36]. It can be categorized into soil-less farming when the plants are grown in the absence of the soil, usually in controlled environmental agriculture, with the aim to boost growth. Due to the controlled environment, unusual sites can be preferred for vertical farming such as buildings, rooftops, shipping containers, tunnels, etc. Some of the vertical farming setups existing around the world are given in Table 1.

As stated in Table 1, the vertical gardening concept uses the current state-of-the-art technologies, in which artificial sunlight is provided using a special LED light [37]. The main benefit of using vertical farming technology is maximizing yields with minimal land requirements [38]. As it allows to grow large varieties of plants at the same time because plants do not share the same soil plots during growth. It also has an additional benefit of reducing the carbon footprint as it allows the local availability of food in urban areas. However, the economic challenges may be faced with the vertical farming setup as compared to traditional agriculture in terms of cost. As these farms have high energy requirements, owing to extra use of light, water circulation, air conditioning, and maintenance [39]. However, in the future, renewable energy may cater to these high energy requirements [40].

Table 1 Vertical gardens around the world

S.NO	Company	Type of crops	Light source	Websites
1	Sky greens (Singapore)	Sky Chinese Cabbage, Sky Lettuce, Shanghai Green, Sky Cai Xin, Sky Spinach	Natural sunlight	https://www.skygreens.com/
2	Crop box (North Carolina, USA)	Herbs, Greens and Lettuce, Strawberries, Fodders, Microgreens	LED	https://cropbox.co/index.php
3	Aero farms (New York, USA)	Lettuce and microgreens	LED	https://aerofarms.com/farms/
4	Nuvege (Kyoto, Japan)	Lettuce	LED	https://agritecture.tumblr.com / post/27440882258/nuvege
5	Plant Chicago (Chicago, USA)	Microgreens, Herbs	Natural sunlight	https://plantchicago.org
6	Freight farms (Boston, USA)	Lettuce, Greens	LED	https://www.freightfarms.com
7	Verti Crop (Vancouver, Canada)	Microgreens, Leafy greens, and Strawberries	Natural sunlight	https://grow.verticrop.com/vertical-farming/
8	Growtainers (Texas, USA)	Lettuces Microgreens, Flowers and leafy greens,	LED	https://www.growtainers.com/
9	Brooklyn grange (Brooklyn, USA)	All varieties of agriculture products, Naturally-managed honey bee hives	Natural sunlight	https://www.brooklyngrangefarm.com/
10	Green girl (Memphis, Tennessee USA)	Microgreens, Lettuces	LED	https://www.greengirlproduce.com/
11	Omega gardens (British Columbia, Canada)	Basil, lettuce, strawberries, tomatoes, marigold flowers	LED	https://omegagarden.com/

3 Role of IoT in Soil-Less Farming

The increase in world population, high food demand along with limited availability of natural resource has posed a threat to food security. In the past few years, IoT has played a significant role in our day-to-day lives, enhancing our perception and ability to change the environment around us. IoT solutions in agriculture are mainly focused on helping farmers to reduce the demand gap, by ensuring high availability, profitability, and environmental preservation [41]. IoT in agriculture includes specialized equipment, wireless communication, software, and IT services. The Business Intelligence (BI) study shows that the acquisition of IoT devices in the agricultural industry is growing 20% annually and has increased from 15 million in 2015 to 75 million by 2020 [42]. Also, the world's largest agricultural market size is predicted to touch 15.3 billion (compared to more than $5 billion in 2016) i.e., triple by 2025 [43]. The emerging IoT-enabled smart farming is used to monitor the agricultural field setup that consists of various sensors such as humidity, light, temperature, soil moisture, etc. [44, 45]. These IoT-based agriculture techniques are useful for the farmer and the agriculture expert to gather the field conditions and plant growth from anywhere. For example, if the soil moisture level has been detected as low, the farmer may send information to start watering.

IoT has the ability to influence agriculture in many areas such as smart greenhouses, animal tracking, and molding, agricultural drones, predictive analysis for crop forecasting, agricultural robots, etc. [46–48]. Therefore, the role of IoT and its integration has been studied by researchers for soil-less agriculture for monitoring and controlling plant growth conditions. This intelligence system will help in maximizing plant yields. Sensors are an integral part of IoT networks. These sensors are used to capture the data that can be further sent to the IoT cloud for providing optimal growing conditions for plants to grow [49–51]. The recorded data can be further used to analyze and improve the system for a higher and healthy yield. Figure 3 shows the complete connectivity of the IoT-integrated system for soil-less farming.

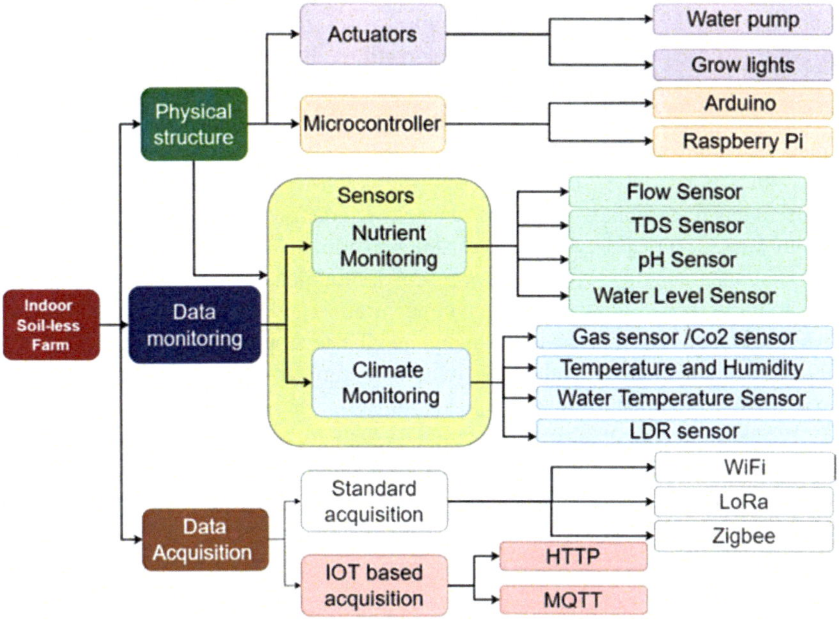

Fig. 3 Major components of IoT enabled indoor soil-less farming

Table 2 present the literature survey of IoT-based hydroponic system carried out by the researcher and academician. In [49] the author compared a hydroponics greenhouse with a conventional system. The main findings of this paper were high precocity in a hydroponic greenhouse, high productivity in fresh and dry matter of aerial part along with more macro and micronutrients. Machine learning algorithm Bayesian network is used in [52] for automated control of hydroponic greenhouse and 66.67% improvement is obtained in crop yield compared to manual control. An IoT-based hydroponic system using Deep neural networks is developed in [50], and it shows an accuracy of 88.50% for the parameters consisting of ppm, temperature, water level, humidity, and LDR.

A Mobile application in integration with AI (Artificial Intelligence) for a smart hydroponic system is implemented in [54]. It also implements an AI framework for predictive analysis of sensor data and disease detection using deep learning convolutional network. The design and implementation of iPONICS system based on Fuzzy logic control which is a low-cost IoT-based intelligent system for a hydroponic greenhouse is reported in [51]. System integration of IoT and ML algorithms i.e., KNN and Lasso Regression is implemented in [55]. A solar-power IoT-based hydroponic system is designed using a Fuzzy association rule mining algorithm for classification based on battery and current weather conditions [56]. Another research for the smart hydroponic system is implemented in [45]. This web interface is designed to monitor and control the parameters with the help of sensors and a controller.

A system to notify the user about the intake of nutrients solution and water is reported in [58] which includes the control of drainage, fan, sprinkler, and water pump. In article [59] the designed system allows users to easily add/remove/exchange sensors as well as actuators and their interactions. It achieved CO_2 concentration reduction 38.54% faster than other systems. A monitoring system for plant growth at different stages, irrigation plans, and water needed by plants is designed with help of an algorithm [60]. A wirelessly controlled system for tomato cultivation in hydroponics is proposed in [61]. It used the 400 MHz band instead of 2.4 GHz because it has zero data loss and good obstacle avoidance. It calculates transpiration. Yildrim et al. designed a system to test the performance of the hydroponic system for Rootstock grapevine [62]. A compact sensor module is developed using an oscillator circuit in [63] to measure nutrient concentration and water level. An automatic nutrition feeding system is designed using the NFT technique [64]. Neural networks and genetic algorithms to conduct the dynamic optimization of water temperature to maximize water content in leaf is reported in [66]. Leaf content is measured using eddy type displacement sensor. An expert system for hydroponics is designed in [67] which the system is able to determine input parameters with output parameters given by the user. Linear regression to implement automatic control of EC and pH is reported in the article [68].

Table 2 Literature study of various hydroponic systems

Sensors crop	Parameters	Crop	Setup/location	User interface/ controller
Domingues [49]	H (pH-meter), EC, LM 35 temp sensor, web cam	Cultivar Vanda (Lettuce)	Greenhouse in center of agricultural sciences, State university of Londrina	ControliHidro App Controller (NTPC-12)
Alipio [52]	pH, EC, RH (relative humidity), LI (light intensity), WT (water temp.)	Lettuce	Manual and Automated (using BN) Local far residence (2.2 m*2.74 m*2.35 m) Laguna, Philippines	ThingSpeak Interface, Raspberry Pi
Ferentinos [53]	EC, pH temperature of solution and controlled pH with metering pump	Lettuce	Kenneth Post Laboratory, Greenhouse at Cornell University in Ithaca, New York	Lab View software
Mehra [50]	Monitored- pH, water conductivity and luminosity, Controlled- pH, temp, light intensity and water level	Tomato	Indoor home-based setup in Chennai, India	Software modules - Raspbian Jessie, Goggle firebase, ARM7, Raspberry PI
Raju [54]	NPK soil, sunlight, turbidity, pH, temperature, water level and camera module	NA	NA	Agri-Hydroponics Raspberry Pi
Tatas [51]	pH, EC, temperature, humidity, motion sensor and camera	NA	Greenhouse operated in lab environment	iPONICS- website Arduino Mega Fuzzy logic control
Srinidhi [55]	EC, DHT 11 and pH sensor	Basil	NA	Mobile application Raspberry Pi Arduino UNO
Zaheer [56]	Light Intensity, Air temp and humidity	NA	NA	Fuzzy association

(continued)

Table 2 (continued)

Sensors crop	Parameters	Crop	Setup/location	User interface/ controller
Palande [57]	EC, pH, water temperature and air temperature	NA	indoor small space	Automation software Domoticz Arduino/ Raspberry Pi
Laxmiprabha [44]	Water flow, humidity, temperature, and motor speed	NA	NA	ThingSpeak FPGA to drive motor and Raspberry Pi
Crisnapati [45]	Ultrasonic sensor, pH, temperature, and EC sensor	Poke Choy lettuce and Kale	Hydroponic system (90*40*20 cm)	Web Interface Raspberry Pi 2 Model D ARM
Aliac [58]	Temperature and humidity sensor, DHT22 and HC-SR04, pH sensor	NA	NA	Google firebase
Wu [59]	CO_2 sensor temperature and humidity sensor	Lettuce	NA	Web-based user interface IPCH-box Arduino- Yun
Pitakphongmetha [60]	DHT temperature and humidity sensor, soil moisture sensor, and ultrasonic sensor	Cantonese	The roof of the building (size 2.4*5.3*2.0)	Arduino IDE
Ibayashi [61]	Temperature, humidity, light intensity, scattered light sensor	Tomato	Greenhouse at Shizuoka Prefectural Research Institute of Agriculture and Forestry, Japan	2.4 GHz and 400 MHz wireless module by Texas instrument
Yildrim [62]	EC by EC59 meter, pH	Grapevine Rootstocks	Pots of 22 cm*72 cm*17 cm Turkey	PIC16/F84
Nishimura [63]	Nutrient Concentration and Water level	NA	NA	Arduino

(continued)

Table 2 (continued)

Sensors crop	Parameters	Crop	Setup/location	User interface/ controller
Eridani [64]	TDS sensor, EC, GPY2YOA21 proximity sensor, MG996R, servo motor	NA	NA	Arduino UNO R3
Peuchpanngarm [65]	Light intensity BH1750, DHT22, Air and temperature, DS18B20 water temperature	NA	NA	Raspberry Pi2
Yumeina [66]	EC, Eddy displacement sensor Air temperature and humidity controller	Tomato	Ehime university, Japan	Neural Networks and Genetic algorithm

4 Results and Discussion

The study of soil-less farming for indoor agriculture is reported as a solution to the tremendously rising population and food demand of the future. As India is going to be the world's highest populated country in the world, soil-less techniques seem to be the future of agriculture due to the lesser use of space and water. Comparing the 3 soil-less techniques hydroponic and aquaponic systems are somewhat similar with only difference in nutrient solution i.e., the aquaponic system uses fish waste as nutrient solution. The aquaponic system seems easier to maintain as there is no need to check the TDS of the nutrient solution throughout the day but it is not suitable for urban areas or households. Comparing hydroponics and aeroponics, Hydroponics uses certain media to support and feed plants whereas the misting system is used in aeroponics. Hydroponics uses 10% of water in comparison to conventional agriculture and aeroponics uses just 5% of the water [38]. The aeroponic system requires constant attention making hydroponic much easier for beginners. However, both of these systems are much more efficient than conventional agriculture. A study has shown that plants grown with the aeroponic technique have 80% stronger root biomass weight compared to ones grown using the hydroponic technique. Also, the aeroponic system uses 65% less water compared to hydroponic systems [69]. The deciding factor depends on ancillary benefits. The study shows that for growing root-based plants such as potatoes, aeroponics is preferred whereas hydroponics is preferred for shoot-based plants.

5 Conclusion and Future Work

The study has shown that modern soil-less farming systems offer the conservation of water and increase in productivity per unit area. As these techniques are based on Controlled environment agriculture, all the plant growth parameters need to be monitored and controlled which requires the use of sensors. Mainly the use of temperature sensors, humidity sensors, LEDs, water pumps, pH, and EC sensors is found in literature, and there is still a scope to use many more sensors for proper monitoring and control of systems like gas sensors, CO_2 sensors, dissolved oxygen sensor, and ORP (Oxidation Reduction Potential) sensor which could further help to increase the yield of the plant. Further, the cost-cutting of deploying these smart agriculture techniques could be further reduced with the help of renewable energy sources.

References

1. World development indicators database (GDP)-Gross Domestic Product, note https://data.worldbank.org/indicator/Ny.Gdp.Mktp.Cd?locations=IN&most_recent_value_desc=true. Accessed 04 September 2022
2. World population prospects 2022: Summary of results., Tech. rep., United Nations Department of Economic and Social Affairs, Population Division (2022),UN DESA/POP/2022/TR/NO. 3. (2022)
3. Pardey PG, Beddow JM, Hurley TM, Beatty TK, Eidman VR (2014) A bounds analysis of world food futures: Global agriculture through to 2050. Aust J Agric Resour Econ 58(4):571–589
4. Bain R, Johnston R, Slaymaker T (2020) Drinking water quality and the SDGs. npj Clean Water 3(1):37
5. Pravalie R, Patriche C, Borrelli P, Panagos P, Rosca B, Dumitrascu M, Nita IA, Savulescu I, Birsan MV, Bandoc G (2021) Arable lands under the pressure of multiple land degradation processes: a global perspective. Environ Res 194:110697
6. Bellard C, Bertelsmeier C, Leadley P, Thuiller W, Courchamp F (2012) Impacts of climate change on the future of biodiversity. Ecol Lett 15(4):365–377
7. Impact assessment of covid-19 on Indian agriculture rural economy, Tech. rep., Department of Economic Analysis Research, National Bank for Agriculture and Rural Development (NABARD), Mumbai, India (2020)
8. Pradhan B, Deo B (2019) Soilless farming—the next generation green revolution. Curr Sci 116:728–732. https://doi.org/10.18520/cs/v116/i5/728-732
9. Beacham A, Vickers L, Monaghan J (2019) Vertical farming: a summary of approaches to growing skywards. J Hortic Sci Biotechnol 94:1–7
10. Moore JW (2010) The end of the road agricultural revolutions in the capitalist world-ecology. J Agrar Chang 10(3):389–413
11. Eliazer Nelson ARL, Ravichandran K, Antony U (2019) The impact of the green revolution on indigenous crops of India. J Ethnic Foods 6(1):8
12. Touliatos D, Dodd IC, McAinsh M (2016) Vertical farming increases lettuce yield per unit area compared to conventional horizontal hydroponics. Food Energy Secur 5(3):184–191
13. Millam S, Sharma SK (2007) "Chapter 32 - soil-free techniques. In: Potato biology and biotechnology. Elsevier Science B.V, Amsterdam
14. Son JE, Kim HJ, Ahn TI (2020) Chapter 20—hydroponic systems. In: Plant factory, 2nd edn. Academic Press, pp 273–283
15. Bradley P, Marulanda C (2001) Simplified hydroponics to reduce global hunger. Acta Hort 554:289–295

16. Jones JBJ (1985) Growing plants hydroponically. Am Biol Teach 47(6):356–358
17. Kerns S, Lee JL (2017) Automated aeroponics system using IoT for smart farming. Eur Sci J (11)
18. Kratsch HA, Graves WR, Gladon RJ (2006) Aeroponic system for control of root zone atmosphere. Environ Exp Bot 55(1):70–76
19. Lakhiar IA, Gao J, Syed TN, Chandio FA, Buttar NA (2018) Modern plant cultivation technologies in agriculture under controlled environment: a review on aeroponics. J Plant Interactions 13(1):338–352
20. Gopinath P, Vethamoni PI, Gomathi M (2017) Aeroponics soilless cultivation system for vegetable crops. 6:838–849
21. Tunio MH, Gao J, Shaikh SA, Lakhiar IA, Qureshi WA, Solangi KA, Chandio FA (2020) Potato production in aeroponics: an emerging food growing system in sustainable agriculture for food security. Chilean J Agric Res 80:118–132
22. Kyaw TY, Ng AK (2017) Smart aquaponics system for urban farming. Energy Procedia 143:342–347
23. Cohen A, Malone S, Morris Z, Weissburg M, Bras B (2018) Combined fish and lettuce cultivation: an aquaponics life cycle assessment. Procedia CIRP 69:551–556
24. Yep B, Zheng Y (2019) Aquaponic trends and challenges—a review. J Clean Prod 228:1586–1599
25. Upendri H, Karunarathna B (2021) Organic nutrient solution for hydroponic system. Acad Lett (07)
26. Waiba KM, Sharma P, Sharma A, Chadha S, Kaur M (2020) Soil-less vegetable cultivation: a review. J Pharmacogn Phytochem 9:631–636
27. Swain A, Chatterjee S, Viswanath M (2021) Hydroponics in vegetable crops: a review. Pharma Innov J 629–634
28. Patil S, Kadam U, Mane M, Mahale D, Dhekale J (2020) Hydroponic growth media (substrate): a review. Int Res J Pure Appl Chem 106–113
29. Geilfus C-M (2019) Hydroponic systems in horticulture. Springer International Publishing, Cham, pp 35–40
30. Moreno MC, Suarez OJ, Garcia AP (2021) IoT-based automated greenhouse for deep water culture hydroponic system. In: 2nd sustainable cities Latin America Conference (SCLA), pp 1–6
31. Harahap MA, Harahap F, Gultom T (2020) The effect of ab mix nutrient on growth and yield of pak choi (Brassica Chinensis l.) plants under hydroponic wick system condition. J Phys Conf Ser 1485(1):012028
32. Al Tawaha AR, Al-Karaki G, Al Tawaha AR, Sirajuddin S, Makhadmeh I, Megat Wahab P, Youssef R, Alsultan W, Massadeh A (2018) Effect of water flow rate on quantity and quality of lettuce (lactuca sativa l.) in nutrient film technique (NFT) under hydroponics conditions. Bul J Agric Sci 24:793–800
33. Daud M, Handika V, Bintoro A (2018) Design and realization of fuzzy logic control for ebb and flow hydroponic system. Int J Sci Technol Res 7:138–144
34. Andaluz VH, Tovar AY, Bedón KD JS, Ortiz, Pruna E (2016) Automatic control of drip irrigation on hydroponic agriculture: Daniela tomato production. In: IEEE international conference on automatica (ICA-ACCA), pp 1–6
35. Despommier D (2011) The vertical farm: controlled environment agriculture carried out in tall buildings would create greater food safety and security for large urban populations. J f'ur Verbraucherschutz und Lebensmittelsicherheit 6(2):233–236
36. Benke K, Tomkins B (2017) Future food-production systems: vertical farming and controlled-environment agriculture. Sustain Sci Pract Policy 13(1):13–26
37. Promratrak L (2017) The effect of using led lighting in the growth of crops hydroponics. Int J Smart Grid Clean Energy 6:133–140
38. Barbosa GL, Gadelha FDA, Kublik N, Proctor A, Reichelm L, Weissinger E, Wohlleb GM, Halden RU (2015) Comparison of land, water, and energy requirements of lettuce grown using hydroponic vs. conventional agricultural methods. Int J Environ Res Public Health 12(6):6879–6891

39. Chowdhury M, Khandakar A (2020) Design, construction and testing of IoT based automated indoor vertical hydroponics farming test-bed in Qatar. Sensors 20. https://doi.org/10.3390/s20195637
40. Al-Chalabi M (2015) Vertical farming: Skyscraper sustainability? Sustain Cities Soc 18:74–77
41. Colizzi L, Caivano D, Ardito C, Desolda G, Castrignanò A, Matera M, Khosla R, Moshou D, Hou KM, Pinet F, Chanet JP, Hui G, Shi H (2020) Chapter 1—introduction to agricultural IoT. In: Agricultural internet of things and decision support for precision smart farming. Academic Press, pp 1–33
42. Misra NN, Dixit Y, Al-Mallahi A, Bhullar MS, Upadhyay R, Martynenko A (2022) IoT, big data, and artificial intelligence in agriculture and food industry. IEEE Internet Things J 9(9):6305–6324
43. Iqbal MZ, Campbell AG (2021) From luxury to necessity: progress of touchless interaction technology. Technol Soc 67:101796
44. Lakshmiprabha KE, Govindaraju C (2019) Hydroponic-based smart irrigation system using internet of things. Int J Commun Syst
45. Crisnapati PN, Wardana INK, Aryanto IKAA, Hermawan A (2017) Hommons: hydroponic management and monitoring system for an IoT based NFT farm using web technology. In: 5th international conference on cyber and IT service management (CITSM), pp 1–6
46. Rayhana R, Xiao G, Liu Z (2020) Internet of things empowered smart greenhouse farming. IEEE J Radio Freq Ident 4(3):195–211
47. Rejeb A, Abdollahi A, Rejeb K, Treiblmaier H (2022) Drones in agriculture: a review and bibliometric analysis. Comput Electron Agric 198:107017
48. Nair AS, Nof SY, Bechar A (2021) Emerging directions of precision agriculture and agricultural robotics. Springer International Publishing, Cham, pp 177–210
49. Domingues DS, Takahashi HW, Camara CA, Nixdorf SL (2012) Automated system developed to control pH and concentration of nutrient solution evaluated in hydroponic lettuce production. Comput Electron Agric 84:53–61
50. Mehra M, Saxena S, Sankaranarayanan S, Tom RJ, Veeramanikandan M (2018) IoT based hydroponics system using deep neural networks. Comput Electron Agric 155:473–486
51. Tatas K, Al-Zoubi A, Christofides N, Zannettis C, Chrysostomou M, Panteli' S, Antoniou A (2022) Reliable IoT-based monitoring and control of hydroponic systems. Technologies 10(1)
52. Alipio MI, Dela Cruz AEM, Doria JDA, Fruto RMS (2019) On the design of nutrient film technique hydroponics farm for smart agriculture. Eng Agricult Environ Food 12(3):315–324
53. Ferentinos KP, Albright LD (2002) Predictive neural network modeling of pH and electrical conductivity in deep-trough hydroponics. Trans ASAE 2007–2015
54. Ramakrishnam Raju SVS, Dappuri B, Ravi Kiran Varma P, Yachamaneni M, Verghese DMG, Mishra MK (2022) Design and implementation of smart hydroponics farming using IoT-based ai controller with mobile application system. J Nanomater (12):4435591
55. Srinidhi HK, Shreenidhi HS, Vishnu GS (2020) Smart hydroponics system integrating with IoT and machine learning algorithm. In: International conference on recent trends on electronics, information, communication technology (RTEICT), pp 261–264
56. Wedashwara W, Jatmika AH, Zubaidi A, Arimbawa IWA (2021) Smart solar-powered hydroponics system using internet of things and fuzzy association rule mining. IOP Conf Ser Earth Environ Sci 712(1):012007. https://doi.org/10.1088/1755-1315/712/1/012007
57. Palande V, Zaheer A, George K (2017) Fully automated hydroponic system for indoor plant growth. Procedia Comput Sci 129(2018):482–488
58. Aliac CJG, Maravillas E (2018) IoT hydroponics management system. In: IEEE 10th international conference on humanoid, nanotechnology, information technology, communication and control, environment and management (HNICEM), pp 1–5
59. Wu TH, Chang CH, Lin YW, Van LD, Lin YB (2016) Intelligent plant care hydroponic box using IoT talk. In: 2016 IEEE international conference on internet of things (iThings) and IEEE green computing and communications (GreenCom) and IEEE cyber, physical and social computing (CPSCom) and IEEE smart data (SmartData), pp 398–401

60. Pitakphongmetha J, Boonnam N, Wongkoon S, Horanont T, Somkiadcharoen D, Prapakornpilai J (2016) Internet of things for planting in smart farm hydroponics style. Int Comput Sci Eng Conf (ICSEC) 2016:1–5
61. Ibayashi H, Kaneda Y, Imahara J, Oishi N, Kuroda M, Mineno H (2016) A reliable wireless control system for tomato hydroponics. Sensors 16(5). https://doi.org/10.3390/s16050644
62. Yildirim M, Dardeniz A, Kaya S, Ali B (2016) An automated hydroponics system used in a greenhouse. Earth Observ Surv Environ Eng V:63–66
63. Nishimura T, Okuyama Y, Matsushita A, Ikeda H, Satoh A (2017) A compact hardware design of a sensor module for hydroponics. In: IEEE 6th global conference on consumer electronics (GCCE), pp 1–4
64. Eridani D, Wardhani O, Widianto ED (2017) Designing and implementing the Arduino based nutrition feeding automation system of a prototype scaled nutrient film technique (NFT) hydroponics using total dissolved solids (TDS) sensor. In: 4th international conference on information technology, computer, and electrical engineering (ICITACEE), pp 170–175
65. Peuchpanngarm C, Srinitiworawong P, Samerjai W, Sunetnanta T (2016) DIY sensor-based automatic control mobile application for hydroponics. In: Fifth ICT international student project conference (ICT-ISPC), pp 57–60. https://doi.org/10.1109/ICTISPC.2016.7519235
66. Yumeina D (2017) Dynamic optimization of water temperature for maximizing leaf water content of tomato in hydroponics using an intelligent control technique. Acta Horticult 1154:55–64
67. Sahin I, Calp MH, Ozkan A (2014) An expert system design and application for hydroponics greenhouse systems. Gazi Univ J Sci 27:809–822
68. Kaewwiset T, Yooyativong T (2017) Electrical conductivity and ph adjusting system for hydroponics by using linear regression. In: 14th international conference on electrical engineering/electronics, computer, telecommunications and information technology (ECTI-CON), pp 761–764
69. Li Q, Li X, Tang B, Gu M (2018) Growth responses and root characteristics of lettuce grown in aeroponics, hydroponics, and substrate culture. Horticulturae 4(4)

Implementation of Sustainable Manufacturing Practices in Indian SME: A Case Study

Naveen Anand Daniel, Ravinder Kumar, Rahul Sindhwani, and K. Mathiyazhagan

Abstract The objective of Sustainable Manufacturing practices (SMP) in manufacturing industries are to minimize the adverse effects of manufacturing operations on the environment and at the same time optimize the production efficiency. In post COVID-19 scenario, customers and manufacturers both are more conscious about the sustainability initiatives. Traditionally the concept of Sustainable Manufacturing (SM) was looked upon by the small-scale industries as the hurdle to the efficiency and profitability. Many researchers have carried out the studies and shown that the adoption of SMP is beneficial in the long term for manufacturing organizations. However, studies on SMP adoption in small and medium enterprises (SMEs) need to be done holistically. Current research presents a case study on adoption of SMP in an Indian small-scale industry. Semi structured interviews, website, and other published information are source of primary and secondary data used for case study. From case study it is observed that technology up gradation, training of employees, formulation of appropriate organizational policies, following government rules and regulations, proper handling of the market competition, creating customer demands are the main factors that case organization considers for the implementation of SMP.

Keywords Sustainable manufacturing practices · SMEs · Indian · Case study

1 Introduction

In modern scenario, sustainability has become very important aspect as it has emerged as an integrating concept [1]. Sustainability is one of the main focus of every entrepreneur, manager, practitioner, customer, and stakeholder in every business [2]. Even though sustainability seems very simple, but its implementation is very complex

N. A. Daniel (✉) · R. Kumar · R. Sindhwani
Amity School of Engineering and Technology, Amity University, Noida, Uttar Pradesh, India
e-mail: nadaniel@amity.edu

K. Mathiyazhagan
Thiagarajar School of Management, Madurai, India

H. Kumar et al. (eds.), *Recent Advances in Intelligent Manufacturing*,
Lecture Notes in Mechanical Engineering,
https://doi.org/10.1007/978-981-99-1308-4_25

and challenging for any organization [3]. The companies are exploring new business models which helps them to face the competition and become more sustainable [4]. Modern organizations have realized the need to implement the mode of productions which are more sustainable such as green manufacturing, green lean implementation, green supply chain management, circular economy, and reverse logistics [5, 6]. Moreover as a result of continuously changing product demand, customer expectations, product life cycle, and new environmental and government guidelines issued from time to time, the SME's are required to plan their manufacturing on sustainable basis so that they are able to face the tough competition and remain in business for a longer period of time. An industry needs a suitable manufacturing strategy to implement the sustainability into its business. Manufacturing strategy is a framework of organizing, managing, and developing the organization resources to compete in the business market" [7]. It is a well co-coordinated plan of action which is used by the organization to achieve its planned goals [8]. Manufacturing should always aim to achieve the strategic goal of an organization [9]. India being a developing nation, manufacturing industries contribute significantly towards the growth of its economy [10]. In the present scenario the manufacturing should be developed in such a way so as to maximize the economic and social benefits and minimize the adverse environmental impacts [11]. In this paper, a case study has been conducted in Indian SME (XYZ Ltd.). The main objective of the case study is to analyze the sustainability issues in manufacturing related to Indian SME in global scenario. Section 2 of the paper highlights the literature related to the study. Section 3 discusses the methodology "SAP–LAP" used in the study. Section 4 discusses the profile of the organization under study. Section 5 discusses case study using "SAP–LAP" methodology. Section 9 will conclude the case study.

2 Literature Review

Due to the present globalized situation it has become very essential for the industry professionals and the academia to collaborate for finding the solution to the issues related to sustainability [12]. There have been studies [13–18] in which the researchers have investigated the enablers and barriers for the implementation of sustainable manufacturing. Most of the studies on this subject have identified the drivers for implementation of SM and ranked them according to their importance. Some of the studies were carried out in India and some were done in other countries like Iran, China, Indonesia, Bangladesh, Taiwan, Malaysia, Egypt, and Uganda [19]. It is evident from the observations that most of the studies carried out in the area of SM are from developing countries of Asia and Africa. It is due to the fact that the developed countries have already adopted the green and sustainable practices owing to the regulatory pressures imposed on them but developing countries of Asia and Africa are still lagging behind in this initiative. Most of the researchers have identified the common drivers of SM. They have applied various tools to identify the essential drivers for implementation of SM and also validated the results with suitable techniques.

3 Research Methodology

The data for the case study has been collected from primary and secondary sources. Semi structured interviews of the concerned manager of relevant department was the main source for primary data whereas the website of the organization and other published information was the source of the secondary data. As per the privacy policy of the company and to maintain the confidentiality of the information the name of the company is not revealed in this paper.

In present study, SAP–LAP methodology [20] is used to analyze the sustainability issues in the Indian SME. The efforts taken by the industry to resolve these issues are also discussed. During the study an in-depth study of the organization was done and there was an interplay between SAP and LAP. Figure 1 shows the SAP–LAP methodology. It is a holistic framework consisting of the elements of the managements along with their function and attributes. SAP–LAP methodology has been used by various researchers in case studies of different sectors [21–24]. Some authors have used SAP–LAP analysis to develop an enquiry model for the Indian energy and power sectors to analyze the status of the supply side of energy security. Some others used the SAP–LAP framework to resolve some critical issues related to adoption of lean manufacturing by Indian MSME's in the context of the scheme of government 'Lean Manufacturing Competitiveness Scheme' (LMCS). Few have analyzed sustainability issues in the context of logistics service providers (LSPs) and also evaluated the sustainability index of an LSP in the Indian scenario by using SAP–LAP methodology while some have carried out a case study on business excellence issues of an Indian automobile manufacturer using the framework of SAP–LAP. The study revealed several critical issues which needed immediate attention.

The case discussed in this study is of an Indian automotive SME. SAP–LAP methodology has been recommended by Sushil (2000) for the critical examination of the case organization. This methodology essentially consists of two steps. SAP analysis is the first step where in the dynamic parameter applicable to the case are identified through the three dynamic interface of any system. These interfaces are situations (S), actors (A), and processes (P). The "situation" represents the present status, organization's environment, and the forces which favors the implementation of Sustainable Manufacturing (SM) in the case organization.

"Actors" who are the individual participants or group of members influence the situation and also evolve business process by defining an organization culture. The examination of the process is done by asking three seminal questions, i.e. what, why, and how? This leads to the generation of the alternatives of the existing process (Sushil 2000). The "Process" is the conversion process that transforms the inputs into output thus recreating the situation.

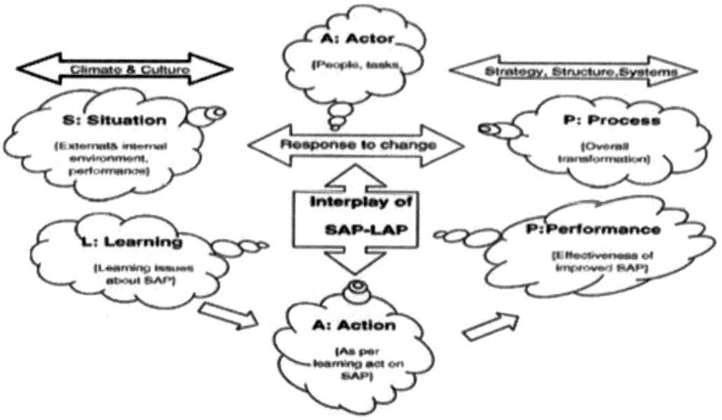

Fig. 1 SAP–LAP framework. *Source* Sushil (2000)

The important points of the emerging situation of the case organization may be understood by understanding situation. These can be top management commitment, technology up gradation, compliance with regulation, proper training system, incentives, and so on. LAP synthesis is the second step. LAP has three components, and these are Learning issues (L), actions recommended (A), and anticipated improvement in performance (P). With the traditional models it is difficult to analyze the entire spectrum in the comprehensive manner. Moreover, the traditional models cannot deal with the changing business situations very effectively. Apart from identifying the sustainability issues the proposed model also explains how to deal with such issues in the changing business scenarios. SAP–LAP is one of the most useful methodologies of analysis and synthesis especially for the organizations adopting the latest technology which is also complex.

4 Profile of the Case Organization

The company XYZ Pvt. Ltd. was established in 1988. The company belongs to a business group which came into existence in 1931 and deals with the manufacturing of large variety of automobile components which are used in two wheelers, three wheelers, passenger vehicles, and commercial vehicles. These components are brake pads, brake shoe, clutch plate, clutch shoe, rubber coupling, filters, lining, and disc pad. XYZ Pvt. Ltd. is in NCR region of Delhi. The total area of the company is 20,000 m^2, and the covered area is approximately 15,000 m^2. Total 300 workers are employed in the organization. Apart from domestic market the company has international market presence in continents like Asia, Africa, Europe, and Middle East. In India all the major automobile manufacturers like Bajaj auto Pvt. Ltd., TVS, Royal Enfield, Tata Motors, Honda cars, Hyundai motors, Maruti Udyog Ltd., Swaraj Mazda, and Eicher are the customers of the case organization. The company operates in 4 shifts. The first shift is from 8:00 AM to 4:30 PM, the second shift is from 4:00 PM to 12:30 AM, the third shift is from 12:00 AM to 8:30 AM, and the general shift is from 9:00 AM to 5:00 PM. The head office dealing with the sales and marketing is located in New Delhi and the regional offices are in Mumbai, Calcutta, Bengaluru, Chennai, and Lucknow. Apart from these the distribution centers are located in 12 other places in India. The main strength of the company lies in the product quality, research and development facilities, high professionalism, effective marketing and sales, and highly motivated manpower.

5 SAP–LAP Analysis

In this section XYZ Pvt. Ltd. is analyzed by using the SAP–LAP model as shown in Fig. 2. For the analysis of the case study SAP–LAP is found to be an effective tool. The analysis of various perspective and the efforts required by the various actors can also be effectively done with this tool. In this study the analysis is conducted in the context of the present scenario of the implementation of SMP in the case organization.

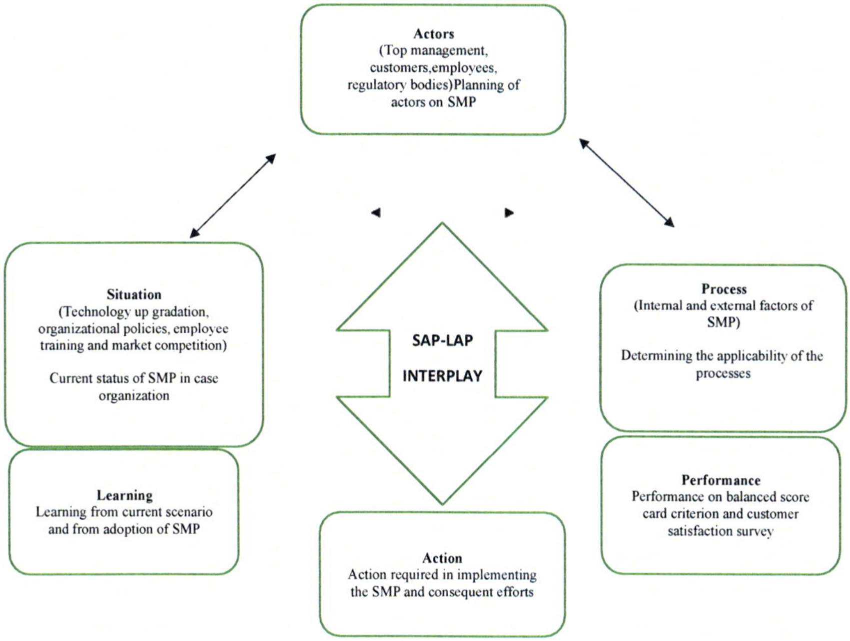

Fig. 2 SAP–LAP framework

5.1 Situation in XYZ Pvt. Ltd.

The present status of the case organization is discussed in this section in terms of the factors of sustainable manufacturing such as technology up gradation, organizational policies, employee training, market competition, customer demand, and pandemic situation.

The mangers of different department of the case organization were interviewed to ascertain the level of implementation of SMP.

- It is interesting to note that since its establishment in 1988, XYZ Pvt. Ltd. is showing a steady progress as per the plan of the company. The total business of the company in the financial year 2015–2016 was of 78 million (US $). This increased to 87 million (US $) in the financial year 2017–2018. But the company faced the tough competition with the entry of foreign competitors in the Indian market. The company realized the challenge to excel in the cut throat competition in the market. The company then adopted the approach of implementing the SMP in its operations. After the efforts of top management which led to the formulation of the suitable organizational policies, technology up gradation, compliance with the regulations, training and skills up gradation of employees, implementation of green supply chain and logistics the case organization has got a total business of 340 million (US $) in the financial year 2020–2021.
- The case organization faces a very intense competition from its competitors in the market. There is a continuous fear of decline in sales and reduction in profits. Moreover there is continuous pressure of cost reduction, waste reduction during production, quality improvement, compliance with the regulations, and also making the work force more productive. More companies are moving towards adopting the green brand image as it creates positive image among the customers.
- The case organization has realized that the most important criteria for the implementation of SMP is the involvement of top management. The authors have observed that the company was strongly influencing the organizational culture to promote sustainable initiative among the employees by providing the required resources. However the authors have observed that the recent situation of pandemic in the global market has created instability leading to obstacles in the implementation of SMP.
- Research and development is a very important function in any business as it leads to technology up gradation by providing powerful knowledge and insights which helps in the development of improved process. The new processes developed are more sustainable and efficient. The case organization has well an equipped R & D centre consisting of chemical laboratory and mechanical laboratory. The chemical laboratory has the facility to perform testing of raw material, wet analysis, and reverse engineering whereas the mechanical laboratory have the modern equipment which can evaluate the physical integrity of the material and carry out physical material testing.

- Due to the shortage of skilled labour in the manufacturing sector the case organization is giving more emphasis on training and skill up gradation of its employees. By doing this the company aims to increase productivity, improve quality, improve efficiency, and ultimately create an effective team. In order to make the training sessions more effective and interactive for the employees the company adopted the approach of hands on training. These type of training program not only captures the attention of the employee but also leaves a lasting impression on him. It also enables the employee to practice the skill learnt during the training in safe and controlled environment.
- Various problems are faced by the XYZ Pvt. Ltd. in the implementation of SMP. Despite that the top management provides all the resources and creates the suitable organizational culture, situations like pandemic can make it difficult to achieve its objectives. The R & D facilities do not have a proper testing facilities to test the prototype. Many workers were not satisfied with the training programs.

5.2 *Actors in XYZ Pvt. Ltd.*

The field visit was undertaken, and the different actors and interlocutors were interviewed. Top level and middle level managers in the production and engineering department were approached and their views were recorded. The following was the views observed.

- In XYZ Pvt. Ltd. the management organizes the lectures for the employees on the sustainable development which gives the employees the right knowledge about SMP. The employees are educated on why the action is important and how to do it. Employees look up to the leaders in the field of sustainability and are more likely to follow them. Group activities on sustainable development are organized which not only explains the social norms but also gives a very positive message to the employees thus creating an organizational culture to promote sustainable initiative among the employees.
- XYZ Pvt. Ltd. lays emphasis on technology up gradation and have developed well equipped chemical and mechanical laboratories. The main materials used in XYZ Pvt. Ltd. are steel, aluminum, plastics, copper, and carbon fiber. The testing and analysis of these material is done in the chemical laboratory. The mechanical laboratory is used for testing the mechanical properties and integrity of the materials.
- The training aims at developing the skills of the employees for operating the latest machines and carrying out the complex operation effectively and efficiently. The common topics of training are
  - On boarding—The employee is introduced to his department and his responsibilities are mapped.
  - Equipment knowledge—Giving instructions about operating the equipment and also repairing it accurately.

- Process training—Explain the complex processes and providing hands on training to develop confidence in the worker.
- Safety training—Teaching the employees the safe and best practices to follow so as to avoid accidents.

5.3 Processes in XYZ Pvt. Ltd.

SMP involves some internal factors which can be handled and decided by the management of the case organization like technology up gradation, organizational policies, and training up gradation of employees. These factors are within the control of the company and can be manipulated according to the current situation.

There are some factors which are external and not within the control of the case organization. These factors are government rules and regulations, customers demand for sustainable products, market competition, and pandemic in the society like COVID 19. The company monitors the external situations very critically. These processes are discussed below.

5.3.1 Internal Factors

- Technology up gradation: The case organization is making efforts in upgrading the technology by its R & D activities. The chemical and mechanical upgradation of the product is the main objective of the company. The company needs to consider the other aspects also like prototype testing and developing a sustainable design of the product.
- Organizational policies: The case organization realizes its responsibility towards society for its impact on society and fulfils its social responsibility by taking steps to reduce carbon footprint and energy consumption.
- Training and upgradation of the employees: The case organization provides training to the employee to get the maximum efficiency because unutilized human skill is also identified as a waste.

5.3.2 External Factors

- Government rules and regulations: Rules and regulations implemented by government have a large impact on the adoption of SMP by the SME's. The case organization follows all the directives issued by the government from time to time.
- Customers demand for sustainable products: Consumer behaviour is the major factor that drives the company to adopt sustainable manufacturing. There is sharp increase in the customers who demand more sustainable product, and the products that are made with sustainable manufacturing practices.

- Market competition: Products manufactured with SMP have an edge over other products as the customers prefer the core values of the company which favours clean environment, human health, fair trade, and social equity.
- Pandemic in the society like COVID 19: Pandemic situation like COVID 19 affects the production and supply chain activities of the company. It disrupts the transportation links and distribution mechanism between suppliers, manufacturing units and consumers. Rise in uncertainty in product demand due to pandemic situation makes the industry to focus on the sales rather than implementation of SM.

6 Learning from XYZ Pvt. Ltd.

The key to successful adoption of sustainable manufacturing by a company is to have a clear and well defined frame work. The XYZ Pvt. Ltd. have already identified the factors of sustainable manufacturing and are working towards their successful implementation. Due to global competition every company is aiming at cost cutting but while doing it the company should also aim at the long term benefits which can be achieved by implementing SMP. Therefore the case organization has to carefully study these factors of sustainable manufacturing and makes efforts to implement them in their processes. The learning related to various internal and external factors of sustainable manufacturing in XYZ Pvt. Ltd. are as follows.

- Technology up gradation: The company was committed to manufacture a quality product and for this reason it has established a well-equipped chemical and mechanical laboratory, but the design aspect was neglected. Due to this the company was not able to provide 100% customer satisfaction.
- Organizational policies: The company is monitoring its energy use and taking steps to eliminate the unnecessary use of energy thus reducing the wastage. The company has also reduced the amount of packaging used for its products. The company should also consider creating positive image in the minds of consumers by contributing to the social causes.
- Training and upgradation of the employees: XYZ Pvt. Ltd. is investing large amount on the employee training as they are targeting happy, challenged, and loyal employees. These employees constantly upgrade their skills and are able to adopt to any changes in the working of the industry. However, the company should evaluate its training program from time to time and make improvements in it so as to make it more useful to the employees.
- Government rules and regulations: Government is passing new laws to address the sustainability problems which is emerging from time to time like waste, pollution, and carbon emission. XYZ Pvt. Ltd. is bound to follow these rules and regulations to maintain legalities and to avoid penalties.

- Customers demand for sustainable products: The highest demand of sustainable products is from the middle class [24]. XYZ Pvt. Ltd. should develop their marketing strategy such that middle level customers are targeted initially to launch their sustainable products.
- Market competition: Sustainability gives a competitive advantage to a company [25]. They provide the strength which is unique to the company. These are the areas where the company excel and makes it different from other companies. The customer values the efforts of the company and start buying its products.
- Pandemic in the society like COVID 19: Pandemic like COVID 19 occurred due to environmental degradation and the manufacturing units are one of the major factor responsible for it. It is a late lesson learned from the early warning. All though sustainability aims at avoiding such situations in the future the companies should also equip itself to face and be stable in such situations in the future.

7 Actions as Per the XYZ Pvt. Ltd.

Based on the situation, actors, and processes the following improvement actions are taken.

7.1 Internal Factors

The internal factors must be analysed critically and implemented in the proper way to take the maximum advantage out of them. The result of the implementation of internal factors can be improved in the following way.

- Technology up gradation: A prototype lab will assist the company to develop a sustainable design of the component. This lab have the facilities like Mixers, Ovens, Raw material storage facility, finished component storage area, and bench marking sample storage. A sustainable product design along with sustainable manufacturing makes the product more marketable.
- Organizational policies: Apart from making policies towards achieving sustainable manufacturing XYZ Pvt. Ltd. should also highlight its positive image in the minds of consumers which will create a larger customer base. The corporate social responsibility makes the company more popular in the market which enhance the profitability and boosts the development of the company.
- Training and up gradation of the employees: Employee listening strategy makes the training and development program more effective and successful. Employees post training survey helps the company to find out the short comings and scope of improvements in the future training programs. It also helps to determine the efficiency of the training program conducted by the company. If it is done in the correct manner it will improve the quality of the training program in real-time with real employee feedback.

7.2 External Factors

The external factors are beyond the control of the organization but they are not unforeseeable. These external factors may change from time to time depending on the current situation. The company should be able to quickly identify and respond to the changes in these factors. The external factors can be successfully implemented as follows.

- Government rules and regulations: The company is bound to follow the government rules and regulations laid down from time to time to avoid penalties. The company also improves its value and sustainable image by taking proactive measures and adopting forward thinking best practices. Proper method of managing, reporting, and improving the environmental, social and governmental (ESG) data creates a positive impact on the investors and stakeholders.
- Customers demand for sustainable products: As the demand for the sustainable products is highest from the middle class, the middle class consumers can be targeted through paid digital campaigns and on site marketing. A right marketing strategy will increase the customers for the products and boost up the sales.
- Market competition: Sustainability gives competitive advantage to the company. The sustainable competitive advantage is the unique strength of the company which are valuable to its clients. The company upgrades its technology and the skills of its employees to get the benefits of the sustainable manufacturing in the long term.
- Pandemic in the society like COVID 19: Pandemic situations in the society makes it difficult for the company to predict future sales. Many restrictions are imposed by the government during these times, and the company needs to adopt the alternatives to keep the business moving. One such alternative is the adoption of digital technology. Another impact of the pandemic is the loss of job for the employees. To tackle this situation the company runs the training programs to provide alternate skills to the workers so that they can be assigned the available job during the pandemic.

8 Performance

The performance measurement and benchmarking of sustainable manufacturing would provide an opportunity to identify the gaps in sustainable manufacturing practices. Performance of case organization is judged on the basis of different criteria related to customer satisfaction, innovation and growth, finance, and internal business (balanced score card method). Table 1 highlights the performance of XYZ Pvt. Ltd. in last 3 financial years.

Table 1 Performance of XYZ Pvt. Ltd. in last 3 financial years

Performance parameter	2018–2019 (millions of US $)	2019–2020 (millions of US$)	2020–2021 (millions of US $)
Sales	110	230	340
Operating profit	12.21	22.40	48.12
Net profit	8.1	16.2	41.5
Net worth	28.01	38.2	130.4
Capital employed	38	62	153

- On the parameter of customer satisfaction, performance of XYZ Pvt. Ltd. was observed and found improvement in the customer satisfaction score. The customer satisfaction survey conducted by the company in 2018–2019 was rated as satisfactory which improved in 2020–2021 with the rating as very satisfactory.
- On the parameter of innovation and growth, performance of XYZ Pvt. Ltd. was observed and found improvement in the design of the product, ability to implement and adopt new technology, increased market share, and ability to compete based on quality.
- On the parameter of finance, performance of the case organization has improved in terms of sales turn over, net profit, and return on investment.
- Export share of the case organization has decreased in the year 2018–2019 but slightly increased in the year 2020–2021.
- Productivity has increased from 62.3 to 81% after the implementation of SMP.
- Internal business performance of case organization was found to be improved in last 3 years on certain parameters such as operator output rates, multiskilling of workers, customer order compliance, increased revenue from operations and level of team work, and coordination among internal departments.

9 Concluding Remarks

The SMP implementation initiative taken by Indian SME has been demonstrated through a case study in this paper. The case organization is one of the largest manufacturer of automobile components which are used in two wheelers, three wheelers, passenger vehicles, and commercial vehicles. This case has provided a good insight about the initiatives for improving the SMP in the SME's. Improved SMP's has led to benefit like improvement in productivity, profit, and reduction in production cost.

Further SAP–LAP analysis of the case study was conducted based on semi structured interview with the experts. This study can be further extended to other cases for cross case comparison. The study was conducted to see how the SMP initiative help SMEs in performance improvement. Some of the concluding observations from the case study are as follows:

- While paying attention to cost reduction and quality improvement, organization should devote resources and efforts for the implementation of SMP.
- The case organization lays more emphasis on technology up gradation, training of employees, formulation of appropriate organizational policies, following government rules and regulations, proper handling of the market competition, creating customer demands as the strategies for implementation of SMP.
- Creating a world class R & D center, creating a demand for the sustainable products among the customers, uncertainty created during pandemic situations and building up a positive green image among the customers are the main challenges faced by the case organization.
- Need of digital technologies such as Artificial intelligence, Block chain enabled technology, Robotics, 3D printing and manufacturing system capacity balancing has been felt by the XYZ Pvt. Ltd. for effective implementation of SMP.
- For analyzing the level of implementation of SMP in the XYZ Pvt. Ltd. response of management was taken on the different aspects of sustainability in the organization. Response were analyzed using different analytical tools.

The research findings will be quite useful for SMEs in making a strategy for the implementation of SMP. By observing various benefits and improvements, SMEs from different sectors will be motivated to implement SMP in their respective organizations. Therefore major implication of this study is that in the present scenario of turbulent global market, fierce competition, and uncertainty due to pandemic situations SMEs should be proactive and dynamic in the implementation of SMP. The manufacturing capacity and the level of expertise of the employees should be examined very carefully before SMP system is implemented. To further validate the findings, study can be extended as an empirical study with certain more number of case studies of different areas for making relative comparison between different sectors.

References

1. Daniel N, Pathak A, Ravichandran P, Adith S, Kumar N (2022) Analysing the role of Industry 4.0 in sustainable lean manufacturing. Int J Serv Oper Manag. https://doi.org/10.1504/IJSOM.2021.10046299
2. Kahupi I et al (2021) Building competitive advantage with sustainable products–a case study perspective of stakeholders. J Clean Prod 289:125699
3. Deshmukh SG (2012) Sustainable business: impressions, implications and impact. In: National conference on emerging challenges for sustainable business. Department of Management Studies, IIT Roorkee, Uttarakhand, India
4. Valdez-Juárez LE, Gallardo-Vázquez D, Ramos-Escobar EA (2018) CSR and the supply chain: effects on the results of SMEs. Sustainability 10(7):2356

5. Piyathanavong V et al (2019) The adoption of operational environmental sustainability approaches in the Thai manufacturing sector. J Clean Prod 220:507–528. https://doi.org/10.1016/j.jclepro.2019.02.093
6. Garg D, Mustaqueem OA, Kumar R (2021) Sustainable circular manufacturing in the digital era: analysis of enablers. In: Advances in industrial and production engineering. Springer, Singapore, pp 541–554. https://doi.org/10.1007/978-981-33-4320-7_48
7. Dangayach GS, Deshmukh SG (2001) Manufacturing strategy: literature review and some issues. Int J Oper Prod Manag. https://doi.org/10.1108/01443570110393414
8. Ocampo LA, Clark EE (2014) Developing a framework for sustainable manufacturing strategies selection. DLSU Bus Econ Rev 23(2):115–131
9. Grant RM (1991) The resource-based theory of competitive advantage: implications for strategy formulation. Calif Manage Rev 33(3):114–135. https://doi.org/10.2307/41166664
10. Bhanot N, Venkateswara Rao P, Deshmukh SG (2015) Enablers and barriers of sustainable manufacturing: results from a survey of researchers and industry professionals. Procedia Cirp 29:562–567
11. Joung CB et al (2013) Categorization of indicators for sustainable manufacturing. Ecol Indic 24:148–157
12. Kwok L (2022) Labor shortage: a critical reflection and a call for industry-academia collaboration. Int J Contemp Hosp Manag. ahead-of-print
13. Aboelmaged M (2018) The drivers of sustainable manufacturing practices in Egyptian SMEs and their impact on competitive capabilities: a PLS-SEM model. J Clean Prod 175:207–221
14. Mutingi M et al (2017) An analysis of enablers and barriers of sustainable manufacturing in southern Africa. In: Proceedings of the world congress on engineering and computer science, vol 2
15. Bhanot N, Venkateswara Rao P, Deshmukh SG (2017) An integrated approach for analysing the enablers and barriers of sustainable manufacturing. J Clean Prod 142:4412–4439
16. Kumar R (2020) Sustainable supply chain management in the era of digitialization: issues and challenges. In: Handbook of research on social and organizational dynamics in the digital era, pp 446–460
17. Kumar R (2021) Sustainable manufacturing in the era of Industry 4.0: a DEMATEL analysis of challenges. In: Research anthology on cross-industry challenges of Industry 4.0. IGI Global, 2021, pp 1807–1815
18. Kumar R, Singh RK, Dwivedi YK (2020) Application of industry 4.0 technologies in SMEs for ethical and sustainable operations: analysis of challenges. J Clean Prod 275:124063
19. Daniel N, Sindhwani R, Kumar R, Mathiyazhagan K (2022) Identifying the drivers and barriers for the implementation of sustainable manufacturing. Int J Logist Syst Manag. https://doi.org/10.1504/IJLSM.2022.10045956
20. Sushil (2000) SAP-LAP models of inquiry. Manag Decis 38(5):347–353
21. Soni V, Singh SP, Banwet DK (2016) Systematic inquiry for energy security and sustainability (a case study approach for India using SAP-LAP framework). J Adv Manag Res
22. Matharu M, Sinha N (2019) Lean implementation in Indian manufacturing MSMEs: a SaP-lap analysis. Manag Prod Eng Rev
23. Gupta A, Singh RK (2020) Study of sustainability issues in an Indian logistics service provider: SAP-LAP approach. Qual Res Org Manag Int J
24. Yan L, Keh HT, Chen J (2021) Assimilating and differentiating: the curvilinear effect of social class on green consumption. J Consum Res 47(6):914–936
25. Lahkar G, Baiju H, Marwah D, Daniel N (2022) Barriers to business growth in pharmaceutical SMEs—a sustainable perspective. Int J Prod Qual Manag. https://doi.org/10.1504/IJPQM.2021.10042581

Hall and Magnetic Impacts on Stream Past a Parabolic Accelerated Vertical Plate with Varying Heat and Uniform Mass Diffusion in the Appearance of Thermal Radiation

M. Aruna, A. Selvaraj, and V. Rekha

Abstract The aim of this work is to investigate and analyse how a stream passing a parabolic accelerating isothermal perpendicular plate is affected by magnetic fields, radiant heat, and the hall effect. The boundary conditions and accompanying nonlinear, related partial derivatives for momentum, mass, and energy are also reduced to a dimensionless form. For precise answers to the velocity, temperature, and concentration, the Laplace transformation is used. The temperature, velocity, and fluid density were represented by their absolute values. The flow phenomena have been described using flow metrics such as the rotation parameter, Hartmann number, hall parameter, Schmidt number, Prandtl number, and Grashof number. When a required rotation parameter is fulfilled, the transverse flow stops, and the fluid only moves along the disk's path. With the associated parameters, graphic representations of the axial and transverse velocity profiles are offered.

Keywords Hall impact · Thermal Radiation · Vertical plates · Mass diffusion · MHD

1 Introduction

The research of MHD edge covering flood with mass transmit effect in the influence of radiant heat and Hall currents is crucial in so many engineering and manufacturing applications, such as chamber heating and cooling, diesel combustion energy

M. Aruna · A. Selvaraj (✉)
Department of Mathematics, Vels Institute of Science, Technology and Advanced Studies, Chennai-600117, India
e-mail: aselvaraj_ind@yahoo.co.in

V. Rekha
Department of Mathematics, Vel Tech Multi Tech Dr Rangarajan Dr Sakunthala Engineering College, Chennai-600062, India

H. Kumar et al. (eds.), *Recent Advances in Intelligent Manufacturing*,
Lecture Notes in Mechanical Engineering,
https://doi.org/10.1007/978-981-99-1308-4_26

processes, and so on. Electromagnetic hydrodynamics is important in serum retention, nuclear reactor solvent refrigerating, magnetic molding, and electromagnetic drug targeting.

The Hall effectuate is defined as the generation on a potential variation between a magnetic anode that is transversal to an electrical area and the current that is vertical to the current. Spinning electronic controls, fluid dynamic sensors, current sensors, and pressure sensors all use Hall Effect transducer. Brushless Direct Current motors located on gesture detection operating on the rule of Hall currents. Hall effectuate sticks are utilized in mining quarrying, carriages, lifts, and scissor miners between different belongings. Unstable hydro magnet flow near an accelerating perpendicular dish: Hall effect involving differential heat and mass dispersion was studied by M. Acharya [1]. Effects of mass transfer and free convection on the flood through a heat perpendicular dish were examined by Basanth Kumar Jha [2]. Heat and mass move down a vertical surface with varying surface tension and concentration, according to E.M.A. Elbashbeshy's [3] presentation. A technique for creating Laplace transformations in inverted exponential form was examined by R.B. Hetnarski [4]. M.A. Sattar [5] created free convective and mass transfer flow across an infinite vertical porous plate through a porous material with time-dependent temperature and concentration. B. Prabhakar Reddy, B. Shanker, and J. Unsteady MHD free convective flow fluid flow embedded in a porous media with heat and mass transfer: Radiation and heat transport effects, according to Anand Rao [6]. Through a vertical plate that is exponentially accelerated in a spinning flood with mass and heat transmit effects, Thamizhsudar [7] investigated the Hall impact on electro dynamic flood. An accurate solution was discussed by Sacheti NC [8] for unstable magnetohydrodynamic MHD free convective with constant heat flux. The effect of Hall currents on the flow of the MHD boundary layer in a semi-infinite flat plate was examined by Katagiri M [9]. The study's title, Hall influences on mhd shear layer circulation over a continuous moving flat plate, was depicted by Pop1 and Watanabe T [10]. The Hall effects on MHD slip flow and heat transfer through a porous material over an accelerating plate in a rotating system were studied by Dileep Singh Chauhan and Priyanka Rastogi [11] jointly. The effects of chemical reaction and Hall currents on the hydro magnetic flow of an extending vertical surface with internal heat generation/absorption were studied by Salem, A. M., and El-Aziz, M. A. [12]. Makinde, O. D. [13] studied the similarity solution of hydro magnetic heat and mass transfer on a vertical plate with a convective surface boundary condition. Ibrahim, S. Y., and Makinde, O. D. [14] described the chemically reacting MHD boundary layer flow of heat and mass transfer across a moving vertical plate with suction simultaneously. In the presence of a variable stream state, Muhaimin, I., Kandasamy, and Hashim I. investigated the effects of thermophoresis and chemical reactions on non-Darcy MHD mixed convective heat and mass transport across a porous wedge [15]. Selvaraj A, Jose S D, Muthucumaraswamy R, and Karthikeyan S examined the MHD parabolic flow past an accelerated isothermal vertical plate with heat and mass diffusion while rotating [16]. The interaction of the heat source with the MHD and radiation absorption fluid flow via an exponentially accelerated vertical plate with variable temperature and mass diffusion across a porous medium was studied by Selvaraj A and Jothi E. [17]

These investigations aim to investigate the effects of thermic irradiation on flood over a parabolic accelerated isotherm plumb dish after the plate has cooled in the presence of an electrical area and Hall currents. These studies aim to examine the effects of thermic irradiation on flood over a parabolal accelerated isotherm plumb dish in the presence of an electrical area and Hall currents when the plate is cooled.

2 Mathematical Formulation of the Problem

At first, the fluid and dish are in a fixed medium that is heated to the same temperature. Here, the x-axis is drawn vertically up the vertical plate, while the y-axis is caught perpendicular to the dish. At first, the dish and fluid rotated at a constant angular velocity about the z-axis, which is normal to the plate and perpendicular to the x- and y-axes. Here, variable mass diffusion and an accelerated parabolical isotherm perpendicular dish are used to model the erratic flood of an incompressible sticky flood. A transverse electrical field with strength B_0 is applied to the plate (Fig. 1).

The unstable stream is regulated by the following by Boussinesq's approximation:

Equation of Momentum:

Fig. 1 Flow model of the problem

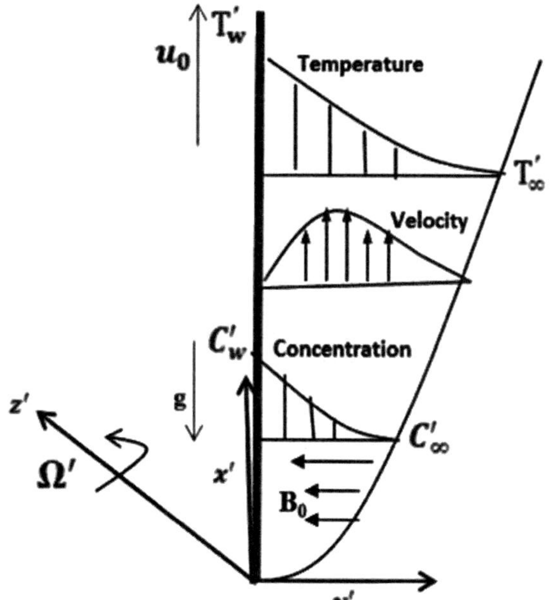

$$\frac{\partial u}{\partial t'} = V\frac{\partial^2 u}{\partial y^2} + 2\Omega' v - \frac{\sigma \mu_e^2 B_0^2}{\rho\left(1+m^2\right)}(u+mv) + g\beta\left(T' - T'_\infty\right) + g\beta^*\left(C' - C'_\infty\right) \tag{1}$$

$$\frac{\partial v}{\partial t'} = V\frac{\partial^2 v}{\partial y^2} - 2\Omega' u + \frac{\sigma \mu_e^2 B_0^2}{\rho\left(1+m^2\right)}(mu+v) \tag{2}$$

Energy Equation:

$$\rho c_p \frac{\partial T'}{\partial t'} = k\frac{\partial^2 T'}{\partial z^2} - \frac{\partial q_r}{\partial z} \tag{3}$$

Equation of mass diffusion:

$$\frac{\partial C'}{\partial t'} = D\frac{\partial^2 C'}{\partial z^2} \tag{4}$$

where the secondary velocity is v and the primary velocity is u. The introductory and extremity terms are

$$t' \le 0 : u = 0, v = 0, T' = T'_\infty C' = C'_\infty, \text{ for all z}$$

$$t' > 0 : u = u_0 t'^2, v = 0, T' = T'_\infty + (T'_w - T'_\infty)At', C' = C'_w \text{ at z} = 0 \tag{5}$$

$$t' > 0 : u \to 0, v \to 0, T' \to T'_\infty, C' \to C'_\infty \text{ as } z \to \infty$$

For the situation of an extremely thin grey gas, the local radiant is stated by

$$\frac{\partial q_r}{\partial z} = -4a^*\sigma\left(T'^4_\infty - T'^4\right) \tag{6}$$

In order to describe 4 T ′ as an analog operate of temperature, it is presumable that variation in temperature inside the stream are sufficiently modest. By omitting higher-degree terms and expanding 4 T ′ in a Taylor series extension about T ′, this is achieved.

$$T'^4 \cong 4T'^3_\infty - 3T'^4_\infty \tag{7}$$

Equation (3) is reduced to using Eqs. (6) and (7).

$$\rho C_\rho \frac{\partial T'}{\partial t'} = k\frac{\partial^2 T'}{\partial y^2} + 16a^*\sigma T'^3_\infty\left(T'_\infty - T'\right) \tag{8}$$

Introduce non-dimensional quantities here:

$U = u\left(\frac{u_0}{v^2}\right)^{\frac{1}{3}}, t = \left(\frac{u_0^2}{v}\right)^{\frac{1}{3}} t', Z = z'\left(\frac{u_0}{v^2}\right)^{\frac{1}{3}}, M = \frac{\sigma B_0^2}{\rho}\left(\frac{v}{u_0^2}\right)^{\frac{1}{3}}, Gc = \frac{g\beta(C' - C'_\infty)}{(v.u_0)^{\frac{1}{3}}},$
$Gr = \frac{g\beta(T' - T'_\infty)}{(v.u_0)^{\frac{1}{3}}}, R = \frac{16a^*\sigma T'^3_\infty}{k}\left(\frac{v^2}{u_0}\right)^{\frac{2}{3}}, C = \frac{C' - C'_\infty}{C'_w - C'_\infty}, T = \frac{T' - T'_\infty}{T'_w - T'_\infty}, Pr = \frac{\mu C_p}{K},$
$Sc = \frac{v}{D},$

$$\Omega = \Omega'\left(\frac{v}{u_0^2}\right)^{\frac{1}{3}}$$

The non-dimensional version of Eqs. (1) through (5) is transformed as

$$\frac{\partial U}{\partial t} = \frac{\partial^2 U}{\partial Z^2} + 2\Omega V - \frac{2M^2}{1+m^2}(U - mV) + G_r\theta + G_c C \tag{9}$$

$$\frac{\partial V}{\partial t} = \frac{\partial^2 u}{\partial Z^2} - 2\Omega U + \frac{2M^2}{1+m^2}(mU - V) \tag{10}$$

$$\frac{\partial \theta}{\partial t} = \frac{1}{Pr}\frac{\partial^2 \theta}{\partial z^2} - \frac{R}{Pr}\theta \tag{11}$$

$$\frac{\partial C}{\partial t} = \frac{1}{Sc}\frac{\partial^2 C}{\partial z^2} \tag{12}$$

The non-dimensional introductory and edge terms are.

$$\begin{aligned} &t \le 0 : U = 0, V = 0, \theta = 0, C = 0 \text{ for all z} \\ &t > 0 : U = t^2, V = 0, \theta = t, C = 1 \ at \ Z = 0 \end{aligned} \tag{13}$$

$$t > 0 : U \to 0, V \to 0, \theta \to 0, C \to 0 \ as \ Z \to \infty$$

The extremity terms (13) and the above mentioned Eqs. (9)–(12) can be coupled as

$$\frac{\partial F}{\partial t} = \frac{\partial^2 F}{\partial Z^2} - aF + G_r\theta + G_c C \tag{14}$$

$$\frac{\partial \theta}{\partial t} = \frac{1}{Pr}\frac{\partial^2 \theta}{\partial z^2} - \frac{R}{Pr}\theta \tag{15}$$

$$\frac{\partial C}{\partial t} = \frac{1}{Sc}\frac{\partial^2 C}{\partial z^2} \tag{16}$$

where $a = \frac{2M^2}{1+m^2} + 2\mathrm{i}\left[\Omega - \frac{M^2 m}{1+m^2}\right]$

With barrier circumstances.

$$t' \leq 0 : F = 0, \theta = 0, C = 0 \text{ for all z}$$
$$t' > 0 : F = t^2, \theta = t, C = 1 \ at \ Z = 0$$
$$t' > 0 : F \to 0, \theta \to 0, C \to 0 \ as \ Z \to \infty$$

Here F = U + i V where U denotes the major velocity and V denotes the minor velocity. The terminology defines each physical variable.

$$\theta(z,t) = \frac{t}{2}\left[\exp\left(2\eta\sqrt{Rt}\right)erfc\left(\eta\sqrt{Pr} + \sqrt{bt}\right) + \exp\left(-2\eta\sqrt{Rt}\right)erfc\left(\eta\sqrt{Pr} + \sqrt{bt}\right)\right] - \frac{\eta\sqrt{Prt}}{2\sqrt{b}}\left[\exp\left(-2\eta\sqrt{Prt}\right)erfc\left(\eta\sqrt{Pr} - \sqrt{bt}\right) - \exp\left(2\eta\sqrt{Prt}\right)erfc\left(\eta\sqrt{Pr} + \sqrt{bt}\right)\right] \quad (18)$$

$$C(z,t) = \frac{1}{2}\left[\exp\left(-2\eta\sqrt{Sc.t}\right)erfc\left(\eta\sqrt{Sc} - \sqrt{t}\right) + \exp\left(2\eta\sqrt{Sct}\right)erfc\left(\eta\sqrt{Sc} + \sqrt{t}\right)\right] \quad (19)$$

$$F = \left\{\left(\frac{\eta^2 t}{a} + t^2\right)S + \left(\frac{1}{4a} - t\right)2\eta\sqrt{t}\,T - \frac{\eta t}{a\sqrt{\pi}}e^{(-\eta^2 - at)} + (g - f)A_1 + e(A_1 - A_2) - g.\exp(ct)A_3 + f.\exp(dt)A_4 - g.A_5 - e(A_6 - A_7) + g.\exp(ct)A_8 + f.(A_9 - \exp(dt)A_{10}\right\} \quad (20)$$

where

$$S = \frac{1}{2}\left[\exp\left(2\eta\sqrt{at}\right)erfc\left(\eta + \sqrt{at}\right) + \exp\left(-2\eta\sqrt{at}\right)erfc(\eta - \sqrt{at})\right]$$

$$T = \frac{1}{2\sqrt{a}}\left[\exp\left(-2\eta\sqrt{at}\right)erfc\left(\eta - \sqrt{at}\right) - \exp\left(2\eta\sqrt{at}\right)erfc(\eta + \sqrt{at})\right]$$

$$A_1 = \exp\left(2\eta\sqrt{at}\right)erfc\left(\eta + \sqrt{at}\right) + \exp\left(-2\eta\sqrt{at}\right)erfc(\eta - \sqrt{at})$$

$$A_2 = \frac{\eta\sqrt{t}}{\sqrt{a}}\left[\exp\left(-2\eta\sqrt{at}\right)erfc\left(\eta - \sqrt{at}\right) - \exp\left(2\eta\sqrt{at}\right)erfc(\eta + \sqrt{at})\right]$$

$$A_3 = \exp\left(2\eta\sqrt{(c+a)t}\right)erfc\left(\eta + \sqrt{(c+a)t}\right) + \exp\left(-2\eta\sqrt{(c+a)t}\right)erfc(\eta - \sqrt{(c+a)t}$$

$$A_4 = \exp\left(2\eta\sqrt{(d+a)t}\right)erfc\left(\eta + \sqrt{(d+a)t}\right) + \exp\left(-2\eta\sqrt{(d+a)t}\right)erfc(\eta - \sqrt{(d+a)t}$$

$$A_5 = \exp\left(2\eta\sqrt{Rt}\right)erfc\left(\eta\sqrt{Pr} + \sqrt{bt}\right) + \exp\left(-2\eta\sqrt{Rt}\right)erfc(\eta\sqrt{Pr} - \sqrt{bt})$$

$$A_6 = t\left[\exp\left(2\eta\sqrt{Rt}\right)erfc\left(\eta\sqrt{Pr} + \sqrt{bt}\right) + \exp\left(-2\eta\sqrt{Rt}\right)erfc(\eta\sqrt{Pr} - \sqrt{bt})\right]$$

$$A_7 = \frac{\eta\sqrt{Prt}}{\sqrt{b}}\left[\exp\left(-2\eta\sqrt{Rt}\right)erfc\left(\eta\sqrt{Pr} - \sqrt{bt}\right) + \exp\left(2\eta\sqrt{Rt}\right)erfc(\eta\sqrt{Pr} + \sqrt{bt})\right]$$

,

$$A_8 = (\exp\left(2\eta\sqrt{Pr}\sqrt{(b+c)t}\right)erfc\left(\eta\sqrt{Pr}+\sqrt{(b+c)t}\right) + \exp\left(-2\eta\sqrt{Pr}\sqrt{(b+c)t}\right)erfc(\eta\sqrt{Pr}-\sqrt{(b+c)t}).$$

$$A_9 = \exp\left(-2\eta\sqrt{Sct}\right)erfc\left(\eta\sqrt{Sc}-\sqrt{t}\right) + \exp\left(2\eta\sqrt{Sct}\right)erfc(\eta\sqrt{Sc}+\sqrt{t}).$$

$$A_{10} = \exp\left(-2\eta\sqrt{Sc.d.t}\right)erfc\left(\eta\sqrt{Sc}-\sqrt{dt}\right) + \exp\left(2\eta\sqrt{Scd.t}\right)erfc(\eta\sqrt{Sc}+\sqrt{dt}).$$

where.

$$\eta = \frac{z}{2\sqrt{t}}, b = \frac{R}{Pr}, c = \frac{R-a}{1-Pr}, d = \frac{a}{Sc-1}, e = \frac{Gr}{2c(1-pr)}, f = \frac{Gc}{2d(Sc-1)}, g = \frac{Gr}{2c^2(1-pr)},$$

$$h = \frac{Gc}{2d^2(Sc-1)}$$

The reciprocal error function is erfc.

With a view to obtain the realistic perception into the position, the quantities of F have been obtained from (20). Given the complex nature of the error function's parameter, which was discovered while evaluating this equation, we divided it into real and unreal components utilizing the conditions down.

$$erfc(a+ib) = erfc(a) + \frac{\exp(-a^2)}{2a\pi}[1-\cos(2ab)+isin(2ab)] + \frac{2\exp(-a^2)}{\pi}\sum_{n=1}^{\infty}\frac{\exp\left(-\frac{n^2}{4}\right)}{n^2+4a^2}[f_n(a,b)+ig_n(a,b)]$$

where

$f_n = 2a - 2a\cosh(nb)\cos(2ab) + n\sinh(nab)\sin(2ab)$ and

$$g_n = 2a\cosh(nb)\sin(2ab) + n\sinh(nab)\cos(2ab)$$

$$|\in(a,b)| \approx 10^{-16}|erfc(a+ib)|$$

3 Result and Discussion

Utilizing complimentary error function, the problem of unstable convective MHD flood gone a parabolical accelerate boundless perpendicular flat panel in the attendance of irradiation parameter has been developed, examined, and solved. Graphs for Pr = 0.71 and t = 0.2 are used to show the impacts of the flood variables on the speed, heat, and absorption outlines of the flood area. These parameters include the magnetic parameter (M), Gasthof no for heat and mass transfer (Gr, Gc), Schmidt no (Sc), Prandtl no (Pr), and radiation parameter (R).

The above Fig. 2 mentioned with a different Schmidt number, the stream's concentration biographies start to change more or less (Sc). Figure 2 shows these options graphically for different Schmidt values Sc = 0.3, 0.6, 1.5, and time t = 0.2. The Schmidt number Sc is defined as the numerical representation of a fact in an air occurrence. The concentration biographies are raised when the Schmidt number is solved. It is discovered that the occurrence of heavy diffusing species is more significantly influenced by the hike down significance of the liquid stream's absorption edge coating. Figure 2 demonstrates how concentration increases as time goes on. Figures 3, 4 demonstrate how significantly the temp of the stream changes when time duration (t) and the radiation component (R) change. The change is shown in the following figures. For air Pr = 0.7, the heat biographies are planned for varied relevance of time t = 0.2, 0.4, 0.6, and heat irradiation variable R = 2,5,10. In heat biographies, the outcome of time duration T as well as the heat irradiation factor R is crucial. It should be observed that the relevance of t and R has been raised in the heat biographies.

With the help of Figs. 4, 5, 6, 7, 8, 9, the speed biographies from liquid steam that differ to a considerable boundary with the impact of steam factors on the steam area are investigated. Figures 4, 5 exhibit the impact of the Grashof no's for mass and heat transfer Gc and Gr upon the speed from that liquid stream. For three different intakes of the Grashof no Gr = Gc = 2,2,2,3 and 4,4 and Gr = Gc = 5,5,5,8,5,10, the speed of the liquid steam is currently displayed against the similarity variable maintaining recurring additional liquid flow variables. For clear relevance of Gr, Gc,

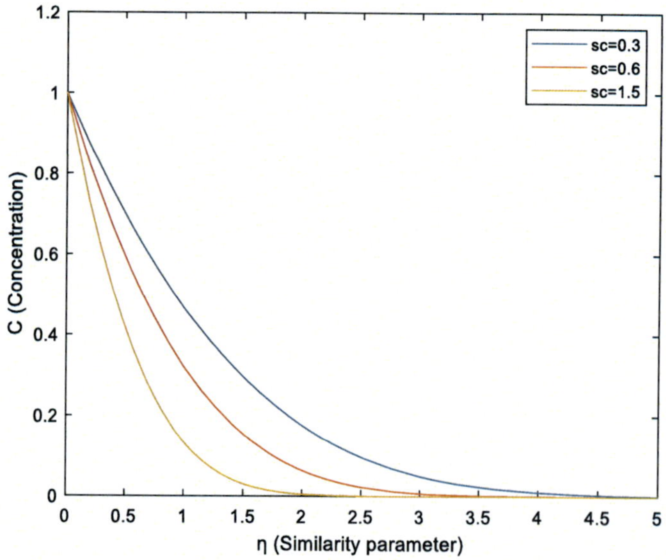

Fig. 2 Profile of a concentration

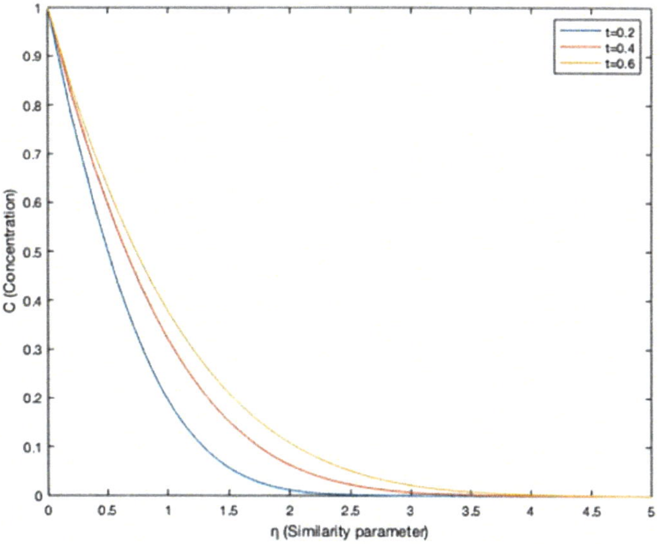

Fig. 3 Profile of temperature based on t

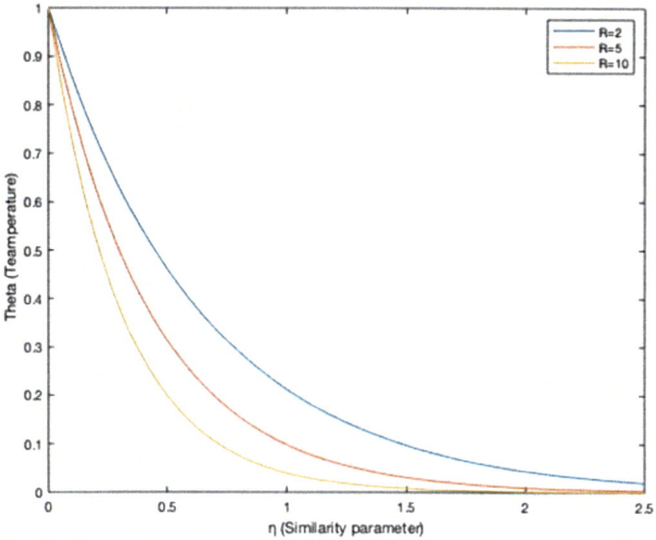

Fig. 4 Profile of temperature based on R

the biographies of the principal velocity and subsequent speed are shown in Figs. 5, 6. It should be noted that, when the heat Grashof number Gr and the mass Grashof value Gc are increased diagrammatically, the fundamental and subsequent velocities increase as well.

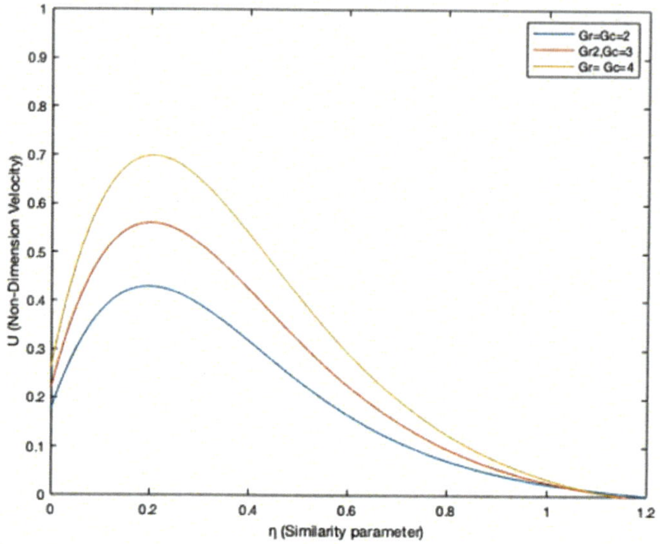

Fig. 5 Profile of Main velocity based on Grash of no

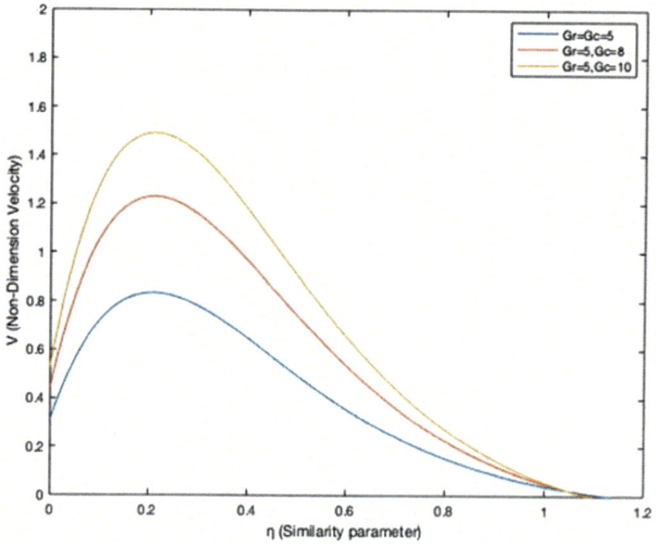

Fig. 6 Profile of velocity based on Grash of no

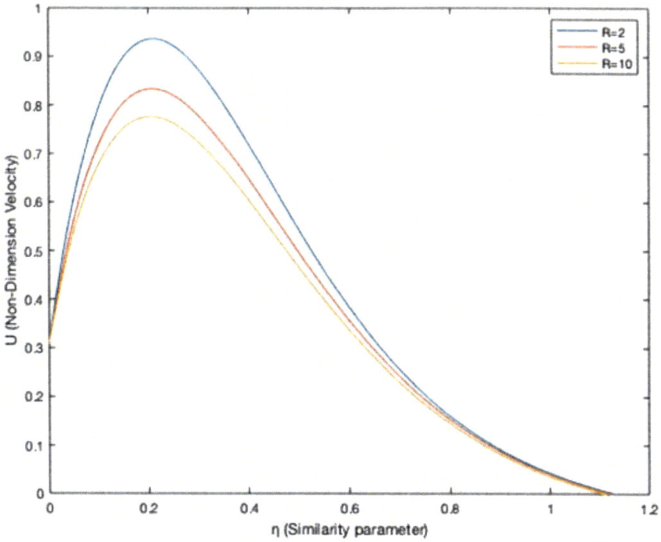

Fig. 7 Profile of main velocity based on R

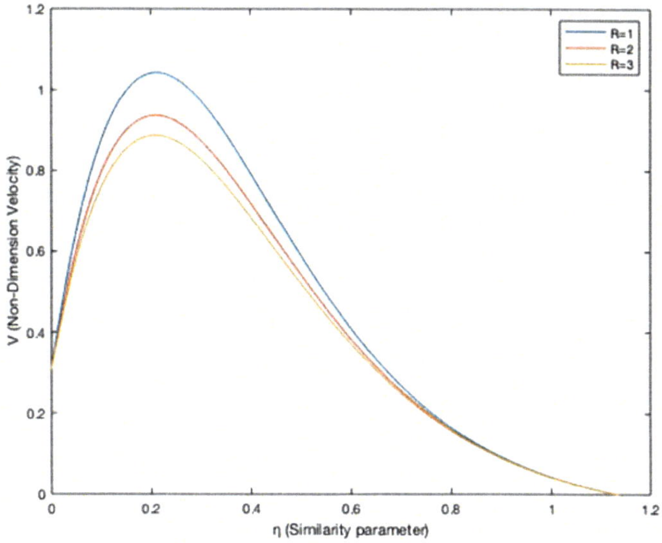

Fig. 8 Profile of velocity based on R

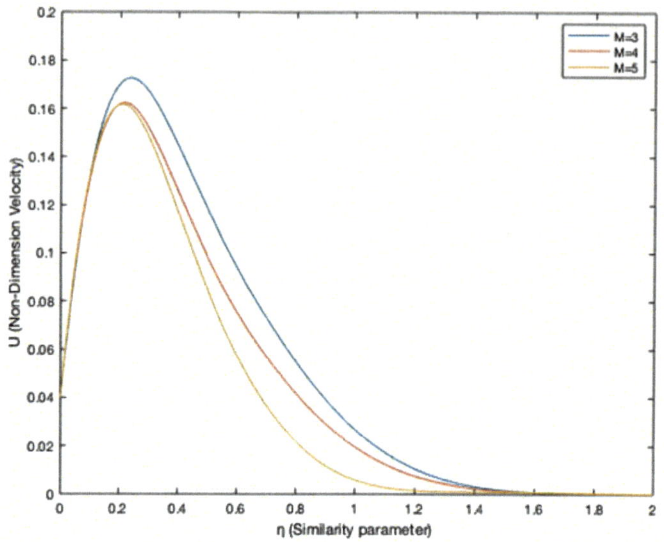

Fig. 9 Profile of velocity based on M

Figures 7, 8 show the main momentum U and second momentum V biographies for various entries of the radiation parameter R = 2, 5, and 10 tends to increase with down grade entries of the heat irradiation variable R. There is also an increase in the minor momentum parts due to a rise in the thermic irradiation variable.

Figure 9 presents the change of the prime momentum U biograph for distinct entries of Hartmann number M = 3,4,5, around here the prime momentum raises about raise down entries of M and the second momentum raises with raising entries Hartmann number.

Figure 10 shows that increasing inputs from that hall parameter m cause the main momentum U and the second momentum V to increase.

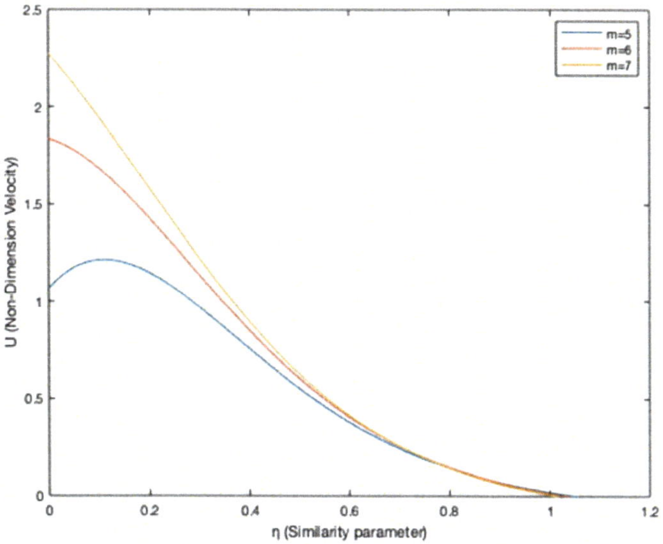

Fig. 10 Profile of Velocity based on m

4 Conclusion

An analytical solution of the of temperature, concentration, primary velocity U, secondary velocity V, and convective flow past parabolic accelerated perpendicular panel in the appearance of electromagnetic region and Hall impact has been carved out. Convection currents are reported to be carried from the plate in the event that the plate cools (Gr > 0, Gc > 0). This is noticed that Concentration (C) declines for increasing Schmidt number relevance (Sc).Temperature biographies rise with increasing time (T) relevance and also decline with increasing thermal irradiation parameter rate (R). The velocity (U) increases due to the growing rate of the hall variable m, the thermal Grashof no Gr, and the mass Grashof no Gc. And Rise down for raising relevance of Thermal irradiation variable R, Magnetic variable M. The Secondary velocity (V) Rises due to the growing rate of the heat irradiation variable R, the magnetic variable M, the mass radiation variable M, and the Grahsof no's Gr and M. and Rise down for raising relevance of hall variable, further analyze and study the Soret and Dufour effect.

References

1. Acharya M, Dash GC, Singh LP (2001) Hall effect with simultaneous thermal and mass diffusion on unsteady hydro magnetic flow near an accelerated vertical plate. Ind J Physics 75B:68–70

2. Basanth Kumar Jha and Ravindra prasad (1990) Free convection and mass transfer effects on the flow past an accelerated vertical plate with heat source. Mech Res Commun 17:143–148. https://doi.org/10.1016/0093-6413(90)90043-B
3. Elbashbeshy EMA (1997) Heat and mass transfer along a vertical plate with variable surface tension and concentration in the presence of the magnetic field. Int J Engg Sci 34:515–522. https://doi.org/10.1016/S0020-7225(96)00089-4
4. Hetnarski RB (1975) An algorithm for generating inverse Laplace transforms of exponential form, *ZAMP26*, pp 249–253, https://doi.org/10.1007/BF01591514
5. Sattar MA (1994) Free convection and mass transfer flow through a porous medium past an infinite vertical porous plate with time dependent temperature and concentration. Ind J Pure and Appl Math 23:759–766. https://doi.org/10.9790/1684-131492100
6. Shanker B, Prabhakar Reddy, Anand Rao J (2010) Radiation and mass transfer effects on unsteady MHD free convective fluid flow embedded in a porous medium with generation absorption, Indian J Pure Appl Phys, 48, pp 157–165
7. Thamizhsudar M, Pandurangan J (2014) Hall effects on magneto hydrodynamic flow past an exponentially accelerated vertical plate in a rotating fluid with mass transfer effects. Elysium J Res Manag 1:143–149. https://doi.org/10.18869/acadpub.jafm.68.225.24418
8. Sacheti NC, Chandran P, Singh AK (1986) An exact solution for unsteady magnetohydrodynamic free convection flow with constant heat flux. Int Commun Heat Mass Transfer 29:1465–1478. https://doi.org/10.1016/0735-1933(94)90090-6
9. Katagiri M (1969) The effect of Hall currents on the magnetohydrodynamic boundary layer flow past a semi-infinite flat plate'. J Phys Soc Japan 27:1051–1059. https://doi.org/10.1143/JPSJ.27.1051
10. Pop1, Watanabe T (1995) Hall effects on magnetohydrodynamic boundary layer flow over a continuous moving flat plate, Acta mech 108:35–47, https://doi.org/10.1007/BF01177326
11. Dileep Singh Chauhan (2012) Priyanka Rastogi Hall effects on MHD slip flow and heat transfer through a porous medium over an accelerated plate in a rotating system. Int J Eng Sci 14:228–236
12. Salem AM, El-Aziz MA (2008) Effect of Hall currents and chemical reaction on hydromagnetic flow of a stretching vertical surface with internal heat generation/absorption. Appl Math Model 32:1236. https://doi.org/10.1016/j.apm.2007.03.008
13. Makinde OD (2010) Similarity solution of hydromagnetic heat and mass transfer over a vertical plate with a convective surface boundary condition'. Int J Physical Sci 5:700
14. Ibrahim SY, Makinde OD (2010) Chemically reacting MHD boundary layer flow of heat and mass transfer past a moving vertical plate with suction'. Sci Res Essays 5:2875
15. Muhaimin I, Kandasamy R, Hashim I (2009) Thermophoresis and chemical reaction effects on non- Darcy MHD mixed convective heat and mass transfer past a porous wedge in the presence of variable stream condition. Chem Eng Research Design 87:1527. https://doi.org/10.1016/j.cherd.2009.04.008
16. Selvaraj A, Jose SD, Muthucumaraswamy R, karthikeyan S MHD parabolic flow past an accelerated isothermal vertical plate with heat and mass diffusion in the presence of rotation, materials today proceedings, vol 46,9, pp 3546–3549. https://doi.org/10.1016/j.matpr.2020.12.499
17. Selvaraj A, Jothi E, "Heat source impact on MHD and radiation absorption fluid flow past an exponentially accelerated vertical plate with variable temperature and mass diffusion through a porous medium. In: Materials Today Proceedings, 46(9), pp 3546–3549 ,https://doi.org/10.1016/j.matpr.2020.11.919.

Implementation of Kaizen in Automotive Industry: A Case Study

Anant Bharat, Duni Chand, Preeti Dahiya, and Sanjay S. Rathore

Abstract As change is the law of nature, humans have evolved from the stone age wheels to today's industry 4.0. This implies that continuous improvement is the habit as well as the need of human beings. This continuous improvement was named Kaizen. The present paper is a case study performed in an automotive industry in Neel Metal Production Ltd.—TVS (JBM) at Hosur, Tamil Nadu, in which the Kaizen improvement activity was performed on the shop floor and the kaizen targets were made. At the end of the kaizen implementation, several objectives have been achieved, such as introduction of work bench, 5S level improvement, identification and traceability system of Dies and Fixture, implementation of automatic spot-welding guns, layout for shop floor, implementation of KANBAN system, standard worksheets on shop floor, Poka-Yoke for critical dimension for critical part made, safety and cost saving. The result and conclusion of the case study includes situations before and after kaizen.

Keywords Kaizen · JBM group · Spot welding · Industry 4.0 · Dies · Fixtures

1 Introduction

Kaizen means a constant effort not only to maintain but also upgrade standards. It means continuous improvement. The word implies improvement that involves everyone. Both managers and workers entail relatively little expense. The kaizen philosophy assumes that our way of life—is it our working life, our social life, or our home life—should be focused on constant improvement efforts [1]. The Kaizen improvement can also be done practically to achieve more productivity.

In industry, we have several goals that we plan to achieve, but our main motive is to increase production ensuring safety with good quality standards. In order to

A. Bharat · D. Chand · P. Dahiya (✉) · S. S. Rathore
Skill Faculty of Engineering and Technology, Shri Vishwakarma Skill University, Palwal, Haryana, India
e-mail: preeti@svsu.ac.in

H. Kumar et al. (eds.), *Recent Advances in Intelligent Manufacturing*, Lecture Notes in Mechanical Engineering,
https://doi.org/10.1007/978-981-99-1308-4_27

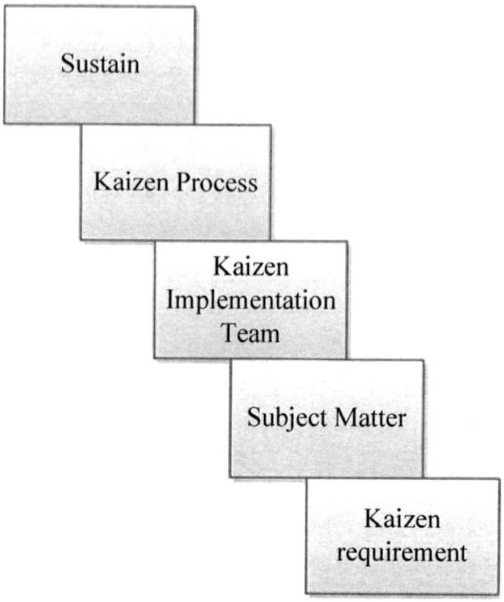

Fig. 1 Steps in Kaizen implementation

increase production, initially we should reduce the wastes in the industry. Before elimination, the type of waste and non-valuable activities (NVA) needs to be identified. The subsequent Sects. 1.1 and 1.2 explains the different stages of identification, elimination of wastes and quality control tools [2, 3] (Fig. 1).

1.1 Elimination of Seven Wastes (Muda)

1. Over production—It means we are producing items which have low or no customer demand, which results in excess cost for transport and inventory.
2. Excessive Inventory—If we procure excess raw material as compared to work in progress and finished goods, this also leads to production imbalance and defects due to long delay time in delivery.
3. Waiting—If due to imbalance the automatic line operator is waiting, stock outs, long process time on the previous stage (bottle neck) will increase the idle time.
4. Motion—Excess operator motion should be avoided so that he gets low fatigue and productivity can be increased.
5. Transportation—Stages having too much distance in between, store to line moment, there should be single flow of material.
6. Reworks/defects—Produced parts sometimes have defects that must be avoided as it increases the efforts and time of the man on the same part which increases our investment in the part.
7. Over Processing—Doing extra process on the product or item for which our customer is not paying us anything.

After finding out different types of wastes in industry we look towards the possible solution. We have several tools [2] which we can ensure while observing our product quality. These tools help us out in monitoring our process which are briefly described in the subsequent subsection.

1.2 Quality Control Tools

1. Check sheet—it helps us in understanding the frequency of a process in a well-structured way.
2. Cause and effect diagram—all the causes which are responsible for a problem are listed on an arrow over a straight line.
3. Control chart—it is taking record of the current data of the process comparing it with previous data. Taking out the conclusion whether the process is controllable or unexpected errors are coming.
4. Histogram—the most commonly used graph for showing frequency distribution in the set of data occurs.
5. Pareto charts—A bar graph showing the significance of the different parameters.
6. Scatter Diagram—comparing two variables with the discrete values on the graph each axis denoting the different variable.
7. Stratification—A technique of comparative analysis of the data gathered from the different sources.

1.3 Driving Force for Kaizen

As every reaction is a response to the particular action, similarly Kaizen is also a reaction towards the problems one face and lack of goal target achievement [4, 5]. Likewise some of the actions are as following:

(i) Customer complains about the process
(ii) End to end process is neither defined or nor understood
(iii) There is a considerable amount of rework or defect in the process
(iv) Process is not standardized
(v) Process inputs/outputs are not error proofed and poor quality
(vi) Process cycle time is lengthy
(vii) Low productivity
(viii) No cost saving

In view of the aforementioned reasons, it is evident that these factors result in decrease of productivity and there is a need for the proper action to be taken, i.e. is the core theme of the Kaizen, continuous improvement. Before implementing any Kaizen, several points need to be considered such as it should be started by people; every level of person should be involved and putting efforts; process improvement will improve results and improved results will improve customer satisfaction. The activity of Kaizen has result in improvement of the overall efficiency in many of the automotive industries [6, 7].

In the present article, the Kaizen has been implemented in Neel Metal Production Ltd.—TVS (JBM) at Hosur, Tamil Nadu, in which the Kaizen has resulted in reuse of scrap shank, use of spot-welding gun arm scrap, isolation of transformer, fixing spot welding gun, and use of scrap to make trollies. The significant change in the plant with the use of Kaizen activities has resulted in improvising the efficiency and working at shop floor of the plant.

2 Kaizen Cases Implemented in JBM Group

This section presents the Kaizen cases implemented in Neel Metal Production Ltd.—TVS (JBM) at Hosur, Tamil Nadu.

2.1 Scrap Shank Reuse

It is well known that in the process of the spot welding, we have shanks but some time due to some problems in robot programming or miss alignment of shanks they got broke down and we scrap them. Since these are very expensive, it is our responsibility as well as need to have some improvement so that these can be used again. While brainstorming, we concluded reusing these broken shanks by processing i.e. machining, grinding, or cutting them and make them reusable as shown in Fig. 2. In this way, this results in saving cost and making the resource management to its full use.

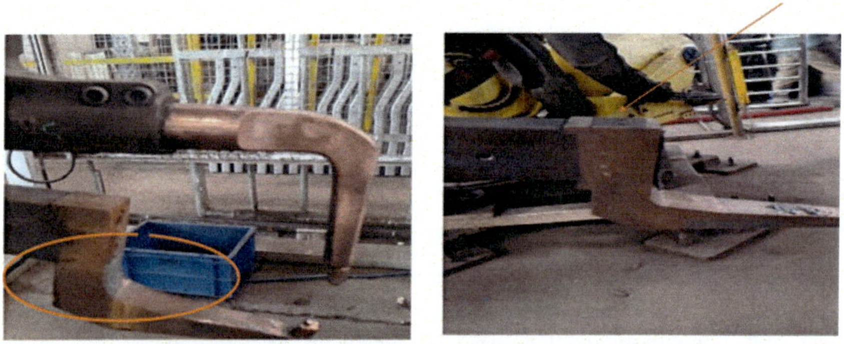

Before Kaizen **After Kaizen**

Fig. 2 **a** Scrap Shank before Kaizen; **b** Scrap Shank after Kaizen

Before Kaizen

After Kaizen

Fig. 3 **a** Spot welding gun arm before Kaizen; **b** Spot welding gun after Kaizen

2.2 *Spot Welding Gun Arm Scarp Used*

While working on a spot-welding robot, it was observed that the spot on part is not that much as much as the length of spot gun arm. This results in breaking of the arm of the guns. These broken-down arms may get distorted or thrown as scrap. Alternatively, it was suggested to use them in other small length arm spot guns or perform some machining and keep these small arms as spears as depicted in Fig. 3. In this way, we reduced our scrap and make a good resource out of that.

2.3 *Transformer Isolation*

In the control panel of the robot, a step-down transformer is used in order to step down three phase supply to single phase supply. Initially, it was not having any isolation due to which there was risk of getting shock, which could be fatal too to the employees on shop floor. So, isolation was provided using an acrylic sheet so that it could not isolated and be away from any touch by the workers while working.

2.4 *Fixing Spot Welding Gun*

There was a hanging spot welding gun in the Gemba. There was a chance that it could strike any person. We decided to fix it on the pillar so that the operator can put the part and just have to press the button. Risk becomes almost zero. The fixed spot-welding gun before and after implementation of Kaizen is shown in Fig. 4.

Before Kaizen **After Kaizen**

Fig. 4 **a** Fixed Spot-welding gun before Kaizen; **b** Fixed Spot welding gun after Kaizen

2.5 Making Use of Scrap

In the press shop, there were plenty of scrap skeletons that were to be thrown out. It was decided to make the trollies out of them for a material moment inside the plant as shown in Fig. 5. This saved sheet and waste was also used in a good way for a fruitful outcome.

The summary of situations before and after Kaizen implementation has been summarized in Table 1. It is evident from the results obtained after implementing Kaizen results in improvising the working area, traceability and identification of dies on shop floor, implementation of KANBAN system, standardization of worksheets, saving cost of scrap material, and optimal use of resources. The plan sheet for implementing Kaizen is given in Table 2.

(a) **(b)**

Fig. 5 **a** Scrap before Kaizen **b** Trolley prepared after Kaizen

Table 1 Summary of situation before and after Kaizen

Subject	Before Kaizen	After Kaizen
Working Area	Maximum work done on floor	Work table were introduced
5S	No 5S implementation	5S implemented
Traceability and identification	No identification marks and no traceability system for dies	Identification and traceability system of dies implemented
Layout	No layout for the shop floor	Proper layouts were demonstrated on shop floor
KANBAN system	Not implemented properly	Implementation of KANBAN system
Worksheets	No standard worksheets and SWCT (Standardised Work Combination Table)	Standard worksheets and SWCT (Standardised Work Combination Table)
Poke Yoke	No Poke Yoke for critical dimension for critical parts	Poke Yoke for critical dimensions of critical parts
Standardisation	Not there	Templates installed

Table 2 Plan sheet for Kaizen

Improvement situation	Before	Target	After Kaizen				Percentage improvement	Remarks
			June					
			26	27	28	29		
Improvement in 5S	10%	100%	10	15	25	20	70%	Bins yet to be made
Inventory control	Not well maintained	Kanban system	10	25	30	10	75%	–
Material handling	–	Proper packing, trollies & racks to be made	5	15	7	28	55%	Plastic racks are yet to be made
Customer complaint reduction	–	50% reduction	–	–	–	–	–	Will be monitored after 6 months
Space utility	–	30%	0	10	10	15	35%	Pillar gun fixed on the pillar

3 Conclusion

Kaizen is a Japanese word, which means to have continuous improvement in order to increase our productivity, quality, and safety. We tend to grow our industry in a thorough way. The bigger improvements in the process have been obtained using the Kaizen implementation on the shop floor. The use of various quality tools has been made to eliminate different kinds of wastes with the implementation of plan-do-check-act (PDCA) in Kaizen cycle. The improvements evident from the introduction of work table on shop floor, implementation of 5S, identification and traceability system of dies, demonstration of proper layouts on shop floor, standard worksheets, saving cost and optimal use of resources in terms of scrap shank reuse, use of spot welding gun scrap, isolation of transformation for protection, fixed spot welding gun, and use of scrap to make trollies on the shop floor are some of the outcomes of implementation of Kiazen in the JBM plant.

Acknowledgements We would like to acknowledge the support and opportunity given by JBM Group to provide opportunities for skilling in the dual education model with Shri Vishwakarma Skill University, Palwal, Haryana.

References

1. García-Alcaraz JL, Oropesa-Vento M, Maldonado-Macías AA (2016) Kaizen control phase models: activities and benefits. Kaizen Plan Implement Control 225–257. https://doi.org/10.1007/978-3-319-47747-3_13
2. Al-Hyari KA et al (2019) The applications of kaizen methods in project settings: applied study in Jordan. TQM J 31(5):831–849. https://doi.org/10.1108/tqm-03-2019-0078
3. Hammerl L, Weber D, Ton AD (2021) Kaizen in automotive innovation: how the Hungarian automotive clusters can profit from the adoption of kaizen principles—a literature review of the Central European automotive industry. Int J Appl Res Bus Manage 2(2):23–38. https://doi.org/10.51137/ijarbm.2021.2.2.3
4. Medinilla Á (2014) Understanding kaizen. Agile Kaizen, pp 3–19. https://doi.org/10.1007/978-3-642-54991-5_1
5. Imai M (1986) In Kaizen: (ky'zen): the key to Japan's competitive success. McGraw-Hill, New York
6. Wittenberg G (1994) Kaizen—the many ways of getting better. Assem Autom 14(4):12–17. https://doi.org/10.1108/eum0000000004213
7. Suzuki M, Sakamaki E (2020) Opportunities for kaizen in Africa: developing the core employability skills of African youth through kaizen. Work Manage Product 141–167. https://doi.org/10.1007/978-981-15-0364-1_7

Implementation of Productivity Improvement in SME by Designing and Developing Assembly Line

Manish Gupta and Rajendra Kumar Shuklal

Abstract Small and medium-sized enterprises (SMEs) have always played a key role in the economies of all large industrial corporations when it comes to gross domestic product. It considerably helps in developing entrepreneurship and generates the second-largest employment prospects. Historically, most new manufacturing concepts and technologies have been implemented in major manufacturing industries. However, SMEs have received insufficient attention in the implementation of these new manufacturing concepts and technologies. In this paper, an attempt has been made to present experiences of implementing new productivity improvement strategies in a medium-sized firm. The work has taken place in Hi-Tech Metal forming Industry, an Indian Company located in Indore, Madhya Pradesh. This firm engages in batch production to generate a wide range of low-volume products for several clients. The implementation is concerned with the effective utilization of the tools, breaking the operations into the elements, Design of the assembly line, and development of the autonomous cell. The article concludes with a Performance analysis and some suggestions for reducing waste and boosting productivity.

Keywords Productivity · Small and medium enterprises (SME) · Assembly line

1 Introduction

The study on the role of micro, small, and medium-sized businesses in inclusive growth concluded that MSMEs are the country's growth engine [1]. The contribution of the Manufacturing Sector, as well as the Service Sector of MSMEs, is insignificant to the total GDP of the Country. The Indian government has taken several steps

M. Gupta
Department of Mechanical Engineering, Shri Vaishnav Polytechnic College, Indore, MP, India

R. K. Shuklal (✉)
Department of Mechanical Engineering, Medi-Caps University, Indore, MP, India
e-mail: rajendrakumar.shukla@medicaps.ac.in

H. Kumar et al. (eds.), *Recent Advances in Intelligent Manufacturing*,
Lecture Notes in Mechanical Engineering,
https://doi.org/10.1007/978-981-99-1308-4_28

and is currently accelerating the development of Micro, Small, and Medium-Sized Enterprises in the nation [2]. So it's important to sustain such types of enterprises by providing low-cost solutions to their problems.

In this paper, a real-life example of the supplier to the automotive industry, a factory of the Indian company "Hi-Tech Metal Forming" in Indore, MP, has been considered to explain the need for change and its implementation. This business is an example of a SME that has numerous issues with output and quality.

To create solid and long-lasting solutions, current trends in industrial firms are concentrating on the continued development of cyber-physical systems and the potential integration of human and machine intelligence [3]. Cost is an additional important factor that must be considered as well. Designing and developing a good physical system for material handling is of prime importance for any industry before going for automation.

In this work, the focus is on the manufacturing area and on cells assembling M&M SWC and Jack manually. It analyses the process and supply chain to see if any improvements can be made, proposes some changes in both the processes (to simplify the tasks) and the flow of material, and implements those changes.

The purpose of the paper is to share some of the lessons learned from using different productivity enhancement strategies and techniques.

2 Background for the Research

The ratio of what is produced to what is needed to produce is known as productivity. Productivity measures the relationship between inputs, such as labor, capital, material, and other resources, and outputs, such as goods and services generated [4].

It is generally acknowledged that increased material handling benefits output. However, it's not just output; it's also how the staff members perceive the brand-new circumstance. Benefits are achievable when perception is positive [5].

As the main method for continuously enhancing everyone's quality of life by emphasizing things like recognition, autonomy, training, and growth of individuals, the Japanese idea of kaizen (Continuous improvement) mixes scientific with humanistic management concepts [6–8]. Therefore, as innovation and production rise, so does the process's natural byproducts. All improvement strategies must be introduced while keeping employee empowerment in mind.

An assembly line normally consists of several workstations, each of which is in charge of carrying out a certain set of duties. According to their order, the product moves along the line from one workstation to the next [9].

Each workstation can become specialized, thanks to the division of the full set of duties into smaller groups that are each carried out in a different workstation. To achieve great efficiency, the jobs may be carried out manually or using a specialized equipment [9, 10].

The use of flexible (and typically automated) machinery, such as robots or flexible machines, to carry out assembly activities has recently been developed as Flexible Assembly Systems (FAS) [11].

This study focuses on manual assembly cells because they have a higher labor cost percentage. Additionally, management wants to lower labor expenses by boosting productivity.

3 Case Study

The Company, Hi-Tech, manufactures various components for the automotive industry. It produces all parts of SWC and then assembles the whole part. Most of the work is done manually; individuals assembled the list of the part as shown in Table 1. So, every person needs all the tools and equipment for assembling all the parts as shown in Table 2, hardly each person can do only 30 assemblies per day/ shift. In assembling every person require all the parts in front of him, so it again creates the storage problem. In discussion with management, were interested in an assembly line, but not complete automation which may incur the cost and trained personnel (Figs. 1, 2, 3).

In marathon discussions, some ideas came up, and all problems can be sorted out by designing an assembly line. But as per the management requirement, we need to design a line in which no external automation is needed. Here the part must be fed from one station to another only by human power and only gravitational force was used for all the movement. Further to improve productivity it is needed to break the whole job into a number of elements. As shown in Table 3, initially, all the operations

Table 1 Part list

Name of parts	Number of parts	Weight of parts
Front shaft	1	840 gm
Rear shaft	1	500gm
Rachet	1	280 gm
Klonic washer	2	–
Lock pin	1	–
Metal washer	2	–
Hex screw	2	–
SWC bracket	1	7.72 kg
Pawl	1	–
Pawl pin	1	–
Hex Nut	2	–
Channel Rope	1	1.40 kg
Total	16	10.44 kg

Table 2 Tool details

Tool name	Quantity	Weight (gm)
Hammer	1	500
Chisel	1	100
Ring spanner	2	20
Socket	2	20
Gauge	1	100
Retapping gauge	1	20
Channel gauge	1	150
Jack nut	1	500

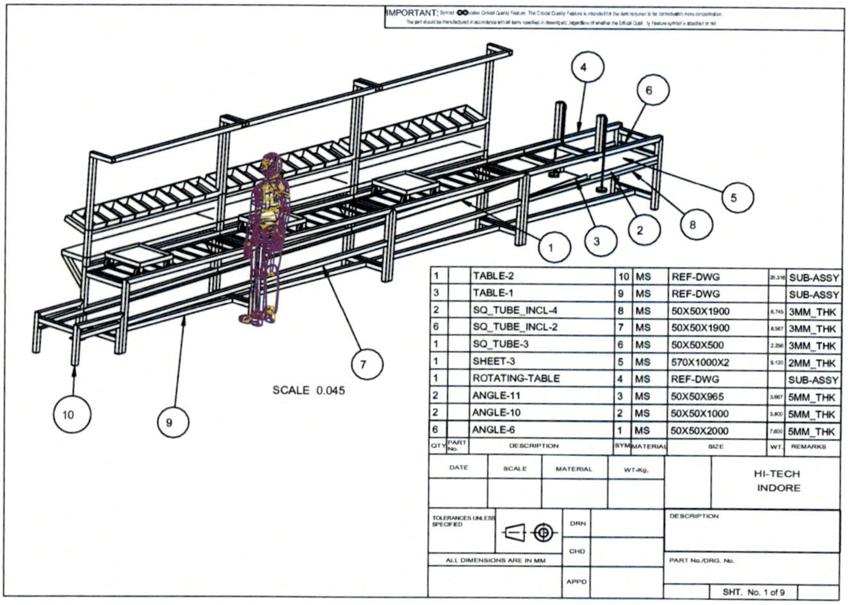

Fig. 1 Assembly line drawing

formed by each individual person need all tools and equipment to perform all the tasks as shown in Table 2. At the end of the day each individual produced, only 20 parts. Further, the company faces another issue, the problem of tool cost and loss of the tools. So, designing an assembly line, as shown he Fig. 1, resolves this problem. In this assembly line, a track has been established, and a trolley was designed as shown in Fig. 3. Each task was broken into elements and groups were made. So each station is designed for their cycle time not to exceed more than 2 min. So, specialization was used to perform each task, and every person needs only that many tools and equipment to perform only that element of the task not the whole work. After performing their element they push the trolley to the next station or are either

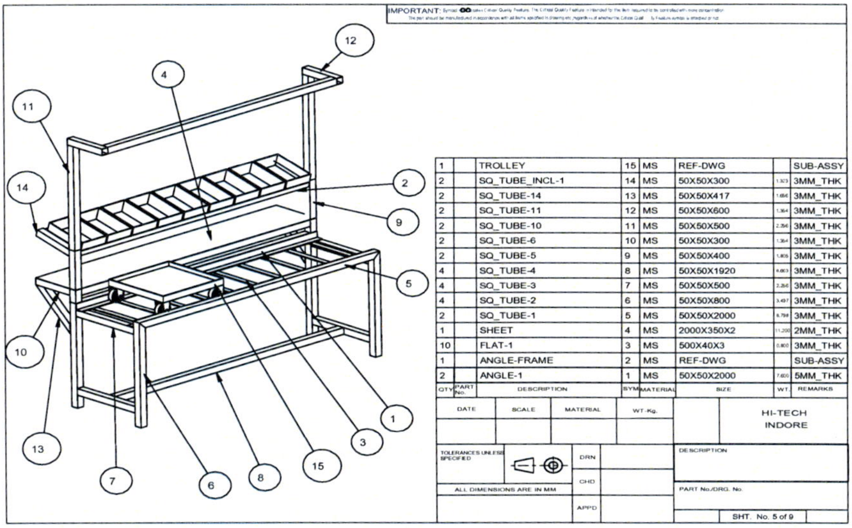

Fig. 2 Single unit of assembly line

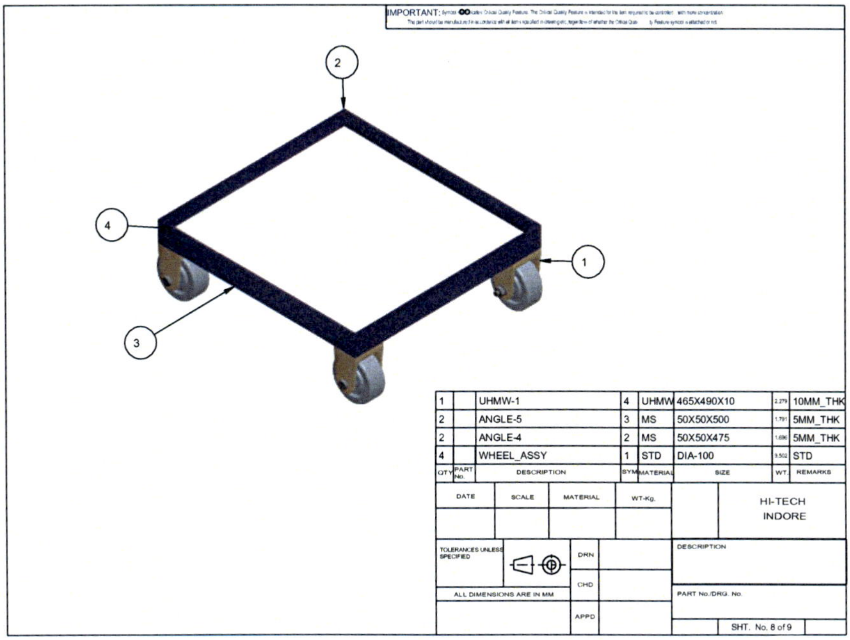

Fig. 3 Line trolley

pulled by the next operator and so on. At the end of the Assembly line, the last worker picks up the part & puts it into a trolley, and a line trolley (as in Fig. 3) is needed to feed station one. So a tab was designed and named as a rotating table, which rotates down due to the load of the trolley and pushes the line trolley into a lower track, and due to the gradient of this track the line trolley moves to the first station, by the gravity force.

By this design, only one set of tools is required, so cost is saved and each person performs the specific task so specialization was used to perform the assembly. Further, the theft or loss of the tools was achieved by providing the hanging arrangement of the tool on each station by the springs. So every time when the tool is needed by the operator to perform the assembly he just pulls it down, and when the work is over as he releases it reaches its fixed station. Further bins were provided to store all materials, which provide a fixed station to each part and cultivate habit formation in the employee to rapid up the assembly. At the end of the experiment to conclude the number of assemblies made was 240 per shift by the team of six persons as shown in Table 4. Initially, they were doing only 25 assemblies per shift by each person, so a team of 6 was achieving only 200 assemblies per day as shown in table-3. With this experiment 240 were achieved by a team of 6 persons in a day, which increases the production by 20%, thus productivity gets improved.

The following are some important considerations for the implementation of productivity development initiatives that are highlighted in the case study:

a. Employee empowerment is crucial to the effective use of productivity development tactics and techniques.
b. Employee education and training in the key areas of the implementation procedures
c. The application of productivity development techniques requires a team made up of key managers from related functional areas
d. Any change process, including the installation of the assembly line and improvement, need the backing of top management.

4 Discussion and Outcome

The economy of the country mostly govern by SMEs but most searchers pay attention to large industries as there are constraints in making investments in the SME sector. To encourage readers to improve productivity in SMEs, the paper has explored the specifics of productivity issues and solutions, implementation procedures, and lessons learned throughout. The following benefits have been realized in the company because of these experiments:

- Improvement in production by 20%
- Design, development, and implementation of the Assembly line.

Table 3 Assembly sequence for SWC (IA207846/IA209706)

Assembly station	1	2	3	4	5	6	7	8	9
Incoming material	Main bracket	Main bracket	Main Bracket from Stn 1 & Front shaft + Rear shaft + Ratchet + Friction plate	Sub assy from stn 2	Sub assy from assy stn 3 & Spring pin	Sub Assy from Assy stn 4 & Pin pawl + Washer and pawl + Hex nut	Sub Assy from stn 5 & wire rope + bolt	Sub Assy from stn 6 + Holding plate + nut + Washer	sub assy from stn 7
Operation description	Re-tapping	Check CD of Main Bracket	Insert Rear Shaft in Main Bracket with friction plate, ratchets and front shaft	Tightening of front shaft and rear shaft with the help of rod and spanner	Inserting spring pin in rear shaft and break it by hammer	Assemble pin pawl, washer and pawl in main bracket with hex nut	Insert wire rope in rear shaft and tight with bolts by air gun	Assemble holding plate with wire rope by nut and tightening by air gun	Winding wire rope in bracket with the help of spanner
Method of operation	Drilling	Manual	Manual	Manual	Manual	Manual	Manual	Manual	Manual

(continued)

Table 3 (continued)

Assembly station	1	2	3	4	5	6	7	8	9
Inspection method	TPG 12X1.25	Gauge should qualify with the cd of Bracket	Visually check all sub parts are assemble in proper manner	Free movement of shaft assembly in Bracket	Check spring pin fitment properly	Tight nut in pin pawl with the help of spanner at both ends	Insure proper fitment of wire rope in rear shaft by pulling wire rope	Manual	Visual
Method of storage	Bin	On table	On table	On table	On table	On table	On table	On table	On table
Number of persons	One	One	One	One	One	One	One	One	One
Average production per day	**For IA207846**	25							
	For IA209706	8							

Table 4 Modified assembly sequence for SWC (IA207846/IA209706)

Assembly station	1	2	3	4	5	6
Incoming material	Main bracket from station1	Main bracket from station 1 & front shaft + rear shaft + Rachet + Friction plate	Sub assembly from station 2	Sub Assy from Station 3 & spring Pin + Pin Pawl + washer + Pawl + hex Nut	Sub Assy station from stn4 and wire rope + bolt	Sub Assy from station 5 + holding plate + nut + washer
Operation description	Check CD of main bracket by Gauge	Inser rear shaft in main bracket with friction plat, rachet and front shaft	Tightening of front shaft and rear shaft with help of rod and spanner	Inserting spring pin in rear shaft and bracket by chisel and hammer & assembly in pawl washer and pawl in main bracket with hex nut	Inser wire rope in rear shaft and tight with bolt by air gun	Assemble holding plate with wire rope by nut and tightening by air gun and winding wire rope in bracket with the help of spanner
Method of operation	Manual	Manual	Manual	Manual	Manual	Manual
Inspection method	Gauge should qualify with the CD of Bracket	Visually check all sub parts are assemble in proper manner	Free movement of shaft assembly in bracket	Check spring pin fitment properly and tight nut in pin pawl with the help of spanner at both ends	Ensure proper fitment of wire rope in rear shaft by pulling wire rope	Manual
Method of storage	Plastic Bin	Plastic Bin	Plastic Bin	Plastic Bin	Plastic Bin	Plastic Bin
Production target (in Min)	2	2	2	2	2	2
Number of persons	1	1	1	1	1	1
Average production per day	**For IA207846**	**240**				
	For IA209706	**240**				

- Development of the transfer line without using any additional external power merely using human power and gravitational force for feeding the material to the next station.
- Breaking the whole cycle of operation into the elements, & thus imparting the specialization for speeding up the operation.
- Reducing the cost of the tooling by using only a single set.
- Implementing the principle of motion economy by the design of tools & equipment and by an arrangement of the workplace.
- Proper positioning and using the hanging option achieve theft and loss of tools.

5 Conclusion

This paper attempts to present the process of implementing productivity improvement strategies. SMEs are struggling with sufficient technology implementation due to limitations of investment. Through limitations of investment these sectors are not capable to make huge investments and lag behind the implementation of higher-end technologies.

So it is required to pay more attention to the SME sector so that they can take advantage of the latest technology and can be more competitive as a large industry. In this work, the low-cost conveyor system is developed as per the need of SMEs which gives the same advantages as the automated conveyor.

In this work, the low-cost conveyor system is developed as per the need of SMEs which gives the same advantages as the automated conveyor. Designing a low-cost setup and where this takes the advantage of industrial engineering, benefits of expertise, and dividing the work into elements caused speed up the work, an advantage of green supply chain management, and reducing the number of tools and dies. This effort causes the same benefits of large-scale industries but in an affordable range.

References

1. Srinivas KT (2013) Role of micro, small and medium enterprises in inclusive growth. Int J Eng Manag Res 3(4):57–61
2. Kumar V (2017) An analysis of the growth of MSMEs in India and their contribution in employment and GDP of the Country. Int J Interdiscipl Multidiscipl Stud 4(2):157–191, ISSN-0348-0343
3. Bohács G, Horváth AM, Gáspár D (2022) Development of a low-cost material handling vehicle solution using teleoperation and marker recognition. In: Shi X, Bohács G, Ma Y, Gong D, Shang X (eds) LISS 2021. Lecture notes in operations research. Springer, Singapore. https://doi.org/10.1007/978-981-16-8656-6_40
4. Hill T (1993) Manufacturing strategy: the strategic management of the manufacturing function, 2nd edn. Open University, Macmillan, London
5. Vieira G, et al (2011) Material handling management: a case study. J Operat Supply Chain Manag 4:19. https://doi.org/10.12660/joscmv4n2p19-30.
6. Ross KW (1994) Assembly-line job satisfaction and productivity. Ind Eng 26:44–45

7. Cheser R (1994) Kaizen is more than continuous improvement, quality progress, 27, 23–26. Drucker, P. (1991) the new productivity challenge. Harv Bus Rev 69:69–79
8. Patel PM (1993) Quality and productivity improvement through effective management of control chart and analysis of statistical data. Quality Forum 19:71–79
9. Ghosh S, Gagnon RJ (1989) A comprehensive literature review and analysis of the design, balancing and scheduling of assembly systems. Int J Prod Res 27:637–670
10. Scholl A (1999) Balancing and sequencing of assembly lines, 2nd edn. Physica-Verlag, Heidelberg, New York
11. Bukchin J, Tzur M (2000) Design of flexible assembly line to minimize equipment cost. IIE Trans 32:585–598. https://doi.org/10.1023/A:1007646714909
12. Drucker P (1991) The new productivity challenge. Harv Bus Rev 69:69–79

Electricity Generation by an Innovative Suspension System in a Two-Wheeler Electric Vehicle

Gurpreet Singh Matharou, Simran Kaur, Vishnu Raj, and Rishabh Dhawan

Abstract The primary goal of a vehicle suspension system is to ease passengers' discomfort caused by road roughness and improve ride handling related to pitching and rolling motions. A modification in the present telescopic suspension system has been implemented with the usage of a spring, rack, pinion setup, and a double-fed induction generator. The pinion is further connected to the doubly-fed induction generator shaft. Due to this configuration, the rotating motion of the pinion rotates the generator. As a result of the generator spinning, electrical energy is generated, which is utilized to provide a charging current sufficient for normal phase mobile charging. The maximum current of (1.98 A) and voltage (3.5 V) have been observed at a suspension vibration of 35 jerks per minute. The paper also addresses the development of a mathematical model of current and voltage concerning vehicle speed and vibrations per minute.

Keywords Shock absorber · Regenerative suspension · Rack and pinion · Regression · ANOVA · Ansys R-16

1 Introduction

Emerging energy regeneration technologies in the automotive region have been examined and developed as energy problems grow into major issues. Such technologies are kinetic energy recovery system (KERS) [1, 2] regeneration system thermal energy wasted in the motor [3], and automotive regenerative suspension systems

G. S. Matharou (✉) · V. Raj · R. Dhawan
School of Engineering and Technology, Manav Rachna International Institute of Research and Studies, Faridabad, India
e-mail: gsmatharou@gmail.com

S. Kaur (✉)
School of Commerce, Manav Rachna International Institute of Research and Studies, Faridabad, India
e-mail: simraunak16k16k@gmail.com

H. Kumar et al. (eds.), *Recent Advances in Intelligent Manufacturing*,
Lecture Notes in Mechanical Engineering,
https://doi.org/10.1007/978-981-99-1308-4_29

[4]. In most conventional vehicles that use fossil fuel, it has been shown that only 16% of the chemical energy of the fuel is used to propel the vehicle [5]. Some of the remaining energy feeds other vehicle subsystems, and, in addition, most of the thermal energy produced by the internal combustion engine is dissipated as heat. The vehicle distributes the remaining percentage of energy that displaces the vehicle to overcome the aerodynamic effects (4%), rolling resistance (4%), and the braking system (4%). The stochastic power spectrum of the road roughness level, which induces vibration energy in the vehicle, mainly generates rolling resistance. They transmit vibrational energy to the suspension system and dissipate into heat form in a viscous fluid damper [6]. There is no interaction between the train drivetrain of the vehicle with the suspension system, which is correct up to a point. The power train meets the demands of the requirement of power depending on the driving mode of the vehicle, while the primary function of the automotive suspension system is to mitigate the vibration generated by the irregular profile of the road and by the movements of unsynchronized chassis. However, the ways of driving directly influence the vehicle's dynamics, allowing a link between the power train and the suspension system [7].

Due to the effectiveness of both energy conversion and the control action of this type of device, several studies have recently been conducted to increase the performance of self-fed active and semi-active suspensions. The first regenerative damper based on linear electromagnetic motors was presented in the work of Karnopp [8]. Later, Suda and Shiiba [9] developed a hybrid suspension composed of a collection system of energy (electromagnetic regenerative damper linear) to improve the dynamic performance of an active suspension and thus reduce the energy consumption of the control system. Furthermore, Zuo et al. [10] designed a regenerative damper composed of a four-phase linear generator to increase the efficiency of regeneration of energy, obtaining a capacity of energy harvesting ranging from 26 to 64 watts of electrical power, considering that the relative speed of the suspension is 0.25–0.5 m/s. Later, Gysen et al. [11] analyzed and fabricated an active suspension composed primarily of a linear electromagnetic regenerative damper and concluded that such a suspension could significantly improve comfort and safety. driving the vehicle, maintaining a minimum consumption power within the range of 150–300 watts. With this in mind, Ebrahimi et al. [12] designed a hybrid electromagnetic damper consuming less energy than an active suspension. They took advantage of the eddy current effect to produce passive electrical damping. In addition, the device reduced up to 70% of the energy consumption of a conventional active suspension. However, the weight of the device was relatively greater than that of traditional shock absorbers, which is a disadvantage in real applications. On the other hand, Asadi et al. [13] designed a damper hybrid regenerative consisting of two parts: a passive one that provides passive damping forces employing a viscous fluid and another electromagnetic, which was used for the regeneration of vibratory energy utilizing a linear electromagnetic generator and also to provide active forces to the system of suspension. Hu et al. [14] theoretically designed a self-powered semi-active regenerative damper, which consists of two parts as well: a linear generator electric that harvests the kinetic energy of the suspension system and the magnetorheological fluid that controls damping levels through the electric current produced by the generator. Chen et al. [15] designed a hybrid semi-active regenerative damper with three adjustable damping levels, consisting mainly of a generator linear and a hydraulic damper. Several studies on semi-active self-feeding suspensions were reported with linear electromagnetic dampers. Gao et al. [16] studied the commitments and dynamic differences between the index of driving comfort and that of energy regeneration in suspensions of active electromagnetic waves through different control strategies. They showed that it is impossible to increase energy regeneration capacity without affecting the comfort index. Ding et al. [17] developed a suspension system hybrid electromagnetic active composed of a linear motor and a magnetorheological fluid damper. The results indicated that the hybrid system minimized the energy consumption of the active electromagnetic suspensions. To reduce the size and weight of the linear electromagnetic dampers, Duong et al. [18] designed and fabricated an 8-phase, 8-slot linear single-phase electromagnetic pickup with permanent magnets and also demonstrated that such a device collected an average power of 96.67 watts. However, the mechanism has adverse effects on the dynamics of the pension system. Linear electromagnetic dampers significantly improve both in active suspensions and semi-active ones. However, the size and weight of these are the main inconveniences for its application in real systems since it is necessary to increase the size of electromagnetic motors to provide large damping forces [19].

It is evident through a comprehensive literature review; that no work has been reported on developing electrical energy through suspension systems using rack and pinion arrangements. The arrangement is sufficient to produce electricity to charge a mobile handset with up to 5000 mAH battery. The paper also outlines the investigation of the usage of a vehicle's suspended mass through a regeneration system using a shock absorber.

2 Regeneration Process of Vibratory Energy

A considerable amount of kinetic energy is dissipated as heat in the fluid dampers that make up automotive suspension's passive. This kinetic energy caused by the speed relative suspension system could be regenerated utilizing intelligent regenerative dampers to reuse later in other vehicle subsystems [20]. Note that only the kinetic energy produced by the relative speed of the suspension system is considered for regeneration. However, tires also dissipate energy that spread by the suspension system is even greater. One-way clutches engage and disengage the regenerative damper rods, to transform the vertical movement from oscillatory to unidirectional rotational motion. The rectified rotational motion drives an electrical generator to produce energy that is later stored in supercapacitors. Ultimately, this energy can be used to supply the electrical power requirements of the vehicle's powertrain [21]. Different regenerative dampers have been designed, manufactured, and analyzed theoretically and experimentally to achieve a high energy conversion density. These kinetic energy harvesters are classified according to how they convert vibrational energy to electricity. Three types of dampers are known as regenerative, which are the following: linear electromagnetic, transmission-based electromechanical mechanical, and electro-hydraulic based on transmission hydraulics. Although these can regenerate vibration energy, some drawbacks arise, mainly in rectifying bidirectional vertical movement of the suspension system to unidirectional rotational movement.

3 Development of the Prototype

The two-wheeled vehicles mostly use telescopic fork-type suspension systems [23]. The suspension systems are employed to absorb vibrations caused due to uneven road surfaces. The vibration is absorbed due to a harmonic motion in the suspension system coil springs and dampers. The paper aims to illustrate how kinetic energy from a car's suspension can be used to maximize the energy that would otherwise be wasted. A rack and pinion setup and a double-fed induction generator has been employed in the project.

The damper oil is compressed due to the shock absorber action, and linear movement of the rack is translated to circular motion owing to the pinion moving as the rack meshes with the pinion. Additionally, the pinion is positioned on the shaft that connects to the doubly-fed induction generator's shaft. Due to this configuration, the generator is rotated by the rotating motion of the pinion. As a result of generator shaft rotation, energy is generated [22]. And this energy is utilised to charge a battery, which is then used to charge mobile handset with up to 5000 mAH battery. Figure 1 shows the block diagram of the prototype. The advantage of the project is that it is simple in construction, and the structure is very inexpensive. It is securely fixed to the two-wheelers mud-guard. It generates between 1 and 3 V on an even street and between 3 and 5 V on an uneven road, The battery stores the electrical charge produces and provides current, which is sufficient for charging the mobile battery while the vehicle is moving on a surface. Rack and pinion transmissions are used to convert rotational movements to linear motions. The rack is flat, and the gear contains teeth, referred to as the pinion.

Rack and pinion systems convert rotational motion to linear motion. The speed gain is used to transform movement in this case. The chain & sprocket system utilizes a gear with 24 teeth (a pair) attached to the chain to enhance the revolutions per minute (rpm). Figure 2 shows the assembled rack and pinion setup. When the vehicle's suspension is operating correctly, the linear motion of the suspension generates friction between the pulleys and the belt. As a result, the pulley begins to rotate. The DC motors' shafts are fitted with pulleys. As the pulleys revolve, the motor shafts also rotate, generating power. The generated power is stored in a battery, which is disseminated to the mobile charging unit for charging a mobile set. Figure 3 shows the rack and pinion arrangement simulation of ANSYS R 16 [25].

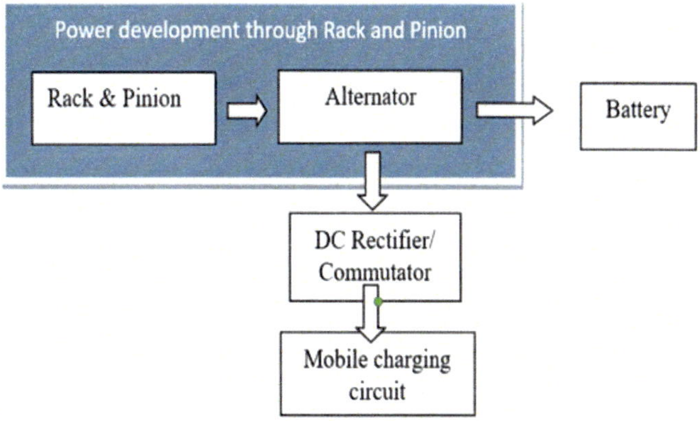

Fig. 1 Block diagram of the prototype

Fig. 2 Rack and pinion setup

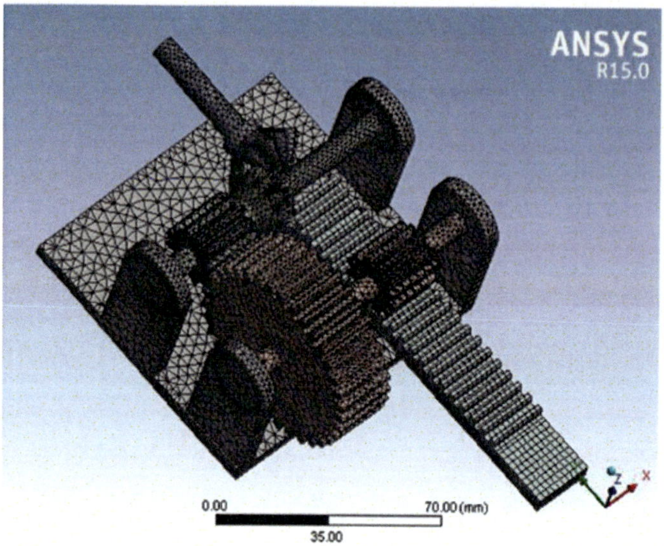

Fig. 3 Rack and pinion arrangement on Ansys 15.0

4 Results and Analysis

4.1 Design of Rack and Pinion

The ANSYS software has been used for design validation. Using ANSYS R 16, the rack and pinion arrangement analysis has been made to check the gear clashes and noise [24]. Figure 4 shows the pinion with a diameter of 50 mm, where the pitch circle of 0.5 mm is found to be sufficient for a selected rack of linear pitch 0.5 mm and pitch line in coordination to pinion pitch radius. This article proposes a system in which the suspension system has a suitably mounted rack and pinion arrangement which gets actuated whenever there is a jerk in the system. When a vehicle is running at a speed of 15–20 km/h due to jerks on road we observe a voltage of 1–3 V with a help of a multi-meter. Likewise, the observations of current, voltage development at various vehicle speeds and corresponding suspension vibration (minutes) are noted and is tabulated in Table 1. A terrain with different potholes and breakers were selected that offers the necessary vibrations. The vehicle was made to run on the different terrains at different speed and output was noted through a multi meter. Figure 5 shows the current development at different vehicle speeds and corresponding vibrations in the suspension system.

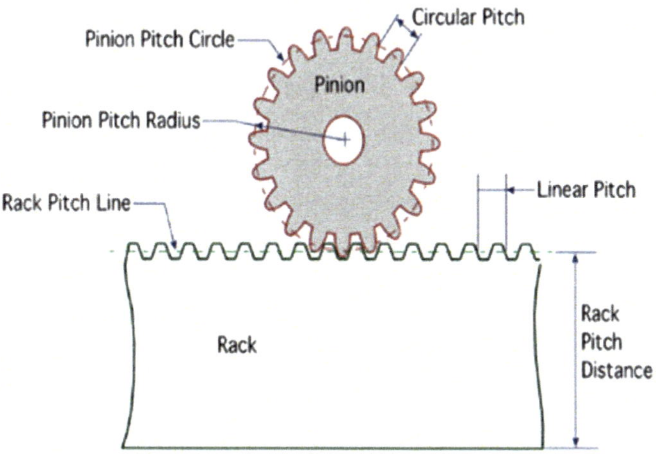

Fig. 4 Rack and pinion configuration

Table 1 Current and voltage development at various vehicle speed and vibrations

S. no.	Vehicle speed km/h	Vibrations per minute	Current (A)	Voltage (V)	Power (W)
1	20	10	0.012	1.590	0.019
2		15	0.035	2.082	0.073
3		20	0.970	2.462	2.388
4		25	1.040	2.769	2.880
5		30	1.330	3.142	4.179
6		35	1.980	3.469	6.869
7	30	10	0.021	1.230	0.026
8		15	0.042	2.134	0.090
9		20	0.860	2.241	1.927
10		25	0.940	2.543	2.390
11		30	1.160	2.982	3.459
12		35	1.840	3.143	5.783
13	40	10	0.038	1.280	0.049
14		15	0.058	2.364	0.137
15		20	0.760	2.186	1.661
16		25	0.870	2.346	2.041
17		30	1.098	2.782	3.055
18		35	1.730	3.236	5.598

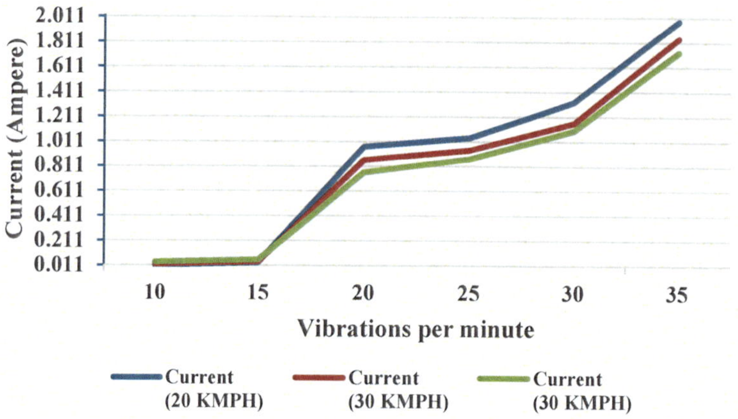

Fig. 5 Current development at different vehicle speed and vibrations

4.2 *Current and Voltage Development*

It can be seen that maximum current of 35 vibrations per minute was obtained at vehicle speed of 20 kmph. It can be seen that the current output is directly proportional to the vibrations. More vibrations in the suspension system causes more lateral movement in the rack causing more rotation in the pinion attached to the electricity generator.

The same trend is also observed in Fig. 6, which shows the voltage development at different vehicle speeds and corresponding vibrations. We can see a direct relation between voltage generation and vibrations. The voltage developed at 20 kmph vehicle speed is seen to be uniform at different suspension vibrations. During 30 kmph vehicle speed and road vibrations of 15 jerks per minute, a sudden surge in voltage development has been seen. This might be due to proper and timely meshing of rack and pinion rotations. A drop in voltage was observed at 20–25 vibrations and after that again a surge in voltage can be witnessed.

4.3 *Mathematical Modeling*

4.3.1 Regression Equation for Current

Regression study (RS) is a collection of analytical techniques employed for the evaluation of relations between a dependent variable and one or more independent variables. It can be appropriated to estimate the vitality of the relation between variables and for designing the future liaison between them. Using RS, a relation between current and voltage as a dependent variable with vehicle speed and suspension vibration is being attempted. Table 2 shows the ANOVA analysis of current

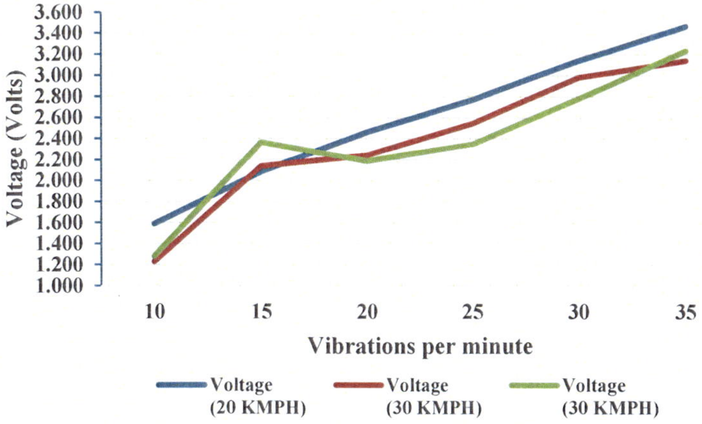

Fig. 6 Voltage development at different vehicle speed and vibrations

Table 2 ANOVA of current with vehicle speed and suspension vibration

Source	DF	Adj SS	Adj MS	*F*-value	*P*-value
Regression	2	6.93634	3.46817	98.41	0
Vehicle speed (kmph)	1	0.05508	0.05508	1.56	0.23
Vibrations per minute	1	6.88126	6.88126	195.25	0
Error	15	0.52866	0.03524		
Total	17	7.46499			
S	R-sq	R-sq (adj)	R-sq (pred)		R-sq
0.187733	92.92%	91.97%	89.89%		92.92%

regarding vehicle speed and suspension vibrations. Equation 1 shows the regression equation for the current. The current produced is directly proportional to the number of vibrations. More vibrations in the suspension develop more current.

$$\begin{aligned}\text{Current} = -0.60 - (0.00678 \times \text{vehiclespeed in kmph}) \\ + (0.07241 \times \text{vibrationsperminute})\end{aligned} \quad (1)$$

4.3.2 Regression Equation for Voltage

Likewise, Table 3 shows the ANOVA of voltage with vehicle speed and suspension vibration. We can see that the regression equation has a p-value = 0, which shows the relationship is valid for a larger sample size. The vibrations per minute as a factor has direct relation with voltage generation.

Table 3 ANOVA of voltage with vehicle speed and suspension vibration

Source	DF	Adj SS	Adj MS	*F*-value	*P*-value
Regression	2	6.4847	3.24236	81.42	0.000
Vehicle speed (kmph)	1	0.1452	0.14520	3.65	0.076
Vibrations per minute	1	6.3395	6.33953	159.20	0.000
Error	15	0.5973	0.03982		
Total	17	7.0821			
S	R-sq	R-sq (adj)	R-sq (pred)	S	R-sq
0.199554	91.57%	90.44%	87.51%	0.199554	91.57%

$$\text{Voltage} = 1.210 - (0.01100 \times \text{vehicle speed in kmph}) + (0.06950 \times \text{vibrations per minute}) \quad (2)$$

For a larger sample size, the relationship is valid in the regression equation. The *p*-values for the coefficients show if the relations are statistically meaningful. The *p*-values further find out whether the relations that have been observed in a small size likewise remain valid in the substantial community. Coefficient of determination (R^2) value of 0.9292, and 0.9157 for current and voltage, respectively, is in line with the respective adjusted coefficient of determination (R^2_{adj}) values 0.9197, and 0.9044. All these values are next to unity, and in both the cases, the R^2 value is higher than the R^2_{adj} value; thus, the data for responses are accurately prepared in the planned designs (Eqs. 1 and 2).

5 Conclusion

In the present paper, a prototype of a two-vehicle suspension system with an arrangement of rack and pinion arrangement has been designed. ANSYS R 16 has been used for data analysis and verification before actual fabrication of the arrangement. The current and voltage developed through rack and pinion arrangement is further analyzed with respect to the vehicle speed and suspension vibration. Thereafter, regression analysis was made and suitable mathematical equation was devised. The model was validated using Analysis of Variance. Based on the results of the study, the following conclusions can be derived.

1. The pinion with a diameter of 50 mm, pitch circle of 0.5 mm is found to be sufficient for a selected rack of linear pitch 0.5 mm and pitch line in coordination to pinion pitch radius.
2. More vibrations in the suspension system causes more lateral movement in the rack causing more rotation in the pinion attached to the electricity generator.
3. Both current and voltage produced have direct relationship with vehicle speed and suspension vibrations.

4. During 30 kmph vehicle speed and road vibrations of 15 jerks per minute, a sudden surge in voltage development has been seen. This might be due to proper and timely meshing of rack and pinion rotations. A drop in voltage was observed at 20–25 vibrations and after that again a surge in voltage can be witnessed.
5. The factor vibration per minute has the maximum f-value with both current and voltage and is the most influential factor.
6. Coefficient of determination (R^2) value of 0.9292, and 0.9157 for current and voltage, respectively, is in line with the respective adjusted coefficient of determination (R^2_{adj}) values 0.9197, and 0.9044. All these values are next to unity, and in both the cases, the R^2 value is higher than the R^2_{adj} value; thus, the data for responses are accurately prepared in the planned designs. The equation for current and voltage thus can be used for prediction on the basis of vehicle speed and vibrations per minute.

6 Future Scope

The innovative suspension system can be designed with different combinations of diameter, pitch circle, and linear pitch of pinion and rack respectively. The combination can further be experimented for variation in voltage and current development along with any new parameter not covered in the study. The electricity output from the innovative suspension system can be used for the necessary energy requirement of automatic braking systems [26] and automated e-carts [27]. Response surface methodology [28] can also be employed for the development of mathematical models of the output parameters along with the usage of artificial intelligence [29].

References

1. Itani K, De Bernardinis A, Khatir Z, Jammal A (2017) Comparative analysis of two hybrid energy storage systems used in a two front wheel driven electric vehicle during extreme start-up and regenerative braking operations. Energy Convers Manag 144:69–87. https://doi.org/10.1016/j.enconman.2017.04.036
2. Pugi L, Pagliai M, Nocentini A, Lutzemberger G, Pretto A (2017) Design of a hydraulic servo-actuation fed by a regenerative braking system. Appl Energy 187:96–115. https://doi.org/10.1016/j.apenergy.2016.11.047
3. Zhao D (2013) Waste thermal energy harvesting from a convection-driven Rijke-Zhao thermo-acoustic-piezo system. Energy Convers Manag 66:87–97. https://doi.org/10.1016/j.enconman.2012.09.025
4. Abdelkareem MAA, Xu L, Ali MKA, Elagouz A, Mi J, Guo S, Liu Y, Zuo L (2018) Vibration energy harvesting in automotive suspension system: a detailed review. Appl Energy 229:672–699. https://doi.org/10.1016/j.apenergy.2018.08.030
5. Zou J, Guo X, Abdelkareem MAA, Xu L, Zhang J (2019) Modelling and ride analysis of a hydraulic interconnected suspension based on the hydraulic energy regenerative shock absorbers. Mech Syst Signal Process 127:345–369. https://doi.org/10.1016/j.ymssp.2019.02.047

6. Zou H-X, Zhao L-C, Gao Q-H, Zuo L, Liu F-R, Tan T, Wei K-X, Zhang W-M (2019) Mechanical modulations for enhancing energy harvesting: principles, methods and applications. Appl Energy 255:113871. https://doi.org/10.1016/j.apenergy.2019.113871
7. Li L, Zhang Y, Yang C, Yan B, Marina Martinez C (2016) Model predictive control-based efficient energy recovery control strategy for regenerative braking system of hybrid electric bus. Energy Convers Manag 111:299–314. https://doi.org/10.1016/j.enconman.2015.12.077
8. Karnopp D (1989) Permanent magnet linear motors used as variable mechanical dampers for vehicle suspensions. Veh Syst Dyn 18:187–200. https://doi.org/10.1080/00423118908968918
9. Suda Y, Shiiba T (1996) A new hybrid suspension system with active control and energy regeneration. Veh Syst Dyn 25:641–654. https://doi.org/10.1080/00423119608969226
10. Zuo L, Scully B, Shestani J, Zhou Y (2010) Design and characterization of an electromagnetic energy harvester for vehicle suspensions. Smart Mater Struct 19:045003. https://doi.org/10.1088/0964-1726/19/4/045003
11. Gysen BLJ, van der Sande TPJ, Paulides JJH, Lomonova EA (2011) Efficiency of a regenerative direct-drive electromagnetic active suspension. IEEE Trans Veh Technol 60:1384–1393. https://doi.org/10.1109/TVT.2011.2131160
12. Ebrahimi B, Bolandhemmat H, Khamesee MB, Golnaraghi F (2011) A hybrid electromagnetic shock absorber for active vehicle suspension systems. Veh Syst Dyn 49:311–332. https://doi.org/10.1080/00423111003602400
13. Asadi E, Ribeiro R, BehradKhamesee M, Khajepour A (2015) A new adaptive hybrid electromagnetic damper: modelling, optimization, and experiment. Smart Mater Struct 24:075003. https://doi.org/10.1088/0964-1726/24/7/075003
14. Hu G, Lu Y, Sun S, Li W (2016) Performance analysis of a magnetorheological damper with energy harvesting ability. Shock Vib 2016:1–10. https://doi.org/10.1155/2016/2959763
15. Chen L, Shi D, Wang R, Zhou H (2016) Energy conservation analysis and control of hybrid active semiactive suspension with three regulating damping levels. Shock Vib 2016:1–14. https://doi.org/10.1155/2016/6196542
16. Gao Z, Chen S, Zhao Y, Liu Z (2019) Numerical evaluation of compatibility between comfort and energy recovery based on energy flow mechanism inside electromagnetic active suspension. Energy 170:521–536. https://doi.org/10.1016/j.energy.2018.12.193
17. Ding R, Wang R, Meng X, Chen L (2019) Energy consumption sensitivity analysis and energy-reduction control of hybrid electromagnetic active suspension. Mech Syst Signal Process 134:106301. https://doi.org/10.1016/j.ymssp.2019.106301
18. Duong M-T, Chun Y-D, Han P-W, Park B-G, Bang D-J, Lee J-K (2020) Optimal design of a novel single-phase 8-slot 8-pole tubular electromagnetic shock absorber to harvest energy. IEEE Trans Ind Electron 67:1180–1190. https://doi.org/10.1109/TIE.2019.2898591
19. Zuo L, Tang X (2013) Large-scale vibration energy harvesting. J Intell Mater Syst Struct 24:1405–1430. https://doi.org/10.1177/1045389X13486707
20. Zhang Z, Zhang X, Chen W, Rasim Y, Salman W, Pan H, Yuan Y, Wang C (2016) A high-efficiency energy regenerative shock absorber using supercapacitors for renewable energy applications in range extended electric vehicle. Appl Energy 178:177–188. https://doi.org/10.1016/j.apenergy.2016.06.054
21. Pu X, Wang ZL (2021) Self-charging power system for distributed energy: beyond the energy storage unit. Chem Sci 12:34–49. https://doi.org/10.1039/D0SC05145D
22. Lv X, Ji Y, Zhao H, Zhang J, Zhang G, Zhang L (2020) Research review of a vehicle energy-regenerative suspension system. Energies 13:441. https://doi.org/10.3390/en13020441
23. Liu J, Li X, Zhang X, Chen X (2019) Modeling and simulation of energy-regenerative active suspension based on BP neural network PID control. Shock Vib 2019:1–8. https://doi.org/10.1155/2019/4609754
24. Matharou GS, Bhatia S, Tanzeel M, Virmani P, Biswas S (2018) Comparative study of SAE 1141 carbon steel, T651–7075 aluminum alloy & 4340 billet steel material for connecting rod using ANSYS 16. Int J Mech Prod Eng Res Dev 8:691–696. https://doi.org/10.24247/IJMPERDFEB201875

25. Matharou GS, Gupta C, Pradhan V, Ahlawat T (2022) The design and development of advanced braking system with microcontroller. ECS Trans 107:2883–2891. https://doi.org/10.1149/10701.2883ecst
26. Tiwari P, Matharou GS, Waghmare N, Sharma B, Shrivastava A (2022) Prototype of operating a vehicle using Arduino micro-controller with wireless communication system. In: International conference on simulation, automation & smart manufacturing (SASM). IEEE Explore, pp 1–6. https://doi.org/10.1109/SASM51857.2021.9841174
27. Waghmare NK, Bhatia S, Matharou GS, Kashyap D, Singh A (2021) Design and manufacturing of automated guided e-cart for stock relocation. In: 2021 international conference on simulation, automation & smart manufacturing (SASM), pp 1–5. https://doi.org/10.1109/SASM51857.2021.9841130
28. Matharou GS, Bhuyan BK (2021) Modelling and combined effect analysis of electric discharge Machining process using response surface methodology. Mater Today Proc © 2021 Elsevier Ltd. 46:6638–6643. https://doi.org/10.1016/j.matpr.2021.04.103
29. Kaur S, Tandon N, Matharou GS (2020) Contemporary trends in education transformation using artificial intelligence. In: Transforming management using artificial intelligence techniques. CRC Press, pp 89–103. https://doi.org/10.1201/9781003032410-7

Ergonomic Assessment of Tractor Seat for Anthropometric Dimensions of Indian Population

Arunesh Chandra, Bibekanad Pathak, Subodh Sharma, Pawan Arora, and Harish Kumar

Abstract Industrial design, apparel design, tool design, and workplace design all heavily rely on anthropometry. For farmers, a tractor is their primary form of power. The design of a contemporary tractor seat must therefore take these human issues into account. When anthropometric dimensions are appropriately considered in the design of seat, these variables enable the operator to carry out a variety of different activities with maximum effectiveness, safety and minimum fatigue. For drivers to have a better driving experience and comfort, the tractor seat's design is crucial. In present study Rapid Upper Limb Assessment (RULA) was used for assessment existing and proposed the design of the tractor seat where it was observed that RULA score (risk factor) was reduced from 5 to 2 when the tractor seat was modified taking into consideration of anthropometric dimensions.

Keywords Anthropometry · Ergonomics · Tractor seat · RULA · Working posture

1 Introduction

In various domains of work, including ergonomics, medicine, and various design disciplines, benefit from the use of anthropometric data. Anthropometric data are used in ergonomics with the goal of optimising daily living by establishing working environments. In order to comprehend the range of human body proportions within

A. Chandra (✉) · S. Sharma
KIET Group of Institutions, Uttar Pradesh, Ghaziabad 201206, India
e-mail: arunesh.chandra@Kiet.edu

B. Pathak
IMS Engineering College, Uttar Pradesh, Ghaziabad 201015, India

P. Arora
Galgotia College of Engineering, Greater Noida 201310, India

H. Kumar
National Institute of Technology, Delhi 110036, India

H. Kumar et al. (eds.), *Recent Advances in Intelligent Manufacturing*,
Lecture Notes in Mechanical Engineering,
https://doi.org/10.1007/978-981-99-1308-4_30

the target audience, designers of a wide range of products rely on anthropometric data. Human factors and ergonomics should play a significant part in population accommodation sizing to improve the product to either get a better fit or widen the range of suitable users in order to accomplish mass customisation and collaborative product design [1]. In an effort to address the conundrum of "fitting the tasks to the human person," anthropometry is regarded as the fundamental ergonomic approach [2]. Inappropriate posture is caused by poorly built systems, which makes operations difficult and increases the risk of MSDs and various physiological illnesses [3]. Anthropometry is now a crucial component of workplace, apparel, tool, and industrial design, among other fields, where statistical information on the distribution of body dimensions in the target population is utilised to optimise products. However, the development of ergonomically good products for usage is lacking significantly in the absence of current anthropometric data sets. The operator benefits greatly from and depends heavily on the ergonomic features of agricultural equipment when used. Both physiological and psychological weariness have an impact on an operator's performance. Ergonomics seeks to boost productivity while also enhancing user comfort and operator safety. Agricultural machineries and equipments will function poorly if ergonomics are not taken into account during design, and will result in a reduction in the amount of time that can be spent actually working. For farmers, a tractor is their primary form of power. The design of a contemporary tractor seat must therefore take human issues into account. When anthropometric dimensions are correctly taken into account during design, the operator is able to complete a variety of difficult tasks quickly, safely and with increased comfort. An efficient strategy for increasing operator comfort is to build tractor operator seats and workstations ergonomically [4].

Tractor seats are available in a huge variety of size and shape today. Common study issues resulting from tractor seat designs include whether the seat is appropriate and better fit for a driver to sit on, as well as whether the seat satisfies tractor drivers' need. Particularly for drivers who frequently use tractors in the field and for other purposes, good seats should match the needs of the customer and take ergonomics criteria and safety measures into consideration. Additionally, the seat needs to be solid and stable enough to support the person using it. Tractor seats must have linear dimensions that correspond to those of their most likely users and it is axiomatic by this point. Since tractor operators are the most likely users, seat characteristics such as seat pan length, width, and tilt, as well as backrest height angle and slope need to be improved with their anthropometric considerations. Rapid Upper Limb Assessment (RULA) analyses the manikin's upper limbs in the present study based on elements including weight, distance, and frequency. When discomfort in the upper extremities is to be recorded, the RULA survey technique is used for assessments of workplaces. The complete body's biomechanical and postural stresses are assessed by the RULA screening tool, with a focus on the neck, trunk, and upper limbs.

2 Methodology

2.1 Subjects and Measuring Techniques

There were no restrictions on the subjects' ethnicity, social class, height, or weight. The subjects' ages range from 26 to 50 years old, with a mean age of 37.4 years. The average height and weight of the participants were found to be 1642.04 mm and 63.62 kg, respectively, with standard deviations of 25.53 mm and 5.14 kg. Farmers from Meerut division of Uttar Pradesh state of India are chosen based on their availability and willingness to participate without payment or any other type of reward. None of the individuals had a history of physical or congenital disabilities or trauma. The farmers are advised to be bare-footed and to wear light clothes so as to reduce the margins of errors. The pilot study was initially carried out in the boy's dormitory of the KIET Group of Institutions, Ghaziabad. The next step is to conduct the measurements in the employees' field area in order to collect the necessary data. The voluntary members of the travelling survey team were made up of four UG students and a trainer. The survey team received the necessary instruction in the operation of the equipment. They only began collecting data while being supervised by a trainer and academic member when their measures were deemed to be precise, dependable and consistent. The measurements were taken in accordance with the methodology discussed by various researchers [5]. Anthropometric measurements were collected usually in open space with ethical guidelines between 10 am to 4 pm.

A high degree of descriptive accuracy and efficient data collecting are priorities when choosing landmarks. Using an eyebrow pencil to draw a dot, the palpated landmarks were discovered and marked on the skin. All measurements for landmarks were then obtained twice, with the mean of the two readings being recorded as the final measurement. These tiny black dots were subsequently used as landmarks. Following anthropometric tools are used for measurement—Hardenpen anthropometer for measurement of stature and arm length (sensitivity of 1 mm), Lafayette (small) anthropometer for measurement of small length (sensitivity of 1 mm), non-stretch, fiberglass, retractable anthropometric Lafayette measuring tape for circumferential measurements (graduation length of 1 mm), portable digital weighing scale for measurement of body weight.

2.2 Anthropometric Measurements

The measurement of body weight and seven anthropometric measurements (stature, sitting height, arm length, buttock popliteal length, popliteal height, and interscye breadth) for 55 male tractor drivers was the extent of this investigation. These measurements are chosen because they are important for seat design and they have been studied in prior similar research projects [6]. The definitions of these 7 anthropometric dimensions are similar to as defined and positioned with land marking of

Table 1 Anthropometric measures of tractor drivers (mm)

Measure	Mean	Std. Dev.	Min	Max	5th Percentile	95th Percentile
Stature (S)	1642.04	25.53	1616	1720	1617.8	1707.8
Sitting height (SH)	1215.65	27.18	1180	1298	1186.8	1286
Arm length (AL)	558.38	22.36	512	598	519.2	597
Buttock popliteal length (BPL)	426.84	8.61	402	444	411.4	442
Functional leg length (FLH)	815	11.97	782	835	789.8	832
Popliteal height (PH)	387.38	10.77	354	409	364.8	406
Interscye breadth (IB)	368.4	19.32	316	390	316	390
Hip breadth (HB)	318.18	14.14	266	342	298.4	339.2
Thigh clearance (TC)	160.78	5.45	120	195	126.7	190.2

all subjects which are considered as per provided definition [7]. Descriptive statistics of these anthropometric dimensions are provided in Table 1.

3 Results and Discussions

Mahindra is the tractor that farmers in the Meerut division of the state use the most frequently out of the ten major suppliers of tractors to farmers in India. Anthropometric measurements should be taken into account while determining the width, depth, and height of the tractor seat. Given that the user spends the majority of their time in the seat while driving, the design is essential to their comfort [8]. Thus, the existing seat dimensions were measured with standard measuring tools and the seat was modeled with existing dimensions in solid works for analysis on CATIA where it is used to create a virtual environment for biomechanical analysis. The study's first part includes gathering anthropometric data and modeling a tractor seat; the study's second phase is used to determine whether farmers' anthropometric data and the design of the current tractor seat are incompatible [9]. RULA evaluates the biomechanical and postural loads on the body, paying special attention to the neck, trunk, and upper limbs. It also takes into account the driver's position with both hands on the wheel. This posture is selected because it is the most potentially used posture while working in the field.

Fig. 1 Driving posture for RULA analysis (without load)

3.1 RULA Analysis of Existing Tractor Seat

This was done without load seat dimensions that are compared with anthropometric dimensions and RULA score for the existing posture as shown in Fig. 1. Overall RULA posture score during driving is obtained as 5 which is critical [10]. Each body part was analysed using RULA and it indicates that leg, neck and trunk are the major contributors for this score that creates this problem regarding MSDs among tractor drivers which may arise due to mismatch between the anthropometric dimensions of the driver and the tractor seat. The measured seat dimensions are seat pan width—320 mm, back rest width—320 mm, seat pan concavity—30 mm, seat pan taperness (vertical)—0°, seat pan taperness (horizontal)—2.1^{0}, back rest concavity—20 mm, back rest taperness—1.7°, steering wheel clearance 175 mm, back rest height 307.82 mm. Therefore it is necessary to adjust the dimensions of the existing tractor seat to increase the comfort during driving.

3.2 Modification in the Seat Dimensions

To achieve a match between the seats and the tractor drivers' anthropometric body dimensions, adjusted seat dimensions were proposed in this study's seat measurements [11]. Figure 2 which is modeled on CATIAV5R20 displays the recently suggested ideal seat measurements for Indian tractor drivers in the northern section of the nation. The modified seat dimensions are estimated to be seat pan width—340 mm, seat pan concavity—37 mm, seat pan taperness (horizontal)—5.7^0, seat pan taperness (vertical)—2^0, back rest concavity—25 mm, back rest width—340 mm, back rest taperness 4.59^0, steering wheel clearance—190 mm, back rest height 356.21 mm [12–14]. These dimensions are based upon the anthropometric dimensions listed in Table 2 against the seat design parameters which are based upon the research contributions from various studies [15, 11].

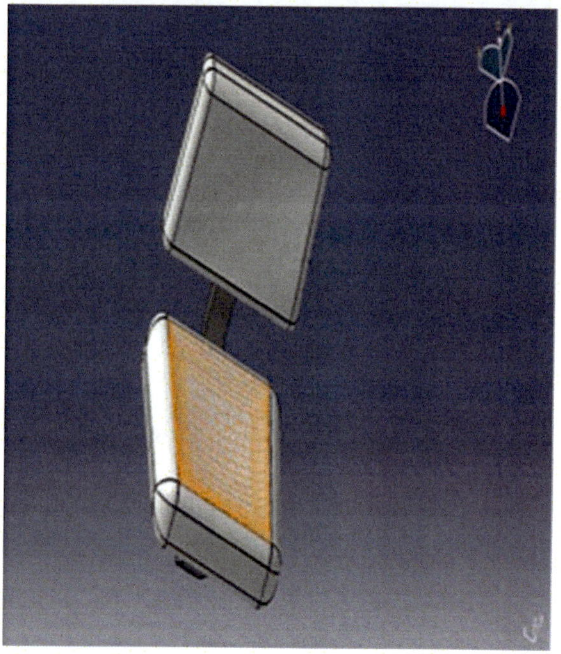

Fig. 2 Modified chair model in CATIA

Table 2 Comparison of major dimensions of existing and modified tractor seat

Anthropometric measures	Mean (mm)	Std. Dev. (mm)	Prediction	Seat existing	Seat modified
Sitting shoulder height (SH)	828.27	16.41	Backrest height	307.82	356.21
Buttock-popliteal length (BPL)	426.84	8.61	Seat length	300	300
Thigh clearance (TC)	160.78	5.45	Steering wheel clearance	175	190
Popliteal length (PH)	387.38	10.77	Seat height	250	222.8
Interscye breadth (IB)	368.4	19.32	Upper back rest width	340	352
Hip breadth (HB)	318.18	14.14	Seat width	320	340

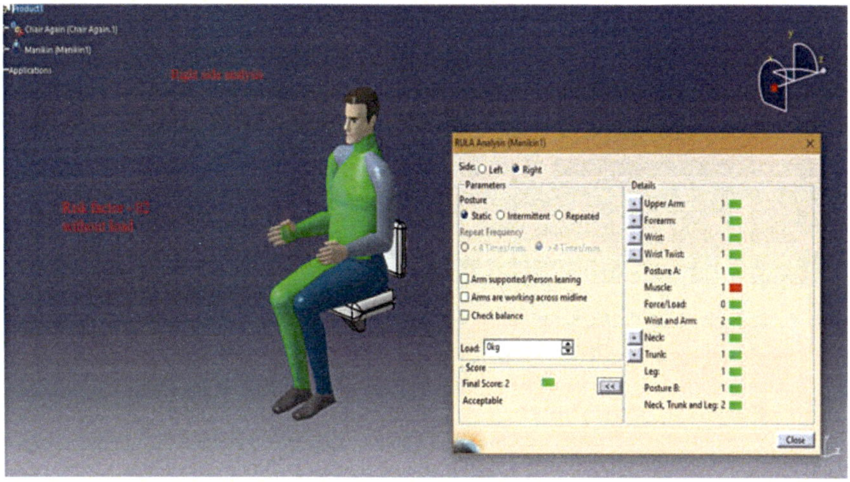

Fig. 3 Driving posture for RULA analysis (without load)

3.3 RULA Analysis of Modified Tractor Seat

The model of the tractor seat was designed on CATIA followed by ergonomic analysis. Figure 3 represents the analysis on the modified seat without load in the CATIA software in which the value of the risk factor comes out to be 2 on the right side of the body and show stress value in different parts of the body. Table 2 shows the comparison of seat dimensions of the existing and modified seat which comes into account after multiple iterations and these iterations will help to reduce the risk factor from 5 to risk factor 2. The lower value of RULA suggested that the comfort level of the tractor will increase and the satisfaction level of the drivers will be higher with improved safety [16, 17].

4 Conclusions

This study's significance lies in its consideration of human variables while adjusting tractor drivers' seats. This research is primarily concerned with assessing tractor seat design and driver postural modifications. Using CATIA, the suggested design of the tractor seat was created. It was then ergonomically assessed using the RULA analysis to get its final score. With certain adjustments to the seat's size, the basic design is then made more efficient. The results revealed that with implementation of modified changes in the seat dimensions the RULA analysis revealed that the risk factor during analysis is reduced from 5 to 2 when manikin is used with recommended changes in the dimensions of the tractor seat. This demonstrated that the tractor driver is operating safely and in the best possible posture. The main drawback of this study was the exclusion of other significant characteristics in favour of simply taking seat adjustment parameters into account. Another drawback of this study is that tractor drivers were not used to evaluate the best seat characteristics in real time. The tractor seat's CAD model does not feature the control panel's buttons or any additional controls.

References

1. Li Z (2006) Anthropometric topography. In: Karwowski W (eds) International encyclopedia of ergonomics and human factors. Taylor and Francis, London, UK
2. Sanders MS, McCormick EJ (1992) Human factors in engineering and design, 7th edn. McGraw-Hill, New York
3. Chandra A, Rathore S, Mallick Z (2021) Ergonomic risk assessment and postural analysis of Indian agricultural workers. In: Part of the design science and innovation book series (DSI). Springer, Heidelberg, pp 73–82
4. Yadav R, Budhrani B, Balani P, Pund S (2017) Anthropometric and ergonomic compatibility of tractor workplace design. J Ergon S6:1–7
5. Pheasant S, Haslegrave CM (2006) Bodyspace—anthropometry, ergonomics and the design of work, 3rd edn. Taylor and Francis, New York
6. Kahya E (2019) Mismatch between classroom furniture and anthropometric measures of university students. Int J Ind Ergon 74(1):102864
7. Chandna P, Deswal S, Chandra A (2010) An anthropometric survey of industrial workers of the northern region of India. Int J Ind Syst Eng 6(1):110–128
8. Aminian NO, Romil FI (2018) Ergonomics assessment of current aircraft passenger seat design against Malaysian anthropometry data. Int J Eng Technol 7(4):18–21
9. Huat NM, Mat S, Dullah AR, Khalil SN (2020) Car seat design using RULA analysis. In: Proceedings of mechanical engineering research. pp 288–290
10. McAtamney L, Corlett EN (1993) RULA: a survey method for the investigation of work-related upper limb disorders. Appl Ergon 24(2):91–99
11. Shukla P, Mehta CR, Agrawal KN, Potdar RR, Kumar M, Singh K (2022) Approach for ergonomic assessment of self- propelled combine harvester seats based on anthropometric body dimensions. Int J Occup Saf Ergon 21:1–12
12. Gite LP, Mehta CR, Agrawal KN, Narwariya BS, Potdar RR (2019) Handbook of ergonomical design of agricultural tools, equipment and work places, 1st edn. Jain Brothers, India
13. Chaffin DB, Andersson GBJ, Martin BJ (2006) Occupational biomechanics, 4th edn. Wiley-Inderscience, United States
14. Halder P, Mahmud T, Sarker E, Karmaker CL, Kundu S, Patel S, Setiwan A, Shah K (2018) Ergonomic considerations for designing truck drivers' seats: the case of Bangladesh. J Occup Health 60(1):64–73
15. Looze MPd, Kuij-Evers LFM, Dieen JV (2003) Sitting comfort and discomfort and the relationships with objective measures. Ergonomics 46(10):985–997
16. Harisinghani MG, Blake MA, Saksena M, Hahn PF, Gervais D, Zalis M, Fernandes LdSD, Mueller PR (2004) Importance and effects of altered workplace ergonomics in modern radiology suites. RadioGraphics 24(2):615–627
17. Vink P, Brauer K (2011) Aircraft interior comfort and design. CRC Press Taylor and Francis Boca Raton

Experimental Investigation on Process Parameter Optimization to Enhance Tensile Strength in FDM—3D Printing Process with PLA Material

Raffik Rasheed, Murugesan Kandasamy, Vijayanandh Raja, Sanjeev Balasubramani, Manoj Kumar Vijayakumar, and Rajavel Mahadevan

Abstract 3D printing technology is employed in many industries for constructing buildings, manufacturing vehicles, and medical devices. Fusion deposition modeling (FDM) is one of the frequently used additive manufacturing (AM) processes to create prototypes and functional parts with geometrically complicated shapes. It is also known as fused filament manufacturing. Due to its ability in shortening product development cycles without using costly instruments and tools, it is becoming more and more popular. But, poor dimensional accuracy, subpar surface quality, part strength, and insufficient mechanical characteristics are restricting the implementation of the FDM technique in numerous industrial applications. Different process parameters, like layer thickness, build orientation, raster width, and print speed, have an impact on the components and part characteristics created by FDM. The optimum and best possible process parameter combinations must be chosen. There are several methods for optimizing and selecting the suitable FDM process parameters. To assess the printed component properties, a thorough overview of pre-processing needs to be done. Depending on the components of FDM machines, the process parameters' range must be specified. The different parameters range depending on the kind of machine, nozzle size, and filament type. A three-dimensional (3D) printed part's quality is thought to be improved by setting the process parameters optimally, which may also mean less post-production effort. This research article investigates the least analyzed process parameter combinations and offers the optimum FDM process parameter range.

R. Rasheed (✉) · M. Kandasamy · S. Balasubramani · M. K. Vijayakumar · R. Mahadevan
Department of Mechatronics Engineering, Kumaraguru College of Technology, Coimbatore, Tamil Nadu, India
e-mail: raffik.r.mce@kct.ac.in

V. Raja
Department of Aeronautical Engineering, Kumaraguru College of Technology, Coimbatore, Tamil Nadu, India

H. Kumar et al. (eds.), *Recent Advances in Intelligent Manufacturing*,
Lecture Notes in Mechanical Engineering,
https://doi.org/10.1007/978-981-99-1308-4_31

Keywords FDM process · 3D printing · Process parameter optimization · Higher tensile strength · Layer thickness · Printing speed · Extrusion temperature

1 Introduction

3D printing technology refers to additive manufacturing processes formerly called Rapid Prototyping (RPT). Additive Manufacturing Technologies are extensively growing in the manufacturing sectors and industrial sectors [1]. Fused Deposition Modeling is one of the emerging bottom-up manufacturing techniques utilized to produce finished products by creating layer-by-layer patterns automatically using CAD-generated design [2, 3]. FDM is the most often used AM method, and it utilizes cost-effective polymeric materials that are simple to print 3D impellers [4] and have a broad variety of applications. But some downfalls such as surface roughness [5], part strength [6], and dimensional inaccuracies [7, 8], occur in FDM components. To overcome these drawbacks, process parameter optimizations are done [9, 10]. Several types of underwater systems [11], robots [12], and drones [13] are designed and commercialized for different purposes by using the FDM process. It is also employed to create medical equipment [14], medical infrastructure coatings [15], weaving loom components [16], and automation system casings [17]. UAVs (Unmanned Aerial Vehicles) and robot components deployed in agricultural fields for pesticide or weedicide spraying processes are also created [18, 19]. The FDM technology is displayed in Fig. 1. Continuously heated filaments are extruded onto a heated bed or deposited over the printed layers. The STL file is generated using the CAD design and it decides the motion of the printer nozzle along the X and Y axes. The functional model is built by adding the semi-molten filament layer by layer [20]. The polymer filament's thermo-plasticity is essential for this technique since it enables the filaments to bond and later solidify at ambient temperature. Identifying the part strength influencing process parameters and the correlation between them is essential [18].

2 Materials and Methods

2.1 FDM Process Materials

The two most extensive and popular FDM process materials are ABS and PLA. Instead of ABS and PLA, other materials like PETG, Nylon [21], Polyphenylsulfone (PPS), polycarbonate (PC), and other polymer composites are also utilized as filament materials to create functional parts, and research projects which provide a greater variety of material options for researchers and makers of functional parts.

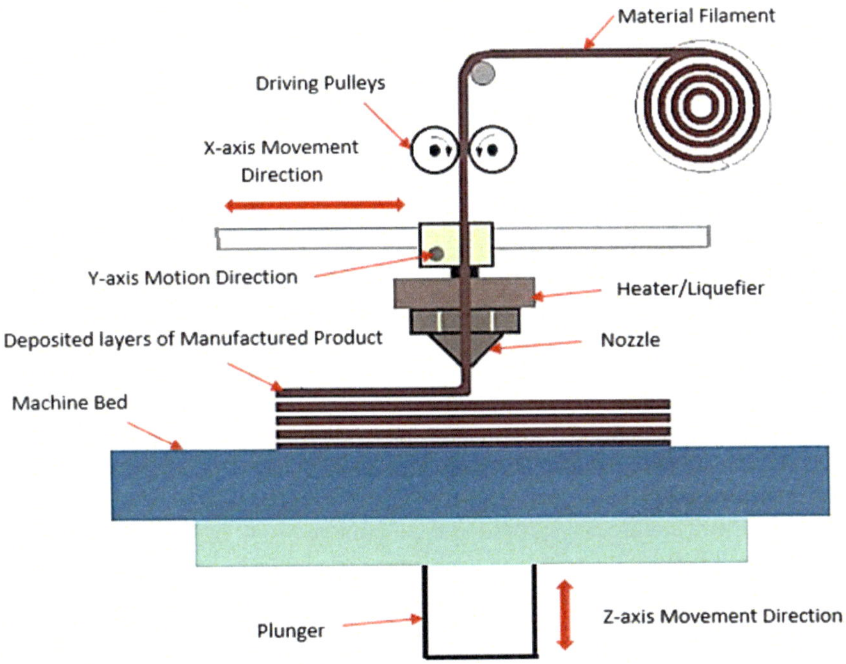

Fig. 1 Fusion deposition modeling process

Investigations on process factors like build orientation, layer thickness, raster orientation, and raster width are vastly analyzed. But process parameters like printing speed, infill pattern, extrusion nozzle temperature, and shell width are less well examined [22]. Little investigations happened and so it simultaneously maximized the features of various sections [23]. The least understood process variables like print speed, infill pattern, and extrusion temperature are taken into consideration in this research investigation process. The FDM process is a complicated 3D printing technique with various process parameters [24]. It requires more investigation in future research with multi-objective process parameter adjustments to improvise the consistency in printed FDM parts during the design and manufacturing process [25–27].

2.2 *Polylactic Acid (PLA) Material*

In this research work, Polylactic Acid is used as the filament material. PLA (Polylactic Acid) is an organic polymer that is biodegradable, renewable, and food safe. It is most often used filament in Fusion Filament Fabrication (FFF) 3D printers due to its versatility and simplicity of use, particularly for non-mechanically or thermally demanding applications. The advantages of using the PLA material are a shiny exterior finish, excellent tensile strength, excellent UV resistance, withstand operational temperatures up to 50 °C, and low humidity resistance [28].

3 Fabrication, Testing, and Optimization

PLA filament with 1.75 mm diameter is used to create ASTM D638 tensile specimens. This study alters the optimization settings, including infill pattern, printing speed, and nozzle temperature, to generate a homogenous sample of nine ASTM D638 tensile specimens [29]. This finds the optimal FDM process parameter combination for producing objects with the highest tensile strength. All nine specimens are prepared by a Creality CR-10 S5 FDM printer. This machine can employ materials with a temperature degradation range of 230–250 °C degrees Celsius. In this study, temperature variations from 200 to 220 °C have been used. In addition, just the ambient temperature is employed as the minimum bed or platform temperature. Tables 2 and 3 provide the selected process parameter ranges and the L9 orthogonal array parameter combinations respectively.

3.1 Test Specimen

The American Society of Testing and Materials (ASTM) has provided standard specimen profiles with dimensional specifications for the tensile testing of materials [20]. The profile and dimensions vary for metals, plastics, ceramics, composites, etc. For plastics and polymers, ASTM D638 should be used. ASTM D68 includes five types of specimens, and the Type I specimen is mostly used. Figure 2 shows the Type I specimen of the ASTM D638 standard. The profile and the dimensional specifications are mentioned in Table 1.

Fig. 2 ASTM D638 specimen

Table 1 ASTM D638—type I specimen dimensions

Dimension	Tolerance
Overall length, mm	165
Width at ends, mm	19 + 6.4
Length of narrow section, mm	57 ± 0.5
Gage length, mm	50 + 0.25
Distance between grips, mm	115 ± 5
Radius of fillet, mm	76 ± 1
Thickness, mm	3 ± 0.4
Width of narrow section, mm	13 ± 0.5
Cross-sectional area of the narrow section, mm^2	35–48.6
Perimeter of the narrow section, mm	30.6–34.2

3.2 Selected Process Parameters

The least analyzed parameters nozzle temperature and infill pattern are considered with layer thickness. The selected parameters are as follows

- Layer Thickness
- Nozzle Temperature
- Printing Speed

Layer Thickness: Layer thickness is the vertical distance covered for one layer of deposition.

Nozzle Temperature: The temperature of the extruding nozzle makes the thermoplastic PLA filament into semi-solid condition to create the parts.

Printing Speed: The extrusion nozzle traveling speed over the part building platform/bed.

3.3 Steps for FDM Process Parameters Optimization

Figure 3 shows the process flow of the FDM process parameter optimization process. Figure 4 shows the tensile testing experimental setup with universal testing machine. Figure 5 gives the images of tested specimens after failure.

3.4 Parameter Range

See Table 2.

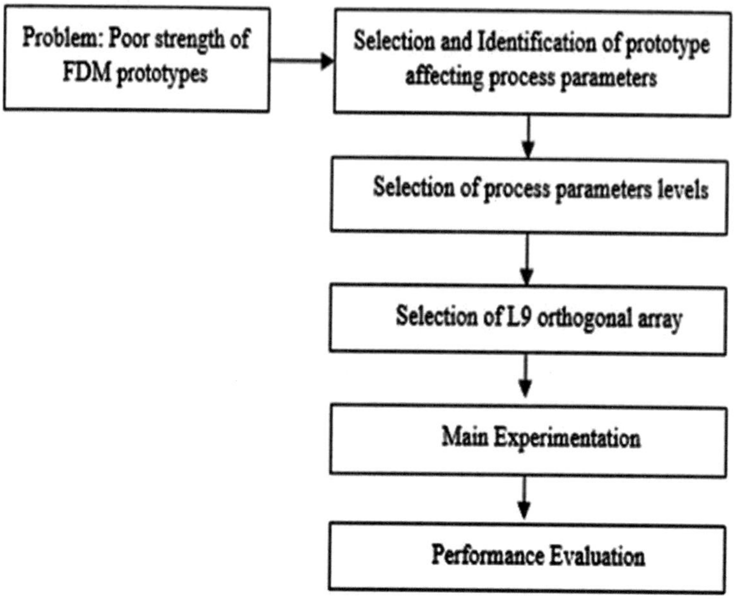

Fig. 3 Steps for optimizing FDM process parameters

3.5 Parameter Combination for L9 Orthogonal Array

Among the selected three parameters, three values were chosen for each parameter between their range as maximum, minimum, and the mean value as tabulated in Table 2. This totally gives 9 values that are used for the L9 orthogonal array combination as shown in Table 3 [29].

3.6 Tensile Testing Process for Optimization of FDM Process Parameters

See Figs. 4 and 5.

4 Results and Discussion

The specimen is loaded and pulled gradually continuously till it fails. The tensile test was conducted to find the ultimate load and the breaking load. The tensile tests are conducted using the Dak Systems Test bench—a universal testing machine. As per the ASTM guidelines, the testing speed has been set as 5 mm/min. The ultimate tensile

Fig. 4 Tensile tested specimen using universal testing machine

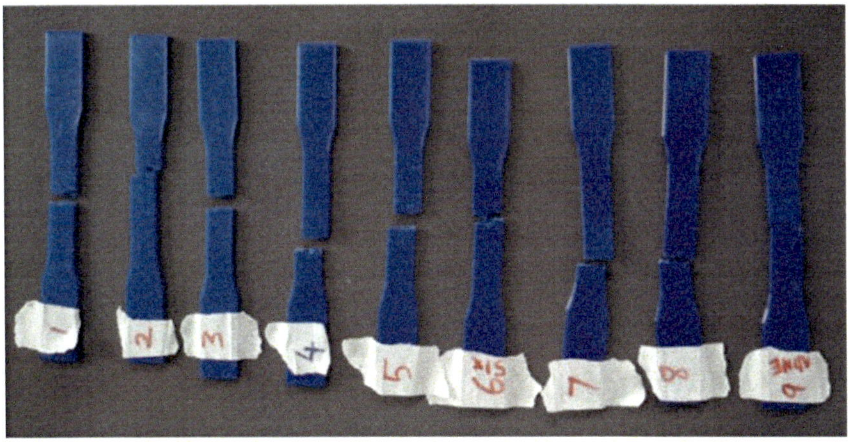

Fig. 5 Tensile tested specimens after failures

Table 2 Selected parameter range

Process parameters	Levels		
	Level 1	Level 2	Level 3
Layer thickness (mm)	0.15	0.2	0.25
Nozzle temperature (°C)	200	210	220
Print speed (mm/sec)	100	75	50

Table 3 L9 orthogonal array with parameter combination

Specimen number	Layer thickness (mm)	Nozzle temperature (°C)	Print speed (mm/sec)
1	0.15	200	100
2	0.15	210	75
3	0.15	220	50
4	0.20	200	75
5	0.20	210	50
6	0.20	220	100
7	0.25	200	50
8	0.25	210	100
9	0.25	220	75

strength, maximum load, and elongation percentage are calculated and tabulated for all orthogonal array settings. Table 4 gives the ultimate tensile load, elongation, ultimate tensile strength, and elongation percentages for all the nine L9 orthogonal specimens The ultimate strength and elongation percentages are calculated using the below formulae and the calculated values are tabulated in Table 4.

$$\text{Ultimate Tensile Strength} = \text{Ultimate load/Cross-sectional Area}$$

$$\text{Elongation percentage} = \text{Linear Strain at failure} * 100$$

4.1 *Tensile Test Results*

See Table 4.

Table 4 Tensile test—experimental results

Specimen no.	Ultimate tensile load (N)	Ultimate strength (N/mm^2)	Elongation (mm)	Elongation %
1	698.230	17.903	2.898	5.759
2	713.009	18.282	4.060	8.119
3	679.434	17.421	3.100	6.201
4	677.748	17.378	4.340	8.680
5	771.407	19.780	4.590	9.179
6	738.185	18.928	3.400	6.800
7	763.655	19.581	4.410	8.819
8	758.226	19.442	4.400	8.800
9	810.452	20.781	3.970	7.940

Table 5 ANOVA analysis of variance for S–N ratios

Source	DF	Seq SS	Adj SS	Adj MS	F	P	Contribution %
Layer thickness	2	1.3654	1.3654	0.68268	2.13	0.32	59.124
Nozzle temperature	2	0.2936	0.2936	0.14681	0.46	0.69	12.715
Printing speed	2	0.0088	0.0088	0.00439	0.01	0.99	0.003
Residual error	2	0.6415	0.6415	0.32077			27.781
Total	8	2.3093					

4.2 ANOVA Analysis Results for Ultimate Tensile Load

See Table 5 and Figs. 6, 7 and 8.

5 Conclusions

5.1 Empirical Analysis

- Part strength is an important characteristic in rapid layered manufacturing for all application purposes.
- Figure 7 shows the stress–strain curve for all the nine specimens and Fig. 8 provides the maximum load in kg, maximum tensile strength in kg/cm^2 and elongation percentage for the specimens.
- The Ultimate load is 810.452 N and Ultimate Tensile Strength is 20.781 MPa.

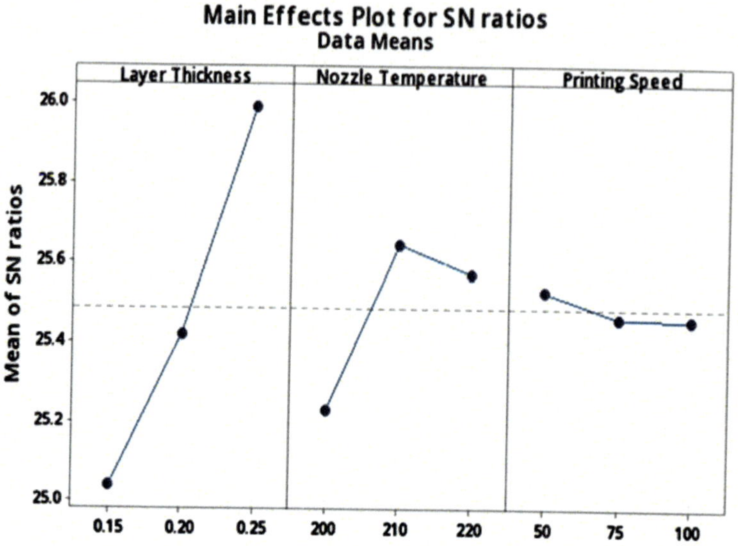

Fig. 6 ANOVA analysis—SN ratio for ultimate tensile strength

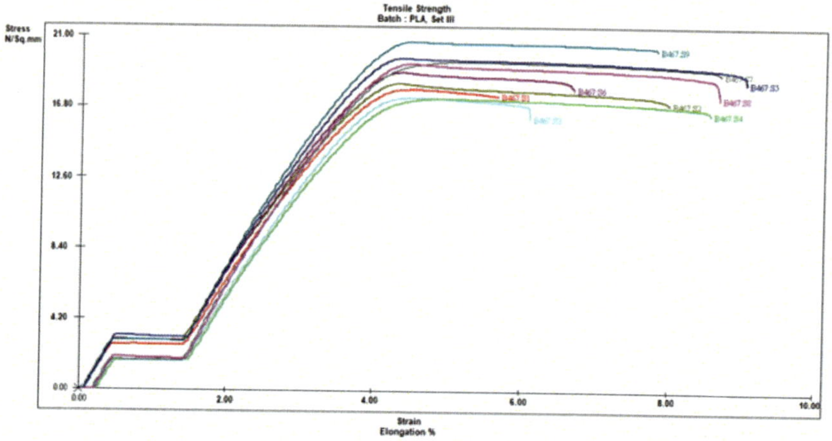

Fig. 7 Stress–strain response curves for specimens

- The parameter combination for the ninth specimen is,
  - Layer thickness—0.25 mm
  - Nozzle Temperature—220 °C
  - Printing Speed—75 mm/sec
- Parts produced by setting these parameter values will yield higher tensile strength.

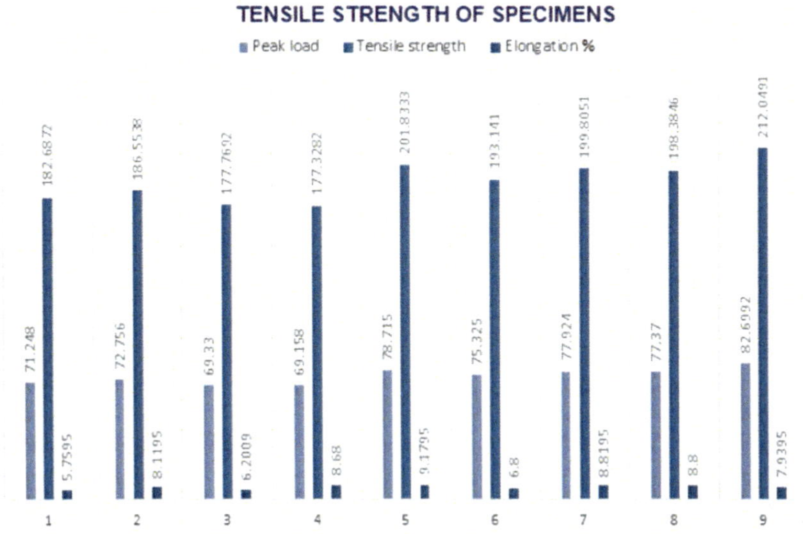

Fig. 8 Tensile test results for L9 orthogonal array specimens

5.2 *ANOVA Analysis*

- According to the ANOVA Analysis, Layer thickness contributes more to the load-withstanding ability of FDM prototypes in tensile loading conditions.
- Table 5 shows ANOVA analysis of variances and Fig. 6 show the mean of S–N ratios details and the process parameter variations and its influencing percentages and impact in obtaining higher tensile strength.
- The better parameter combination to obtain a higher tensile strength is,
  - Layer thickness—0.25 mm
  - Nozzle Temperature—210 °C
  - Printing Speed—50 mm/sec
- Layer thickness is the most influential process parameter followed by nozzle temperature and the least influencing factor is printing speed.
- Layer thickness contributes 59.12%, Nozzle temperature affects about 12.72%, Printing speed provides the least impact about 0.03% and other parameters and errors contribute about 27.78% in the ultimate tensile load handling capability of 3D printed PLA prototypes.

References

1. Dey A, Yodo N (2019) A systematic survey of FDM process parameter optimization and their influence on part characteristics. J Manuf Mater Process 3(3):64. https://doi.org/10.3390/jmmp3030064
2. Patel R, Jani S, Joshi A (2022) Review on multi-objective optimization of FDM process parameters for composite materials. Int J Interact Des Manuf. https://doi.org/10.1007/s12008-022-01111-9
3. Rehman RU, Zaman UKU, Aziz S, Jabbar H, Shujah A, Khaleequzzaman S, Hamza A, Qamar U, Jung DW (2022) Process parameter optimization of additively manufactured parts using intelligent manufacturing. Sustainability 14:15475. https://doi.org/10.3390/su142215475
4. Zhou H, Zhao G (2021) Error analysis and process parameter optimization of impeller model manufactured by FDM process. In: 8th annual international conference on geo-spatial knowledge and intelligence, 693012069. IOP conference series: earth and environmental science. IOP Publishing Ltd., China
5. Saad MS, Mohd Nor A, Abd Rahim I et al (2022) Optimization of FDM process parameters to minimize surface roughness with integrated artificial neural network model and symbiotic organism search. Neural Comput Appl 34:17423–17439
6. Panda SK, Padhee S, Anoop Kumar SOOD, Mahapatra SS (2009) Optimization of fused deposition modelling (FDM) process parameters using bacterial foraging technique. Intell Inf Manag 1(02):89
7. Equbal A, Sood AK, Ansari AR, Equbal A (2017) Optimization of process parameters of FDM part for minimiizing its dimensional inaccuracy. Int J Mech Prod Eng Res Dev 7(2):57–65
8. Deswal S, Narang R, Chhabra D (2019) Modeling and parametric optimization of FDM 3D printing process using hybrid techniques for enhancing dimensional preciseness. Int J Interact Design Manuf (IJIDeM) 13(3):1197–1214
9. Gao G, Xu F, Xu J (2022) Parametric optimization of fdm process for improving mechanical strengths using Taguchi method and response surface method: a comparative investigation. Machines 10(9):750
10. Shankar AN, Jagota V, Jamadon NH, Raffik R, Suneetha VL, Samori IA, Karnan L (2022) An AHP-TOPSIS approach for optimizing the mechanical performance of natural fiber-based green composites. Adv Mater Sci Eng
11. Raffik R, Kamal SS, Arun S, Raja PP, Kumar RM (2018) Automatic tank cleaner. Int J Mech Prod Eng Res Dev 8(113)
12. Raffik R, Mayukha S, Hemchander J, Abishek D, Tharun R, Kumar SD (2021) Autonomous weeding robot for organic farming fields. In: 2021 international conference on advancements in electrical, electronics, communication, computing and automation (ICAECA). IEEE, pp 1–4
13. Ezhil VS, Sriram BR, Vijay RC, Yeshwant S, Sabareesh RK, Dakkshesh G, Raffik R (2022) Investigation on PID controller usage on unmanned aerial vehicle for stability control. Mater Today Proc 66(3):1313–1318
14. Rakesh D, Keerthivaasan RKV, Mohan A, Samvasan P, Ganesan P, Raffik R (2021) Automated public screening and health vitals monitoring station. In: 2021 international conference on advancements in electrical, electronics, communication, computing and automation (ICAECA). IEEE, pp 1–6
15. Zhang J, Liu T, Hu X, Raffik R, Bhatt MW, Ofori I (2022) Anti-cracking performance test of thick steel structure fireproof coating under vibration fatigue load for critical health infrastructure. J Eng 1086–1094
16. Raffik R, Naveen C, Siyaguru J, Vijayanandh R (2022) Performance analysis of modified shuttle mechanism in polymerized bag weaving circular loom. In: AIP conference proceedings, vol 2446, no 1. AIP Publishing LLC, p 180028
17. Sharma BB, Raffik R, Chaturvedi A, Geeitha S, Akram PS, Natrayan L, … Sathyamurthy R (2022) Designing and implementing a smart transplanting framework using programmable logic controller and photoelectric sensor. Energy Rep 8:430–444

18. Senthilkumar S, Anushree G, Kumar JD, Vijayanandh R, Raffik R, Kesavan K, Prasanth SI (2021) Design, dynamics, development and deployment of hexacopter for agricultural applications. In: 2021 international conference on advancements in electrical, electronics, communication, computing and automation (ICAECA). IEEE, pp 1–6
19. An Z, Wang C, Raj B, Eswaran S, Raffik R, Debnath S, Rahin SA (2022) Application of new technology of intelligent robot plant protection in ecological agriculture. J Food Quality
20. Soundararajan R, Raffik R, Karthikeyan S (2022) Mechanical properties enhancement of A356 alloy hybrid composite by FSP and SP route (No. 2022-28-0541). SAE technical paper (2022)
21. Ramesh M, Panneerselvam K (2021) Mechanical investigation and optimization of parameter selection for Nylon material processed by FDM. Mater Today Proc 46:9303–9307
22. Rayegani F, Onwubolu GC (2014) Fused deposition modelling (FDM) process parameter prediction and optimization using group method for data handling (GMDH) and differential evolution (DE). Int J Adv Manuf Technol 73(1):509–519
23. Jin L, Zhou Z, Li K, Zhang G, Liu Q, Yao B, Fang Y (2022) Applying evolutionary multitasking for process parameter optimization in polymerization process of carbon fiber production. Appl Sci 12(18):9378
24. Wang Z, Li J, Wu W, Zhang D, Yu N (2021) Multitemperature parameter optimization for fused deposition modeling based on response surface methodology. AIP Adv 11(5):055315
25. Malviya M, Desai K (2019) Build orientation optimization for strength enhancement of FDM parts using machine learning based algorithm. Comput Design Appl 17:783–796
26. Alam N, Alam M, Ahmad S (2021) Optimization of fused deposition modelling process parameters using teaching learning based optimization (TLBO) algorithm. In: IOP conference series: materials science and engineering, vol 1149, no 1. IOP Publishing, p 012014
27. Yadav D, Chhabra D, Garg RK, Ahlawat A, Phogat A (2020) Optimization of FDM 3D printing process parameters for multi-material using artificial neural network. Mater Today Proc 21:1583–1591
28. Raja S, Agrawal AP, Patil P, Thimothy P, Capangpangan RY, Singhal P, Wotango MT (2022) Optimization of 3D printing process parameters of polylactic acid filament based on the mechanical test. Int J Chem Eng
29. Tura AD, Mamo HB, Rao DK (2021) Study on the effect of fused deposition modelling (FDM) process parameters on tensile strength and their optimal selection. Int J Eng Artif Intell 2(2):81–91

Empirical and Experimental Analysis on Influencing FDM Process Parameters Optimization with PLA Material on Tensile Strength Using ANOVA Analysis

R. Raffik, P. Magudapathi, R. P. Roshan, C. Subash, B. Subashini, and D. K. Anusha

Abstract Fused deposition modeling (FDM) is the first 3D printing technique that most people and consumers are getting introduced. Globally, 3D printing with the FDM method is most often used with 3D printers. Among the seven primary various additive manufacturing techniques, usually preferred material extrusion technology is FDM or Fused Filament Fabrication (FFF). In the FDM technique, a continuous thermoplastic or composite material thread is employed to make 3D components. Plastic filament is fed through an extruder, melted at the extruding nozzle, and then deposited layer by layer onto the build platform automatically according to the part structure. In the investigation of process parameters, the specimens are made of PLA (polylactic acid) material. With Taguchi method, the design of the experiments is prepared, and the specimens are created. There are nine specimens made with different parameters, namely the nozzle temperature, infill pattern, and layer thickness. ANOVA analysis is employed to identify the optimum process parameters in FDM process.

Keywords FDM (Fusion Deposition Modeling) · Process parameter optimization · Tensile strength · PLA (polylactic acid) · ANOVA analysis

1 Introduction

The FDM process is utilized to create functional models, parts and components. Due to the flexibility of component fabrication and the lack of costly tooling, FDM is particularly beneficial for a variety of industrial applications. A fascinating and

R. Raffik (✉) · P. Magudapathi · R. P. Roshan · C. Subash · B. Subashini · D. K. Anusha
Department of Mechatronics Engineering, Kumaraguru College of Technology, Coimbatore, Tamilnadu, India
e-mail: raffik.r.mce@kct.ac.in

H. Kumar et al. (eds.), *Recent Advances in Intelligent Manufacturing*,
Lecture Notes in Mechanical Engineering,
https://doi.org/10.1007/978-981-99-1308-4_32

expanding field in many industries is 3D printing. Despite the severe number of problems with product strength and quality, they have persisted. Fused deposition modeling processing parameter optimization is a difficult task owing to the vast quantity of variables that impact the degree of the manufactured components [1]. An experiment-based study is carried out to determine the variables that affect the quality and to establish an acceptable amount of these variables that would improve the FDM process's quality. As a result, improving the mechanical characteristics, surface roughness, or dimensional accuracy typically calls for a required number of test samples [2]. The method of "design of experiments" was applied in the study that was just given to reduce the number of tests required to optimize the factor. The Taguchi Method is a useful technique that has been employed for improving the quality of a manufactured product. It is a straightforward yet effective method. The Taguchi approach was used to construct experiments utilizing the orthogonal array method to get the best performance out of the 3D-printed parts. The performance of prototypes is susceptible to changes in process parameters, which is the main constraint of the FDM technique [3]. The FDM processing method is shown in Fig. 1. The impact of printing process settings on an object's mechanical characteristics is one of these concerns. In general, the elastic modulus and tensile strength increase as well as the layer thickness rise. The strength and stiffness of FDM components are discovered to be affected by interactions and second-order effects. In this investigation, the consequences of print layer thickness, infill pattern and nozzle temperature on the tensile strength of the FDM samples are considered [4].

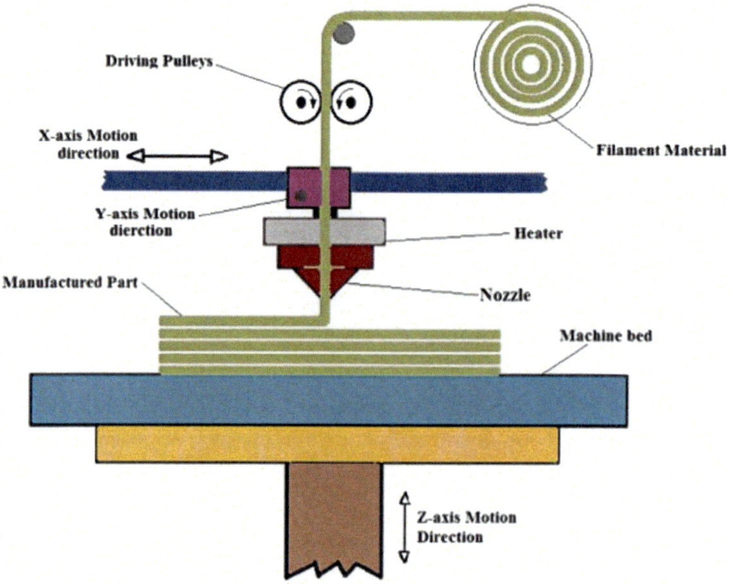

Fig. 1 Fusion deposition modeling process

2 Equipment and Procedures

2.1 *Process Materials for FDM*

One of the key characteristics of FDM is the diversity of materials available. From simple engineering polymers like PA, TPU, and PETG to thermoplastics like PLA and ABS thermoplastics with high performance, such as PEEK and PEI are used for prototyping throughout the embodiment phases of new product design [5].

2.2 *Material Made from Poly-Lactic Acid (PLA)*

A thermoplastic monomer made of sustainable, organic ingredients like sugar cane or corn starch is called polylactic acid, or PLA, and is produced from biomass resources, in a contrast to most plastics, which are created by distilling and polymerizing petroleum. After thermoplastic starch, PLA is the next frequently made bioplastic. It shares many characteristics with polypropylene (PP), polyethylene (PE), and polystyrene (PS), and it is biodegradable [6]. Compared to conventional plastics production, PLA utilizes 65% less power, creates 68% less carbon emissions, and is toxin-free. There are several unique varieties of polylactide because of the nature of lactic acid. These include poly-L-lactide (PLLA), which is created when L-lactide is polymerized (also known as L-lactide) [7].

Additionally, while preparing PLA from many biomass materials, the inclusion of different compositions helps in producing the additional 3D printed components with PLA filaments with enhanced properties [8]. Although adding material compositions to PLA can make 3D printing more challenging and change the material's characteristics. Different blends also need to be understood to change the extrusion temperature to melt the printing material [9].

2.3 *3D Printing Applications in Industrial Sectors*

Several types of underwater systems [10], robots [11], and drones [12] are designed and commercialized for different purposes by using the FDM process. It is also employed to create medical equipment [13], medical infrastructure coatings [14], weaving loom components [15], and automation system casings [18]. UAVs (Unmanned Aerial Vehicles) and robot components deployed in agricultural fields for pesticide or weedicide spraying processes are also created [16, 17].

3 Testing and Evaluation

To evaluate the tensile strength of L9 Orthogonal array specimens, ASTM D638 type IV specimens as shown in Fig. 2 are made with the PLA material according to the design of experiments, and with different process parameters. Table 1 provides the dimensions of the ASTM D638—Type IV specimen.

3.1 Steps for Optimization of FDM Process Parameters in PLA Materials

See Fig. 3.

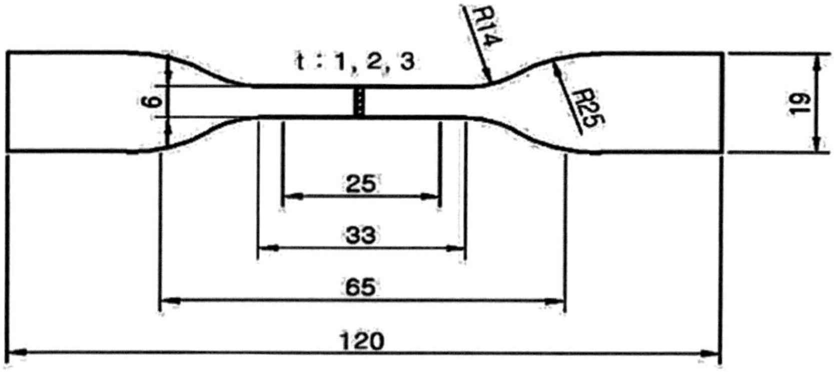

Fig. 2 ASTM D638 type IV specimen

Table 1 ASTM D638—Type IV sample dimensions

Size	Type IV
Full extent, l_3	115 mm
Equivalent extent, l_2	33 mm
Index extent, l_1	25 mm
Equivalent segment breath, $sturdy_1$	6 mm
Depth, h	4 mm or less
Clutch segment breath, $sturdy_2$	19 mm
Separating the clutches	65 mm

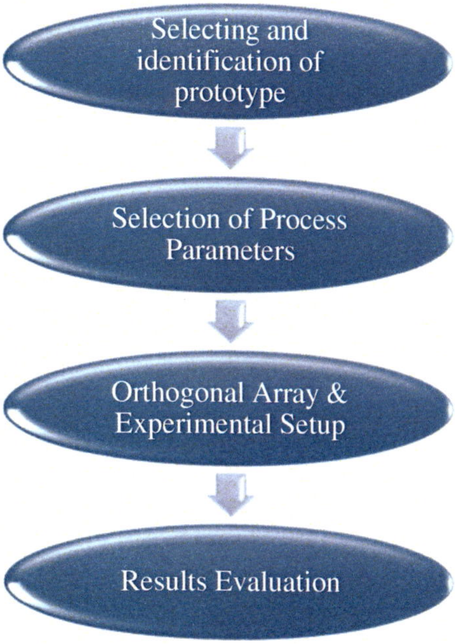

Fig. 3 Steps for optimization of FDM process parameters in PLA materials

3.2 Selected Process Parameters

The above-mentioned parameters are the FDM process parameters. We selected three parameters for evaluating the specimens, the process parameters are as follows,

- Layer Thickness
- Nozzle Temperature
- Infill Pattern

Layer Thickness: In FDM, the parts are produced layer by layer. Layer thickness is the vertical thickness covered for one layer.

Nozzle Temperature: The temperature of the extruding nozzle which makes the thermo-plastic PLA filament into semi-solid condition to create the parts.

Infill Pattern: The semi-liquified filament deposition procedure over the part building platform/bed.

Table 2 Parameter range

Factors	Levels		
	Level 1	Level 2	Level 3
Layer thickness (mm)	0.15	0.2	0.25
Nozzle temperature (°C)	200	210	220
Infill pattern	Grid (1)	Tri-hexagon (2)	Concentric (3)

Table 3 PLA Specimens process parameter ranges

S. no	Layer thickness (mm)	Nozzle temperature (°C)	Infill pattern
1	0.15	200	Concentric
2	0.15	210	Tri-Hexagon
3	0.15	220	Grid
4	0.20	200	Tri-Hexagon
5	0.20	210	Grid
6	0.20	220	Concentric
7	0.25	200	Grid
8	0.25	210	Concentric
9	0.25	220	Tri-Hexagon

3.3 Parameter Range

3.4 Design of Experiments for PLA's Specimens

From the selected parameters, three values were selected for each parameter between maximum, minimum, and mean values. Tables 2, 3 show the process parameter range and its L9 orthogonal array combinations. With these values, L9 orthogonal array was designed to conduct the experiments. The process parameter influence on part strength has been examined with ANOVA analysis method [19, 20].

3.5 Steps for Optimizing FDM Process Parameters Testing on the Universal Testing Machine

Figure 3 shows the testing process method, Fig. 4 exhibits the specimen testing process in the UTM machine. The universal testing machine used for evaluating the FDM 3D printed samples is also known as a tensile testing machine.

Type of testing machine: Tensile Testing Machine.

Test speed: 5 mm/min.

Fig. 4 Specimen testing in a universal testing machine (UTM)

4 Results and Discussion on the Findings

In this study, the tensile test determined the ultimate load and rupture load. Tensile tests are conducted using the Dak Systems test bench—a universal testing machine. The specimen with the best strength value was chosen as the best specimen with optimum process factors. According to the ASTM guidelines, the test speed was set at 5 mm/minute. Figure 5 shows the specimens after testing. The final tensile

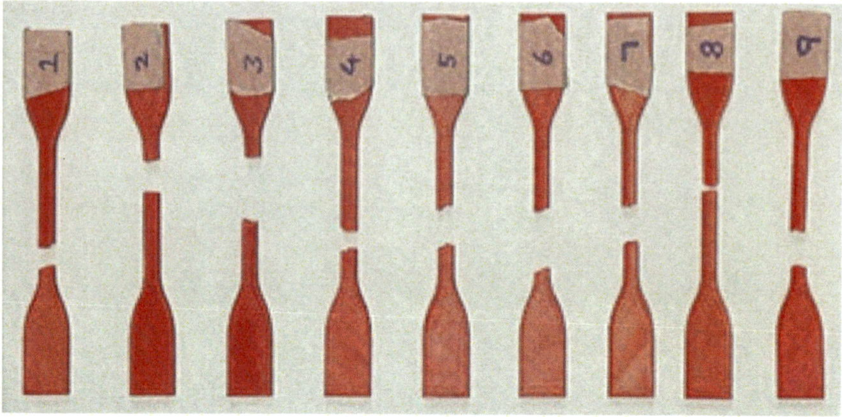

Fig. 5 Sample specimens after testing on universal testing machine

Table 4 Tensile testing results

Specimen No	Ultimate load (N)	Tensile strength (N/mm^2)	Elongation %	Elongation (mm)
1	410.649	22.8139	9.0391	2.259775
2	357.504	19.8613	6.4791	1.619775
3	378.595	21.0297	8.0791	2.019775
4	390.488	21.6938	7.8400	1.95
5	411.389	22.8549	10.2791	2.569775
6	396.929	22.0516	10.7591	2.689775
7	383.288	21.2938	12.4400	3.11
8	452.5245	25.1403	15.3991	3.849775
9	407.0904	22.6161	7.4800	1.87

strength, maximum load, and percentage elongation are calculated and tabulated in Table 4 for all orthogonal array parameters. To compute the values, the used formulas are,

Ultimate Tensile Strength = Ultimate load / Cross-sectional Area.

Elongation percentage = Linear Strain at failure * 100.

4.1 Test Results

4.2 ANOVA Analysis Results for Ultimate Tensile Load

Tensile Test Results for L9 Orthogonal Array Specimens.

5 Conclusions

5.1 Empirical Analysis

In this study, the maximum elongation, load, and tensile strength for all nine specimens were observed, and tabulated. The eighth specimen exhibits the best tensile strength response from the test results. Figure 7 shows the stress–strain curve for all nine specimens, and Fig. 8 provides the specimens' maximum load in N, maximum tensile strength in N/mm^2 (or) MPa and elongation percentage.

- Maximum load = 452.5245N
- Maximum tensile strength = 25.1403 MPa
- The parameter combination for the eighth specimen is,
- Layer thickness –0.25 mm
- Nozzle Temperature –210 °C
- Infill Pattern—Concentric Pattern
- These parameter settings will result in parts with increased tensile strength.

5.2 ANOVA Analysis

According to ANOVA Analysis, Layer thickness contributes more to the load-withstanding ability of FDM prototypes in tensile loading conditions. Figure 6 and Table 5 show the show the results of the ANOVA analysis of Variance for the S–N ratio for ultimate strength.

The parameter combination for the better product with higher tensile strength is,

- Layer thickness—0.25 mm
- Nozzle Temperature—210 °C
- Infill Pattern—Grid Pattern
- Infill Pattern is the most significant process parameter followed by layer thickness and the least influencing factor is nozzle temperature.

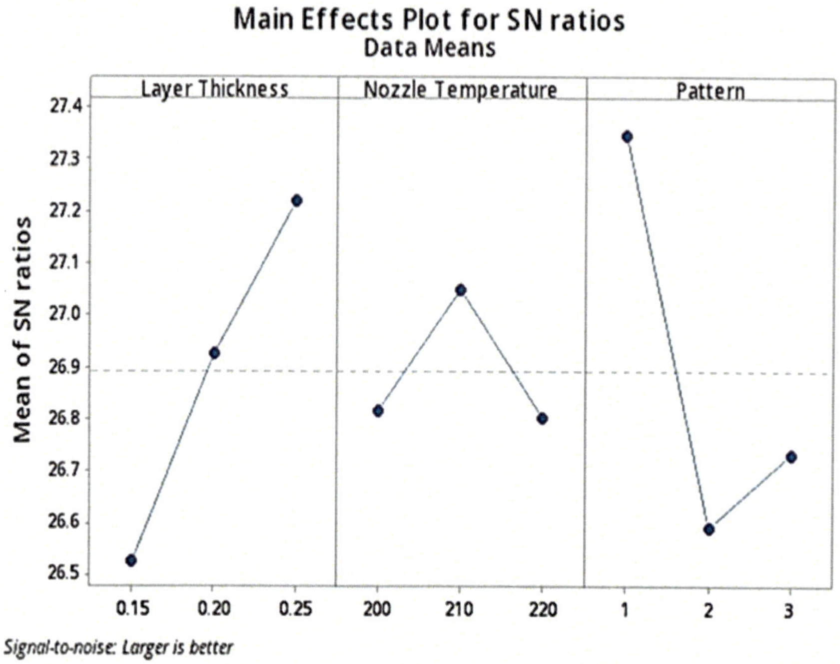

Fig. 6 ANOVA Analysis—SN ratio for ultimate tensile strength

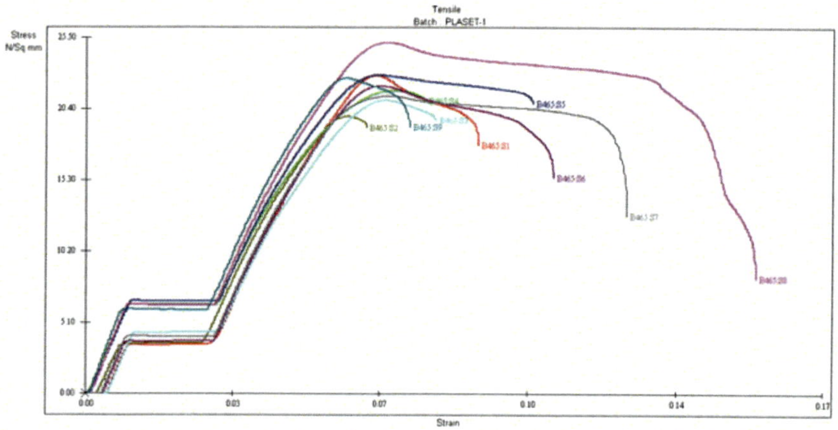

Fig. 7 Stress–Strain response curves for specimens

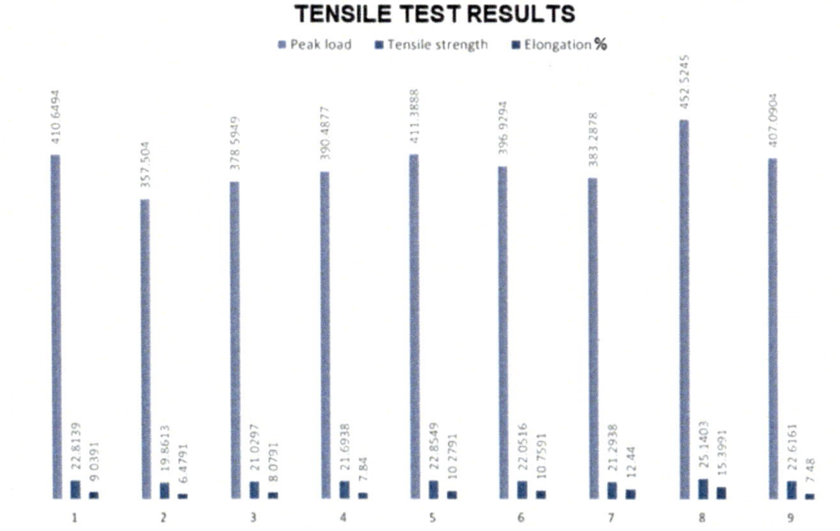

Fig. 8 Tensile test results for L9 orthogonal array specimens

Table 5 S–N ratios ANOVA analysis of variance

Source	DF	Seq SS	Adj SS	Adj MS	F	P	Contribution %
Layer Thickness	2	0.726	0.726	0.3632	0.88	0.532	27.60
Nozzle Temperature	2	0.113	0.113	0.0565	0.14	0.880	4.30
Infill pattern	2	0.966	0.966	0.4830	1.17	0.461	36.70
Residual error	2	0.827	0.827	0.4133			31.40
Total	8	2.632					

- Infill Pattern contributes 36.70%, Layer thickness contributes 27.60%, and Nozzle temperature provides the least impact, and it affects about 4.30% in the ultimate tensile load handling capability of 3D printed PLA prototypes.

References

1. Qattawi A, Alrawi B, Guzman A (2017) Experimental optimization of fused deposition modelling processing parameters: a design-for-manufacturing approach. Procedia Manufacturing 10:791–803

2. Ahmad, M. N., Ab Rahman, M. H., Maidin, N. A., Osman, M. H., Wahid, M. K., Mohamed Saiful Firdaus, H., & Abd Aziz, N. A. Optimization on Surface Roughness of Fused Deposition Modelling (FDM) 3D printed parts using Taguchi approach. In Symposium on Intelligent Manufacturing and Mechatronics (pp. 230–243). Springer, Singapore. (2019).
3. Farashi, S., & Vafaee, F. Effect of printing parameters on the tensile strength of FDM 3D samples: a meta-analysis focusing on layer thickness and sample orientation. Progress in Additive Manufacturing, 1–18. (2022).
4. Alafaghani, A. A., Qattawi, A., Truong, J., Jaman, M. S., & Sun, J. Q. Optimizing Fused Deposition Modeling Processing Parameters Using Response Surface Method for Strong Parts. In International Manufacturing Science and Engineering Conference (Vol. 84256, p. V001T01A030). American Society of Mechanical Engineers. (2020).
5. Ansari AA, Kamil M (2021) Analysis of dimensional quality in 3D printed polylactic acid parts fabricated by fused deposition modeling. Materials Today: Proceedings 47:2281–2287
6. Banjanin, B., Vladić, G., Pál, M., Dimovski, V., Adamović, S., & Delić, G. (2018). Production factors influencing mechanical and physical properties of FDM printed embossing dies. In 9th international symposium on graphic engineering and design. (2018).
7. Bazin, M. M., Othman, M. Z. M., Padzi, M. M., & Ghazali, F. A. Optimisation of 3D printing parameter for improving mechanical strength of ABS printed parts. International Journal of Mechanical Engineering and Technology, 225–260. (2019).
8. Eryildiz M (2021) Effect of Build Orientation on Mechanical Behaviour and Build Time of FDM 3D-Printed PLA Parts: An Experimental Investigation. European Mechanical Science 5(3):116–120
9. Ouballouch A, Ettaqi S, Bouayad A, Sallaou M, Lasri L (2019) Evaluation of dimensional accuracy and mechanical behavior of 3D printed reinforced polyamide parts. Procedia Structural Integrity 19:433–441
10. Raffik, R., Kamal, S. S., Arun, S., Raja, P. P., & Kumar, R. M. Automatic Tank Cleaner. International Journal of Mechanical and Production Engineering Research and Development, 8(113). (2018).
11. Raffik, R., Mayukha, S., Hemchander, J., Abishek, D., Tharun, R., & Kumar, S. D. Autonomous Weeding Robot for Organic Farming Fields. In 2021 International Conference on Advancements in Electrical, Electronics, Communication, Computing and Automation (ICAECA) (pp. 1–4). IEEE. (2021).
12. Ezhil VS, Sriram BR, Vijay RC, Yeshwant S, Sabareesh RK, Dakkshesh G, Raffik R (2022) Investigation on PID controller usage on Unmanned Aerial Vehicle for stability control. Materials Today: Proceedings. 66(3):1313–1318
13. Rakesh, D., Keerthivaasan, R. K. V., Mohan, A., Samvasan, P., Ganesan, P., & Raffik, R. Automated Public Screening and Health Vitals Monitoring Station. In 2021 International Conference on Advancements in Electrical, Electronics, Communication, Computing and Automation (ICAECA) (pp. 1–6). IEEE. (2021)
14. Zhang, J., Liu, T., Hu, X., Raffik, R., Bhatt, M. W., & Ofori, I. Anti-cracking performance test of thick steel structure fireproof coating under vibration fatigue load for critical health infrastructure. The Journal of Engineering, 1086–1094. (2022).
15. Raffik, R., Naveen, C., Siyaguru, J., & Vijayanandh, R. Performance analysis of modified shuttle mechanism in polymerized bag weaving circular loom. In AIP Conference Proceedings (Vol. 2446, No. 1, p. 180028). AIP Publishing LLC. (2022).
16. Senthilkumar, S., Anushree, G., Kumar, J. D., Vijayanandh, R., Raffik, R., Kesavan, K., & Prasanth, S. I. Design, Dynamics, Development and Deployment of Hexacopter for Agricultural Applications. In 2021 International Conference on Advancements in Electrical, Electronics, Communication, Computing and Automation (ICAECA) (pp. 1–6). IEEE. (2021).
17. An, Z., Wang, C., Raj, B., Eswaran, S., Raffik, R., Debnath, S., & Rahin, S. A. Application of New Technology of Intelligent Robot Plant Protection in Ecological Agriculture. Journal of Food Quality, (2022).
18. Sharma, B. B., Raffik, R., Chaturvedi, A., Geeitha, S., Akram, P. S., Natrayan, L., ... & Sathyamurthy, R. (2022). Designing and implementing a smart transplanting framework using programmable logic controller and photoelectric sensor. Energy Reports, 8, 430–444. (2022).

19. Nyiranzeyimana G, Mutua JM, Mose BR, Mbuya TO (2022) A grey-based Taguchi method to optimize fused deposition modelling process parameters for manufacture of a hip joint implant. Materialwiss Werkstofftech 53(1):89–108
20. Patil P, Singh D, Raykar SJ, Bhamu J (2022) Multi-objective optimisation and analysis of fused deposition modelling parameters: best infill patterns. International Journal of Six Sigma and Competitive Advantage 14(1):18–31

Optimization of Machining Parameters for Drilling on EN-24 Material Using the Taguchi Technique

Raj Kumar and Kedar Narayan Bairwa

Abstract High-quality CNC machining depends on proper operator skills as well as optimum factor selection for the machine. The current study aims to establish the relationship between response parameters such as tool wear rate (TWR) and input variables such as spindle speed, feed rate, and depth of cut. Analysis was done using the DOE technique named 'Taguchi' and optimized parameters were determined by using ANOVA. The validity of the trials can be confirmed by displaying the predicted and experimental values of the optimal solution's TWR response range. This study came to the conclusion that when drilling EN-24 steel, the depth of the cut is the most important component, whereas spindle speed is the factor that contributes the least to effective machining.

Keywords CNC drilling · EN-24 · Optimization of TWR · DOE · Taguchi · ANOVA

1 Introduction

In production engineering where material removal is one important industrial process, CNC drilling is an important industrial requirement, like other machining work CNC drilling has also issues like product quality, surface finish production time, and many more. These issues are taken by researchers to improve CNC drilling machining for different types of material objects. In CNC manufacturing especially in CNC drilling conditions to reduce the involvement of human interface to reduce errors and quality issues, researchers try to work on find out suitable conditions for manufacturing of CNC drilling by selection of machining parameters for selective object materials

R. Kumar (✉)
Swami Keshvanand Institute of Technology, Management and Gramothan, Jaipur 302017, India
e-mail: raj.kumar@skit.ac.in

K. N. Bairwa
Regional College for Education Research and Technology, Jaipur 302022, India

H. Kumar et al. (eds.), *Recent Advances in Intelligent Manufacturing*,
Lecture Notes in Mechanical Engineering,
https://doi.org/10.1007/978-981-99-1308-4_33

with proper cutting tool selection. In this study, the Taguchi method is utilized to achieve optimal results for TWR.

Ozler et al. [1] examined the influence of response parameters such as washer geometry, length of bush and petal geometry on drilling of square shaped AISI 10 steel tube. According to the findings, the length of the bushing is distorted when the feed rate is increased, but it is improved when the bushing is longer. Ku et al. [2] utilized the ANOVA method to investigate the influence of spindle speed, friction angle, friction area, and feed rate during drilling of the 2 mm thick plate of SS 304 steel. According to the findings, the element that had the greatest impact on the surface roughness was the spindle speed, followed by the feed rate. On the other hand, the friction area had a considerable bearing on the length of the bushing. Tyagi et al. [3] evaluated the effects of speed, feed and depth of cut on SR and MRR during drilling of mild steel through HSS drill. The study reported that SR and MRR were most significantly influenced by speed and feed respectively. Sharma et al. [4] proposed Taguchi technique to investigate the performance of CNC drilling for ovality of drilling hole and SR on AISI 304 steel. Results revealed that feed rate and depth played a significant role for SR and ovality of hole. SR is most significantly influenced by feed rate followed by depth of cut whereas ovality is affected by depth of cut followed by feed rate. Rajaparthiban et al. [5] suggested the Taguchi approach for finding the optimal solution for various machining parameters. Results revealed the good agreement between predicted results and experimental results. Kalyankumar et al. [6] examined the influence of cutting speed, feed rate and cutting depth on TWR by using grey relational analysis (GRA). The results revealed that GRA technique can effectively improve the response speed of the turning process.

Chow et al. [7] performed an investigation on the sintered carbide drill that was used for drilling the sheet of the AISI 304 steel. During the course of the analysis, a Taguchi, L-18 orthogonal array was utilized. The drilling speed, the friction contact area ratio, the friction angle, and the feed rate were the input parameters, while surface roughness was the output parameter. For demonstration of surface roughness microstructure investigation was performed. Outcomes revealed that high hardness along with fine grain structure were found in the area near the drilled hole.

Hýþman Çelik [8] investigated the influence of the process parameter such as feed (0.05–0.15 mm/rev), cutting speed (12.5–25 m/min) and twist drill (900–1400 point angles) effects on drilling of Ti-6Al-4 V alloy. The main response during the investigation were chip formation, hole quality, surface roughness and burr height. The results showed that with higher speed a decrement was observed in surface roughness, and with higher point angle and feed rate a decrement was observed in burr height. Increase in burr height was observed at high speed. At higher feed and point angles increase in hole diameter was observed and was closed to nominal size at high speed. Experiments on drilling the AISI 304 stainless steel were carried out by El-Bahloul et al. [9], who employed the Taguchi method and the fuzzy logic approach in order to optimize the drilling process. It was suggested that these approaches were simple and capable for multi-objective optimization. To minimize the delamination factor in drilling operation of CFRPs. Gaitonde et al. [10] provided evidence that demonstrated the approach of Taguchi technique optimization. Experiments were conducted utilizing an L-27 orthogonal array, and the process parameters speed, feed, and point angle were taken into consideration. For finding the optimum process parameters for delamination, factor ANOVA was used. Finally it was revealed that for minimum delamination factor, the point angle, then feed and speed were the notable factors. Somasundaram et al. [11] investigated the roundness errors of the drilled holes under varying feed rate, spindle speed, work-piece thickness and percentage weight of reinforcement particles on aluminum matrix composite sheets using high speed steel drill. They proposed RSM to prove the effect of process parameters to minimize the roundness error. The outcomes showed that with the increase in the percentage of reinforcement in aluminum metal matrix composites, roundness error decreased whereas other input process parameters increased the roundness error. Han et al. [12] carried out an experiment on Ti6Al4V to improve the quality of hole exit like burr formation and surface finish. The alpha- beta titanium alloy Ti6Al4V was used as the work-piece. Drilling operations were performed on BJXK5070 five axis high speed machining center. Input parameters were taken as the cutting speed (30 rpm, 60 rpm, 90 rpm and 110 rpm), feed (0.07 mm/rev, 0.08 mm/rev, 0.09 mm/rev and 0.10 mm/rev) for hole diameter of 8 mm. According to the findings, the burr's width and height increased as the feed rate was increased, even though the spindle speed remained the same. If the feed rate remains the same, the burr height will grow proportionally with the spindle speed. Heisel et al. [13] carried out a study on the angle of the drill point while drilling on the carbon fiber reinforced plastic. Experiments were performed using cutting speed (21–513 m/min) and feed (0.05–0.40 mm/rev). Cemented carbide drill having varying point angles (155°, 175° and 185°) was used in experiments. The results revealed that with increase in point angle and feed improved the quality of drill hole at entrance but decreased the quality at exit. An investigation was done on the drilling operation by Miller et al. [14] on the AISI 1020 steel sheet. The surface contact that existed between the work piece and the drill served as the basis for the mathematical modeling of torque and axial force. The findings of the FEA simulation and the experiments were found to be in good agreement.

2 Materials and Methods

2.1 Materials

A thick sheet of the industrial steel alloy EN-24 that was 50 mm × 50 mm × 12 mm in size served as the work piece in the current set of studies. This industrial alloy is a hot-worked, special version of EN-24 that has good hardness and abrasion resistance properties. The seller who provided the test sample was able to ascertain the chemical makeup of this product, and that data is included in Table 1.

Table 1 Chemical composition of industrial alloy EN-24 (Kannan et al. [15])

C	Si	Mn	S	P	Cr	Mo	Ni
0.36–0.44	0.10–0.35	0.45–0.70	0.040	0.035	1.00–1.40	0.20–0.35	1.30–1.70

2.2 Experimental Setup

The trials are performed on a vertical CNC-drilling machine (MAXIMILL MTAB) manufactured by Batliboi Machine Tools Ltd. and installed in the advanced manufacturing laboratory of the Mechanical Engineering Department at CIPET in Jaipur (Rajasthan, India). The pictorial view of the CNC-drilling machine shown in Fig. 1 having the following specifications as shown in Table 2

The industrial alloy EN-24 plate of 100 mm x 100 mm × 12 mm size is mounted on the Wire EDM machine and four specimens of 50mmx50mm size were cut which were mounted on CNC Drilling to perform the experiments as shown in Fig. 2a. Due of its convenience and simplicity, a 50 mm × 50 mm × 12 mm square cut was chosen for this study. The cut is made beginning at the edge of the working piece, as depicted in Fig. 2b. Center to center distance of the hole is 32 mm for each hole. A flat end drill was used during the drilling operation and 6 mm diameter holes were cut. The TWR measurements in micrometer were carried out three times using Mitutoyo's surf tester, a portable TWR tester; the average result was taken into account as the TWR value for the purpose of analysis (Table 2).

The trials' levels and contributing factors are calculated using Taguchi's method. The nominal MRR and best TWR have been determined using the following formulas after all trials have been completed using the data from MINITAB software of L16 OA.

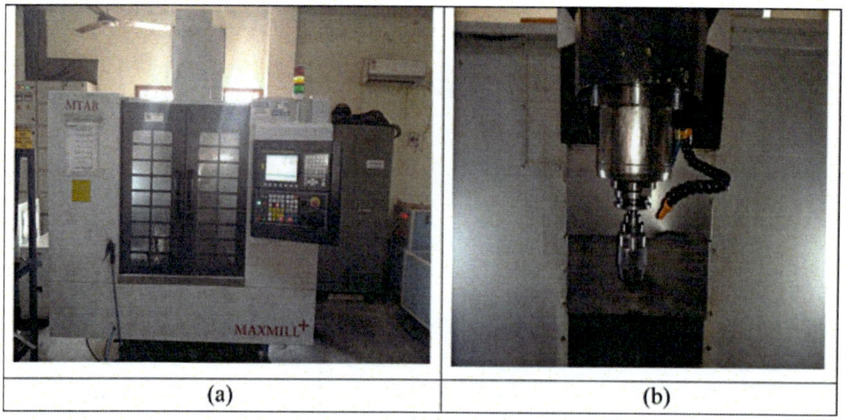

Fig. 1 Pictorial view and cutting operation on CNC drilling machine

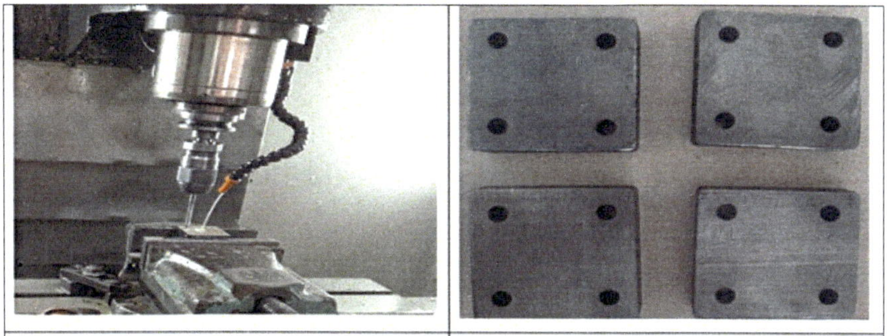

(a) (b)

Fig. 2 Working zone and specimens of work material after drilling

Table 2 Specification of CNC-drilling machine (MAXMILL MTAB)

Sr. No	Description	Specification
01	Travel (mm × mm × mm)	640 × 600 × 500
02	Clamping area (mm × mm)	850 × 600
03	Spindle speed (rpm)	8000
04	Spindle motor (kW)	5.5/7.5
05	Maximum loading capacity (kg)	500
06	Air source (bar)	6–8
07	Machine net weight (kg)	3000
08	Power requirement (kVA)	15
09	Tool magazine (No. of tools)	30
10	Max. tool dia. /weight (mm/kg)	100/7
11	Position accuracy XYZ (microns)	0.8 × 2 × 2

2.3 *Selection of Levels*

The levels and components of the process parameters, by which the machining processes are carried out, are shown in Table 3.

Table 3 Factors and levels for CNC drilling operation

Designation	Factors	Level			
		I	II	III	IV
A	Feed (mm/min)	10	11	12	13
B	Spindle speed (RPM)	800	900	1000	1100
C	DOC (mm)	1.0	1.25	1.50	1.75

3 Results and Discussions

This section uses Taguchi's technique to calculate the experimental findings for optimal TWR using three unique process parameters: feed, spindle speed, and depth of cut. The results are shown in Table 4.

3.1 Effects on TWR

The S/N ratio for TWR Smaller is better option is selected because TWR should be reduced to improve productivity of the machine of the CNC drilling process. The **equation for** calculating S/N ratios for "smaller is better"

Smaller is Better:

$$\left(\frac{S}{N}\right)_{LB} = -10\log(MSD)_{LB}$$

where

Table 4 Experimental results of TWR

S.No	A: Feed (mm/min)	B: Speed (RPM)	C: Depth of cut (mm)	TWR (gm)
1	1	1	1	0.0037
2	1	2	2	0.0034
3	1	3	3	0.0030
4	1	4	4	0.0027
5	2	1	2	0.0033
6	2	2	1	0.0031
7	2	3	4	0.0029
8	2	4	3	0.0026
9	3	1	3	0.0033
10	3	2	4	0.0031
11	3	3	1	0.0028
12	3	4	2	0.0025
13	4	1	4	0.0032
14	4	2	3	0.0030
15	4	3	2	0.0028
16	4	4	1	0.0025

$$(MSD)_{LB} = \frac{1}{R}\sum_{j=1}^{R}(y_j^2)$$

S/N ratios for all 16 experiments are presented in Table 5 for the TWR response parameter.

The most interesting result from S/N ratio analysis is rank identification of factors for selective response variable. Tables 6 and 7 show the rank of factors for TWR using S/N ratio and raw data analysis. The depth of the cut is the most ranked factor among other factors for response variable TWR. The least ranked factors are spindle speed, the reason behind this task is the narrow range of the spindle speed.

Table 5 S/N ratio analysis for TWR

S. No	A: Feed (mm/min)	B: Speed (RPM)	C: Depth of cut (mm)	TWR (gm)	S/N ratio
1	1	1	1	0.0037	48.6360
2	1	2	2	0.0034	49.3704
3	1	3	3	0.0030	50.4576
4	1	4	4	0.0027	51.3727
5	2	1	2	0.0033	49.6297
6	2	2	1	0.0031	50.1728
7	2	3	4	0.0029	50.7520
8	2	4	3	0.0026	51.7005
9	3	1	3	0.0033	49.6297
10	3	2	4	0.0031	50.1728
11	3	3	1	0.0028	51.0568
12	3	4	2	0.0025	52.0412
13	4	1	4	0.0032	49.8970
14	4	2	3	0.0030	50.4576
15	4	3	2	0.0028	51.0568
16	4	4	1	0.0025	52.0412

Table 6 Rank identification of factors for response TWR "**S/N ratio**"

Level	A	B	C
1	49.96	49.45	50.48
2	50.56	50.04	50.52
3	50.73	50.83	50.56
4	50.86	51.79	50.55
Delta	0.9	2.34	0.08
Rank	**2**	**1**	**3**

Table 7 Rank identification of factors for response TWR **"Raw data option"**

Level	A	B	C
1	0.0032	0.003375	0.003025
2	0.002975	0.00315	0.003
3	0.002925	0.002875	0.002975
4	0.002875	0.002575	0.002975
Delta	0.000325	0.0008	0.00005
Rank	**2**	**1**	**3**

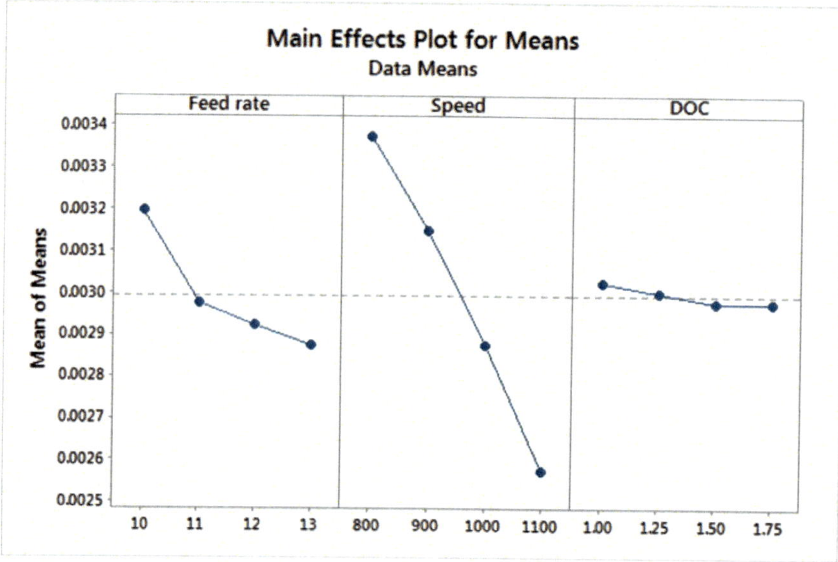

Fig. 3 Main effect plots for response TWR "raw"

In both cases either S/N ratio and raw data analysis results are the same of rank identification for response TWR. Figures 3 and 4 show the effect of factors on response TWR in a graphical manner.

3.2 Interaction Plots for TWR

These plots are generated among two factors for selective response to show interaction among them. Full matrix interaction plot for TWR is presented in Fig. 5 that shows the full scale interaction plots for all three input parameters with response parameter TWR.

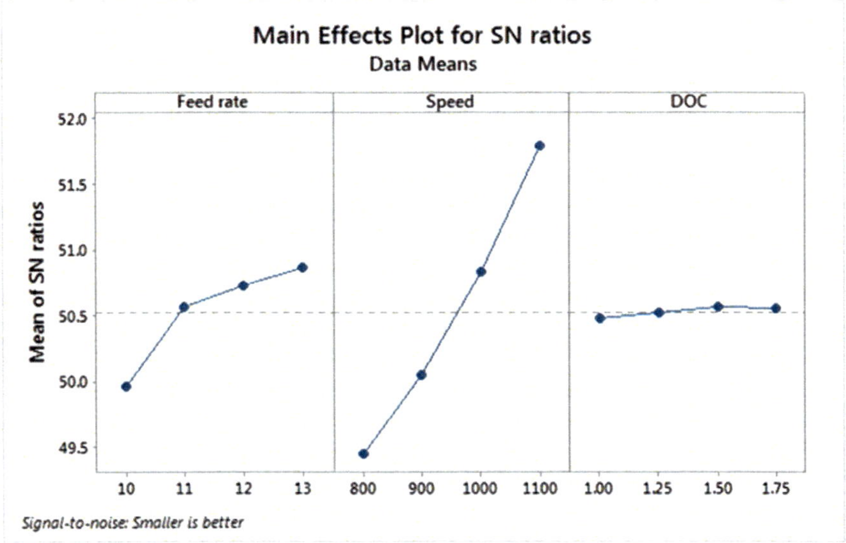

Fig. 4 Main effect plots for response TWR "S/N ratio"

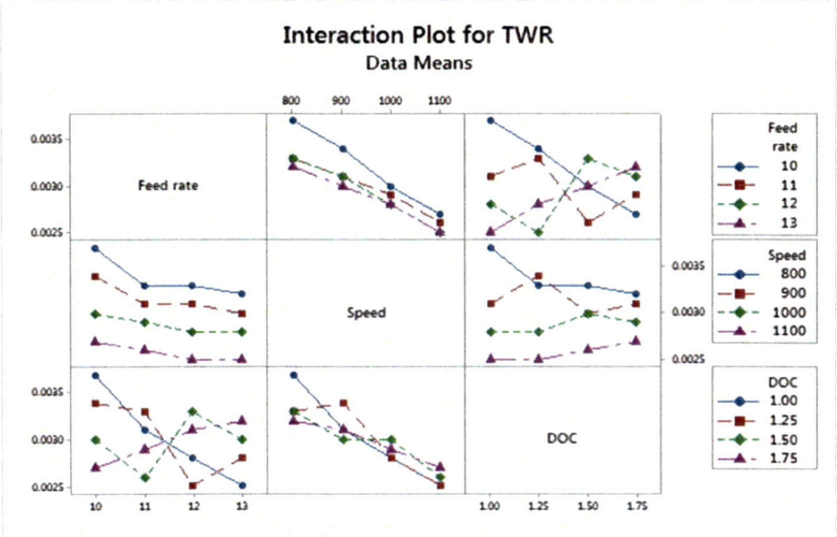

Fig. 5 Full scale interaction plot for TWR

Table 8 Optimal solution for both responses

Response	A	B	C
TWR	4	4	3

3.3 Optimal Solution Prediction

After completing the ranking of the selective replies, the next step is to determine the ideal ranking of the amounts of variables for both of the responses, which are shown in Table 8. The results are reported in a coded value according to the Taguchi method of the DOE approach. The maximum values in rank identification table are considered as optimum levels of each factor for TWR. For response TWR, maximum values (S/N ratio Analysis) for factors A(feed), B(speed) and C(DOC) are at level 4, level 4 and level 3 respectively.

3.4 Estimation of Optimal Solution

As can be seen in Table 8, the top levels for both responses have been generated; the next step is to determine the estimated values of the responses so that they may be verified. Within this section, the Roy estimate technique is utilized in order to do this assignment.

3.5 Estimation of Optimum Response Characteristics for TWR

$$\mu_{TWR} = \overline{A}_4 + \overline{B}_4 + \overline{C}_3 - 2TWR$$

where

TWR is average of TWR = 0.00299.

A_4 represents mean value of factor A for level 4 which is = 0.0028.

B_4 represents mean value of factor B for level 4 which is = 0.0025.

C_3 represents mean value of factor C for level 3 which is = 0.0029.

Substituting the values of A, B, C and TWR in the above equation,

$$\mu_{TWR} = 0.0028 + 0.0025 + 0.0029 - 2 * 0.0031 = 0.0020$$

Using the Equations, one can determine the confidence intervals for the population (CI_{POP}) and the confirmation experiments (CI_{CE}) at a level of 95%.

$$CI_{CE} = \sqrt{F_{\alpha}(1, f_e)V_e\left[\frac{1}{n_{eff}} + \frac{1}{R}\right]}$$

And

$$CI_{POP} = \sqrt{\frac{F_{\alpha}(1, f_e)V_e}{n_{eff}}}$$

where, F_{α} $(1, f_e)$ = The F ratio against DOF 1 and error degree of freedom (f_e) at the confidence level of $(1 - \alpha)$.

$$n_{eff} = \frac{N}{1 + [\text{DOF associated in the estimate of mean response}]}$$

$$\eta_{eff} = \frac{16}{1+3} = 4$$

R = Sample size for confirmation experiments = 1.
V_e = Error variance = 0.00001.
f_e = error DOF = 12.
$F_{0.05}$ (1, 12) = 4.
Substitute all values in formulas

$$CI_{CE} = \sqrt{F_{0.05}(1, 12) * V_e * \left[\frac{1}{n_{eff}} + \frac{1}{R}\right]} = \sqrt{4.75 * 0.0001 * \left[\frac{1}{4} + \frac{1}{1}\right]} = \pm 0.00012$$

$$CI_{POP} = \sqrt{\frac{F_{0.05}(1, 12) * V_e}{n_{eff}}} = \sqrt{\frac{4.75 * 0.0001}{4}} = \pm 0.000177$$

The estimated confidence interval for confirmation experiments is as follows:

$$Mean\mu_{TWR} - CI_{CE} < \mu_{TWR} < Mean\mu_{TWR} + CI_{CE}$$

$$= 0.00124 < \mu_{TWR} < 0.00324$$

Confidence interval for the entire population at 95% level is (Table 9):

$$Mean\mu_{TWR} - CI_{POP} < \mu_{TWR} < Mean\mu_{TWR} + CI_{POP}$$

$$= 0.0026 < \mu_{TWR} < 0.00331$$

ANOVA analyses for TWR after the elimination of insignificant terms have been presented in the Table 9.

Table 9 ANOVA for TWR

Source	DF	Seq. SS	Contribution	Adj. SS	Adj. MS	F-value	P-value
Model	3	0.000002	95.26%	0.000002	0.000001	80.36	0
Linear	3	0.000002	95.26%	0.000002	0.000001	80.36	0
Feed rate	1	0	12.15%	0	0	30.75	0
Speed	1	0.000001	82.75%	0.000001	0.000001	209.43	0
DOC	1	0	0.35%	0	0.00001	0.9	0.362
Error	12	0	4.74%	0	0.00001		
Total	15	0.000002	100.00%				

Table 10 Confirmation of TWR results

Response parameter	Optimal set of parameters	Predicted optimal value	Predicted confidence intervals at 95% confidence level	Experimental value
TWR	$A_4B_4C_3$	0.0020	CI_{POP}: 0.0026 < μ_{TWR} < 0.0033 CI_{CE}: 0.00124 < μ_{TWR} < 0.00324	0.0023

3.6 Confirmation of Experiment for TWR

Final result (Table 10) confirms the validity of the trials by displaying the predicted and experimental values of the optimal solution's TWR response range. A TWR authorization is granted in this instance.

4 Conclusion

The preceding discussions make it abundantly evident that the TWR (0.0023 gm) that is optimal can be attained at A4, B4, and C3 respectively. It has been determined, after a large number of calculations and analyses for optimum TWR, that the optimum TWR value for the EN24 tool steel can be attained by using the DOE to experiment with varying the different levels of input parameters such as speed, feed rate, and depth of cut. This has led to the formulation of the following conclusion: Therefore, the conclusion that can be drawn from this body of work is that the three process parameters at levels 4, 4, and 3 contribute equally to the ideal value of TWR (feed 13 mm/min, speed 1100 rpm and depth of cut 1.50 mm). According to the results of an analysis of variance (ANOVA), the speed is the most influential factor in determining the TWR, accounting for 82.75% of the total variance. The feed rate came in second, accounting for just 12.15%. Nonetheless, it is generally accepted that the depth of cut is the most important factor in determining the TWR.

Acknowledgements The authors would like to extend their gratitude to CIPET, Jaipur for providing the CNC drilling machine (MAXMILL MTAB) that was used to evaluate the tool wear rate of EN-24 steel plates having input variables feed, speed and depth of cut.

References

1. Ozler L, Dogru N (2013) An experimental investigation of hole geometry in friction drilling. Mater Manuf Process 28(4):470–475
2. Ku WL, Hung CL, Lee SM, Chow HM (2011) Optimization in thermal friction drilling for SUS 304 stainless steel. Int J Adv Manuf Technol 53(9–12):935–944
3. Yogendra T, Vedansh C, Jyoti V (2012) Parametric optimization of drilling machining process using Taguchi design and ANOVA approach. Int J Emerg Technol Adv Eng 2(7):339–347
4. Sharma K, Jatav A (2015) Optimization of machining parameters in drilling of stainless steel. Int J Sci Rec Tech 4(8)
5. Rajaparthiban J, Saravanavel S, Ravichandran M, Vijayakumar K, Stalin B (2020) Investigation on effect of machining parameters using TGRA approach for AISI 316 steel. Mater Today Proc 24(2):1282–1291
6. Kalyankumar S, Chandy ST, Muhammed KTA, Rohith PS (2021) Multi-response optimization of machining parameters of turning operation with green environment in EN24T using grey relational analysis in Taguchi method. Mater Today Proc 75(7):6193–6197
7. Chow HM, Lee SM, Yang LD (2008) Machining characteristic study of friction drilling on AISI 304 stainless steel. J Mater Process Technol 207(1–3):180–186
8. Hýþman Çelik Y (2013) Investigating the effects of cutting parameters on the hole quality in drilling the Ti-6Al-4V alloy. Mater Tehnol 48(5):653–659
9. El-Bahloul SA, El-Shourbagy HE, El-Midany TT (2015) Optimization of thermal friction drilling process based on Taguchi method and fuzzy logic technique. Int J Sci Eng Appl 4(2):54–59
10. Gaitonde VN, Karnik SR, Rubio JC, Correia AE, Abrão AM, Davim JP (2011) A study aimed at minimizing delamination during drilling of CFRP composites. J Compos Mater 45(22):2359–2368
11. Somasundaram G, Rajendra Boopathy S, Palanikumar K (2012) Modeling and analysis of roundness error in friction drilling of aluminum silicon carbide metal matrix composite. J Compos Mater 46(2): 169–181
12. Han Q, Wan H (2019) Through-hole exit characteristics in drilling of TIAL4V and quality improvement of hole exit. Int J Recent Adv Mech Engg 8(1)
13. Heisel U, Peifroth T (2012) Influence of point angle on drill hole quality and machining forces when drilling CFRP. Procedia CIRP 1(1):471–476
14. Miller SF, Tao J, Shih AJ (2006) Friction drilling of cast metals. Int J Mach Tools Manuf 46(12–13):1526–1535
15. Kannan TTM, Pavedhan R, Ajith R, Yuvasiri S (2016) Design optimization of drilling process parameters of EN-24 steel plates using design of experiments. Trans Innov Sci Technol 1(3):60–64

Flexural Analysis of Hand Layup Made Multi-Layer Fiber Reinforced Plastic Composite

Zahoor Ahmad, Waquar Ahmad, Yatin Gautam, Shubham Gupta, and Yogesh Shrivastava

Abstract The evolution of the world is taking place at a great pace with the evolution of science, and every conventional material is getting an upgraded and advanced form of itself with the help of research and development in their respective field. With these advancements, fiber-reinforced plastic (FRP) composites came into play with their property better than most of the metals used for rigid structures. It is widely available and requires very less processing work. This work deals with FRPs and their sequencing to produce or create a new and better material to replace conventional materials. The performance of this material shall be analyzed through flexural testing.

Keywords Composites · Flexural · Fiber · Polymer · Hybrid · Reinforced

1 Introduction

When two or more different materials combine, the combination is called composite. The new material is formed to increase its individual properties. All composite materials show two phases primary phase and secondary phase [1, 2]. Fiber-reinforced composites (FRC) have initiated widespread use innovatively. The use of many engineering fields due to their high durability / weight of scale and high performance of the structure, is often the result of precise design and production techniques aimed at improving the response of these composite structures to specific operating conditions. Studies have shown that optimal fiber formations can increase structural strength, load failure, and compression pressure over fiber distribution without increasing the weight [3–5]. Combined compositions are made by combining two or more different reinforcing elements in a shared matrix. In doing so, the new thing is made with new and added materials. Carbon fibres are widely known for their durable materials [6–9]. Different types of short or continuous fiber bond together by the organic polymer matrix called polymer matrix composites. PMC is described to transfer loads

Z. Ahmad · W. Ahmad · Y. Gautam · S. Gupta · Y. Shrivastava (✉)
Galgotias College of Engineering and Technology, Greater Noida, UP, India
e-mail: yogeshshrivastava90@gmail.com

H. Kumar et al. (eds.), *Recent Advances in Intelligent Manufacturing*,
Lecture Notes in Mechanical Engineering,
https://doi.org/10.1007/978-981-99-1308-4_34

between fibers of a matrix. A polymer matrix composite is a component that consists of a polymer (resin) matrix composed of a dispersible reinforcing material. Polymer Matrix Composites are very popular because of their low cost and simple production methods. Polymer matrix compounds (PMCs) as highly efficient engineering tools are the subject of scientific research worldwide. PMCs as a combination of polymers and other organic or inorganic with high concentrations of heavy metal and additional water contaminants can be used to treat contaminated water applications such as adsorbents, practically [10–13]. Fiber-reinforced (FRP), also identified as Fiber-reinforced plastic, is an amalgamated quantifiable product prepared with a reinforced polymer matrix. Fibers are usually basalt, carbon, or aramid, although other fibres such as paper, wood, or asbestos have sometimes been used. Following the zero, there are two procedures in which the polymer is enhanced—(a) the phase development polymerization and (b) the accumulation of polymerization. Blending plastics are made when two identical materials with different properties come together to harvest the final product with the anticipated mechanical properties and materials. These amalgamated constituents can be of two types, fiber reinforced and particle reinforced. Fiber reinforced plastic is the stage at which mechanical strength and resistance of plastics are improved by the enclosure of fiber ingredients. The matrix, which is the core of non-fiber reinforcing materials, is strong but weak in comparison and should be reinforced with the accumulation of strong reinforcing fibres or fibres. It is the most important fiber in separating the parent polymer from the FRP. Blending plastic refers to those types of plastics that are the result of assembling two or more identical materials with different properties to obtain the final product with the desired properties and mechanical properties. Fiber-reinforced plastics are a class of composite plastics that specialize in using fiber materials to enhance the strength and elasticity of plastics [14–16]. The carbon fibers are 5 to 10 μm wide and are self-possessed mainly of carbon atoms. Carbon fiber is a new type of high strength and high modulus fiber material with more than 90% carbon content, with frivolous properties, high strength, corrosion resistance, high modulus, low density, no crawling, and high temperature confrontation in the Nontarnishinghing atmosphere, good resistance to fatigue, good electrical and thermal flexibility, etc. Not only does it have the natural features of carbon materials, but it also has the faintness and texture of fabric fibers, and is a new group of strengthening fiber [17–19]. Kevlar is a heat resistance and strong synthetic fiber. Aramid also called Kevlar is one of the significant organic fibres with a perfumed polyamide planning which is widely used as reinforcement in composites. This fiber is generally made by compression of aromatic diacid by-products with aromatic diamines. Basalt fiber is a material prepared from the extremely fine fiber of glass which is composed of the "mineral plagioclase pyroxene" in addition to olivine.

Table 1 Properties of fiber

Properties	Unit	Carbon	Kevlar	Basalt
Tensile strength	MPa	3450	2920	4840
Elastic modulus	GPa	230	86	89
Weight	GSM	240	200	300
Elongation of break	%	3.8	3.5	3.15
Fiber thickness	mm	0.24	0.2	0.22

2 Methodology, Material Fabrication and Experiment

In the present work, Carbon, Kevlar and basal have been used for the fabrication of the composite. Table 1 shows the mechanical properties of different fibers.

The motorized belongings of the innumerable ingredients at room temperature are abridged below. In this effort, plaited basalt, carbon fibres of thick 200 gsm, and each Kevlar were used to make mix FRP using a hand-lay-up method. The Hardener CT / AH-60 with Epoxy resin CT / E-120 has been used as a polymer matrix due to its attractive motor properties and adhesive construction. Epoxy and Hardener are mixed in a ratio of 100: 18 respectively. As part of the volume of the fire sand their arrangement plays a correspondingly important role in significantly enhancing the mechanical characteristic of any mixture FRPs, therefore, in this study, basalt and carbon, as well as Kevlar encloses were equally enclosed in shape [20]. The dimensions of the composite are equal to 200 × 250 × 3 mm and we test other mechanical features namely; flexural strength and tensile strength are determined using a universal computer-assisted measuring machine. Initially, a layer of gel / grease is coated on the skin, after which the woven fiber is spread over a gel duct inspected by an epoxy layer that spreads over the pre-laid fiber. A brush was used to extract the liquid epoxy and another layer was placed on top of the epoxy and a roller was used to take the air around the layers to proceed further, until the required size is obtained. Finally, the mould was wrapped in C-Clamp and left to cool in a cool place for 24 h. In Fig. 8 the process follows the practical steps using the hand lay-up method [20–24] (Fig. 1).

3 Possible Sequences

In the present work a total of six samples have been made and these samples contain eight layers each, Carbon fibers are broadly known for emphasizing the material due to their great structural strength such as high-performance capacity and modulus pliability, low compactness, and decent blaze confrontation. It makes carbon fiber unique in a varied range of engineering technologies such as automotive. Mixing carbon fiber with basalt and Kevlar fibers will eventually reduce the cost and increase the use in different fields of composite. Basalt and Kevlar have been used differently.

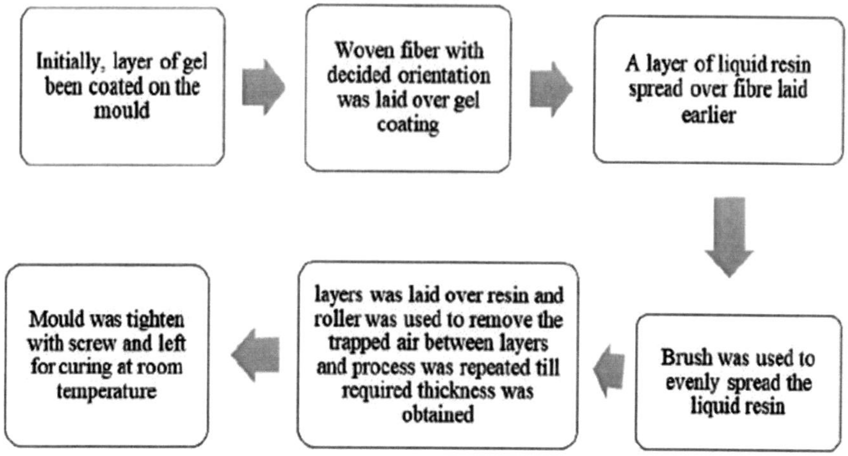

Fig. 1 Stepwise procedure for fabrication through hand-lay-up method

Table 2 Sequence of fabricated material

Sample 1	CBKBBKBC
Sample 2	CBBKKBBC
Sample 3	CKBKKBKC
Sample 4	CBKBBKBC
Sample 5	CKBKBKBC
Sample 6	CBKBKBKC

It is a sample of refined carbon above and below. Kevlar and basalt have been used differently [25–28] (Table 2).

4 Flexural Test

According to ASTM D790, the specimen has a rectangular dimension 127 × 22 mm. The testing of the specimens was prepared according to ASTM standard of flexural testing. There were six specimens of different sequences taken. Tensile testing was completed because of the area of application of the material and its properties. Other test such as compressive shear testing remained not done because of its low implication. The dimension of the specimen was according to ASTM D790. The machine used for the testing was 100 series UTM manufactured by Ban brose engineering Pvt Ltd. (Taiwan). The 3-point rotation test was performed following the ASTM D790 process. A standard 10 kN standard electro mechanical test machine with a constant loading rate of 2 mm/min was used to test the curve [29, 30]. The dimension is 12.8 × 3 mm^2 having length of 100 mm. a three-point load setup up was done on the

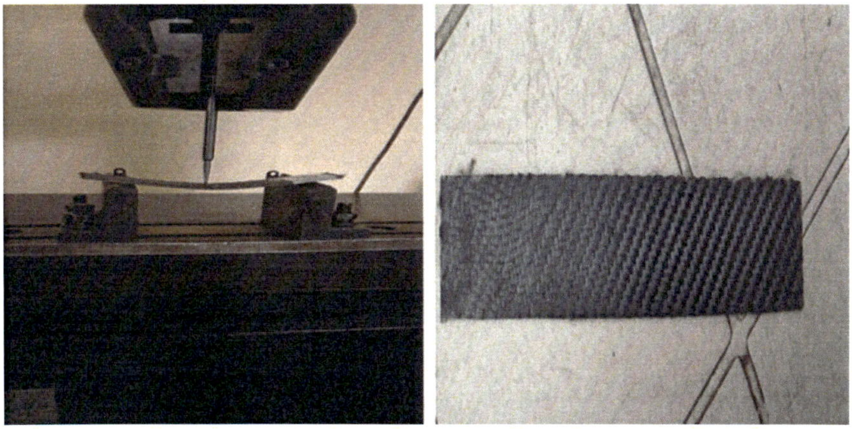

Fig. 2 Specimen and specimen on the universal testing machine

UTM. The nature of the load was point load, which ranged from 0.2 KN to 10 KN. The load was applied until the specimen was deformed to attain a peak force. Peak force was attained after deformation to calculate further Flexural Strength (Fig. 2).

5 Result and Discussion

Flexural strength determines the capacity of the laminated compound to carry the bend before reaching the fracture point (failure). The Flexural strength of the composite compounds were determined according to the following calculations.

The relation between the parameters and flexural strength can easily be understood as:

$$\sigma = \frac{3 \times \boldsymbol{F} \times \boldsymbol{L}}{2 \times \boldsymbol{b} \times \boldsymbol{d}^2}$$

where,

σ = flexural strength (MPa)

F = Peak load (N)

L = span length (mm) = 127 mm

b = width of the beam (mm) = 22 mm

d = thickness of beam (mm) = 3 mm

The flexible structures of the templates are short in Table 3. The magnitude of the curve is expressed by a carbon-based template in the casing, followed by a growing

Table 3 Data of flexural testing

Specimen	Peak force (N)	Maximum deformation (mm)	Flexural Strength (Mpa)
CBK B_2 KBC	610.6	31.604	587.47
CKB K_2 BKC	314.2	42.56	302.29
C B_3K_3 C	374.2	7.353	360.02
C $B_2K_2B_2$ C	426.6	20.278	410.44
C $K_2B_2K_2$ C	210.7	0.127	202.71
CBKBKBKC	342.6	35.45	329.62

sequence of Kevlar and Basalt laminates while the carbon-based template on the skin provides a high flexible modulus. Since, by having high flexibility, high-strength solids should be placed on the outer layers. In addition, laminates above the neutral axis are pressed, while laminates below the neutral axis face tension under bending (Fig. 3).

The above graph states the deformation and load curve of sample 1. The sample (CBKB_2KBC) which was the first sample withstood maximum bearable load it can resist. The sample started deformation was brittle in nature. That is why the sample started breaking directly after 4 mm. This sample reached the yielding point at 0.16 Kn (approx.) load. The fracture point or point of fracture is achieved at a maximum load which is 610 N. The region under displacement between 4 and 19 mm is known as elasticity region. The region after the displacement from 19 mm is known as the

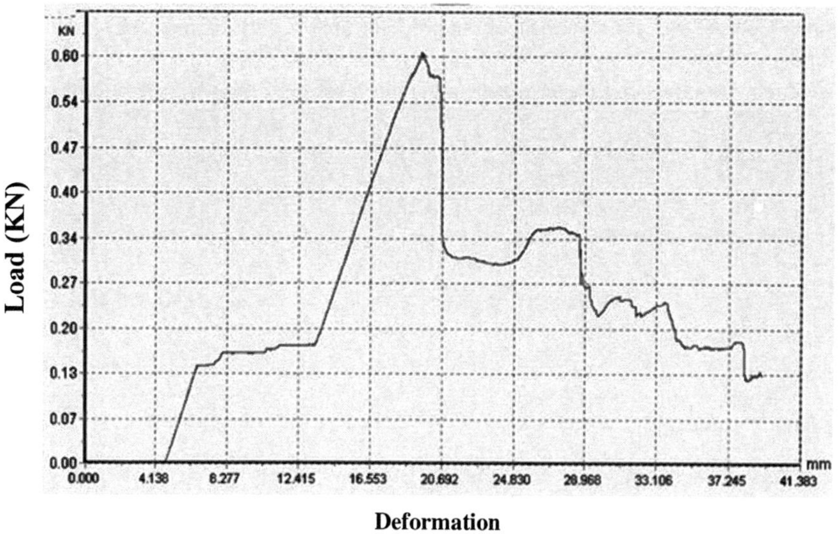

Fig. 3 Load versus deformation graph of sample 1 (CBKBBKBC)

plastic region. The maximum force applied at fracture is also called the fracture force or ultimate force.

6 Conclusions

The work focuses on the development of a three-component FRP compound using the hands-on approach and analysis of its dynamic capabilities. Six hybrid compounds were prepared, namely, CBKB_2KBC, CKBK_2BKC, CB_3 K_3C, CB_2 K_2 B_ 2C, CK_2 B_2 K_2C, CBKBKBKC. The idea behind the experiment is that in combination the FRP provides a better and more efficient hybrid composite than that of an object made using one or two parts. The following observations from the study were noted: -

- Template with Carbon and Kevlar together in the casing, followed by another series of Kevlar and Basalt laminates with high value. However, the results in the study also say that there is a loss of energy after conversion. However, it should be noted that the hybrid FRP sample CBKB_2KBC is the strongest among the other five samples.
- Sample 5 which is C $K_2B_2K_2$ C having carbon at casing following Kevlar and basalt in between showed least amount of resistance, basalt at core was least supportive for the sample in flexural testing.
- The difference between the highest load resisting sample and the least load resisting sample is very much noticeable as the difference is 3 times between the least and most resistive samples.
- The reason behind sample 5 having the least flexural strength is because of no alteration between different fibres and stacking their double layer between the carbon casing.
- The test also showed that sample 1 indicated good flexural strength compared to other samples because of the fact that carbon was used as casing and Kevlar and Basalt was used alternatively. Also the double layer of the basalt gave core strength and Kevlar resisted complete breakage of the sample in the flexural test.

Due to the significant values of durability and flexibility, sample 1 CBKB_2KBC can effectively substitute the metal used when the required load is less than 6 KN. The former can be used effectively in the field of automotive, energy, and space equipment. The project can also be researched further to improve hybrid FRP properties using different types of fibers. It has a wide range of hybrid FRP and has the ability to modify the existing materials with similar properties or uses.

References

1. Gurdal Z, Olmedo R (1993) In-plane response of laminates with spatially varying fiber orientations-variable stiffness concept. AIAA J 31:751–758
2. Dehkordi MT, Nosraty H, Shokrieh MM, Minak G, Ghelli D (2010) Low velocity impact properties of intra-ply hybrid composites based on basalt and nylon woven fabrics. Mater Des 31:x3835-3844
3. Rahman M, Ramakrishna S, Prakash J, Tan D (1999) Machinability study of carbon fiber reinforced composite. J Mater Process Technol 89:292–297
4. Whitney J, Riley M (1966) Elastic properties of fiber reinforced composite materials. AIAA J 4:1537–1542
5. Borenstein A, Hanna O, Attias R, Luski S, Brousse T, Aurbach D (2017) Carbon-based composite materials for supercapacitor electrodes: a review. J Mater Chem A 5:12653–12672
6. Clyne TW, Hull D (2019) An introduction to composite materials. Cambridge University Press
7. Vinson JR, Chou T-W (1975) Composite materials and their use in structures
8. Mehndiratta A, Bandyopadhyaya S, Kumar V, Kumar D (2018) Experimental investigation of span length for flexural test of fiber reinforced polymer composite laminates. J Market Res 7:89–95
9. Srivathsan A, Vijayaram B, Ramesh R (2017) Investigation on mechanical behavior of woven fabric glass/Kevlar hybrid composite laminates made of varying fibre inplane orientation and stacking sequence. Mater Today Proc 4:8928–8937
10. Caminero M, Rodríguez G, Muñoz V (2016) Effect of stacking sequence on Charpy impact and flexural damage behavior of composite laminates. Compos Struct 136:345–357
11. Mouritz AP, Gibson AG (2007) Fire properties of polymer composite materials, vol. 143. Springer Science & Business Media
12. Garoushi S, Säilynoja E, Vallittu PK, Lassila L (2013) Physical properties and depth of cure of a new short fiber reinforced composite. Dent Mater 29:835–841
13. Friedrich K (2012) Application of fracture mechanics to composite materials. Elsevier
14. Rajesh S, Ramnath BV, Elanchezhian C, Abhijith M, Riju RD, Kishan KK (2018) Investigation of tensile behavior of Kevlar composite. Mater Today Proc 5:1156–1161
15. Soutis C (2005) Fibre reinforced composites in aircraft construction. Prog Aerosp Sci 41:143–151
16. Karataş MA, Gökkaya H (2018) A review on machinability of carbon fiber reinforced polymer (CFRP) and glass fiber reinforced polymer (GFRP) composite materials. Def Techn 14:318–326
17. Ashcroft I, Hughes D, Shaw S (2000) Adhesive bonding of fibre reinforced polymer composite materials. Assembly Automation
18. Puck A, Schürmann H (2002) Failure analysis of FRP laminates by means of physically based phenomenological models. Compos Sci Technol 62:1633–1662
19. Little JE, Yuan X, Jones MI (2012) Characterisation of voids in fibre reinforced composite materials. NDT and E Int 46:122–127
20. Jain A, Singh B, Sharma KK, Shrivastava Y (2021) Fabrication, testing and machining of hybrid basalt-glass fiber reinforced plastic composite
21. Jain A, Kumar CS, Shrivastava Y (2021) Fabrication and machining of fiber matrix composite through electric discharge machining: a short review. Mater Today Proc
22. Gholizadeh S (2016) A review of non-destructive testing methods of composite materials. Proc Struct Int 1:50–57
23. Hannon A, Tiernan P (2008) A review of planar biaxial tensile test systems for sheet metal. J Mater Proc Technol 198:1–13
24. Gallo E, Schartel B, Acierno D, Cimino F, Russo P (2013) Tailoring the flame retardant and mechanical performances of natural fiber-reinforced biopolymer by multi-component laminate. Compos B Eng 44:112–119
25. Fiore V, Scalici T, Di Bella G, Valenza A (2015) A review on basalt fibre and its composites. Compos B Eng 74:74–94

26. Tsai SW, Hahn HT (2018) Introduction to composite materials. Routledge
27. Khan A, Ko D-K, Lim SC, Kim HS (2019) Structural vibration-based classification and prediction of delamination in smart composite laminates using deep learning neural network. Compos B Eng 161:586–594
28. Bunsell AR, Joannès S, Thionnet A (2021) Fundamentals of fibre reinforced composite materials. CRC Press
29. Dong C, Davies IJ (2015) Flexural strength of bidirectional hybrid epoxy composites reinforced by E glass and T700S carbon fibres. Compos B Eng 72:65–71
30. Jones RM (2018) Mechanics of composite materials. CRC Press